Orioles, Blackbirds, and Their Kin

Orioles, Blackbirds, and Their Kin

A Natural History

Alexander F. Skutch

Scratchboard Illustrations by Dana Gardner

The University of Arizona Press: Tucson

The University of Arizona Press

⊗ This book is printed on acid-free, archival-quality paper.
Manufactured in the United States of America
01 00 99 98 97 96 6 5 4 3 2 1
Library of Congress Cataloging-in-Publication Data

Skutch, Alexander Frank, 1904–
Orioles, blackbirds, and their kin : a natural history / Alexander F. Skutch ; illustrations by Dana Gardner.
p. cm.
Includes bibliographical references and index.
ISBN 0-8165-1584-0 (acid free). — ISBN 0-8165-1601-4 (pbk. : acid free)
1. Icteridae. 2. Icteridae—Behavior. I. Title.
QL696.P2475S56 1996 95-4431
598.8'81—dc20 CIP

British Cataloguing-in-Publication Data
A catalogue record for this book is available from the British Library.

To Rafael Guillermo Campos-Ramírez
my former student and continuing friend

Contents

Illustrations

Preface

Oriole, Troupial, oropendola, cacique, grackle, blackbird, meadowlark, marshbird, Bobolink, cowbird—the array of names suggests the great diversity of forms and habits in the most varied family confined to the Western Hemisphere, if not in the world. Some are as small as sparrows, others as big as crows. Some are gorgeous in yellow or orange and black; others wholly black, but not without the iridescence that makes them handsome. They sing in the mellowest of voices or call with shrill notes. Some lurk in pairs amid dense thickets; others fly in great flocks across the open sky. Some live throughout the year in the same place; others migrate over vast distances. They breed in cooperating monogamous pairs, in harems attracted by polygynous males, or in promiscuous colonies. Their nests are shallow cups hidden beneath meadow grasses, bulky open bowls, domed structures with a side entrance, or long pouches hanging in clusters from lofty tropical trees, among the finest examples of avian weaving. At the other extreme, numerous cowbirds build no nest but foist their eggs upon other birds, who hatch them and rear the intruded young.

Not only the family as a whole but single genera or lineages present great diversity. Among both the grackles and the marsh-nesting blackbirds are monogamous and polygamous species. Among cowbirds we can trace an instructive series from a species that rears its own young in nests built by other birds through one that parasitizes only this relative, then on through brood parasites with wide ranges of hosts to the more narrowly selective Giant Cowbird. Surprisingly, among such competent nest weavers as the orioles we find one, the Troupial, that habitually pirates closed nests of other birds instead of building its own, thereby reminding us of the Bay-winged Cowbird. All these diversities of habit stimulate thought about their origins and the circumstances that promote them.

To my regret, the tropical valley where I have spent most of my life contained, until recently, only two moderately abundant members of this family, the Yellow-billed Cacique throughout the year and the Baltimore (Northern) Oriole during the months of the northern winter. But in travels and long sojourns in other regions, I have learned all that I could about orioles, blackbirds, and their kin, including several little-known species. By supplementing my own observations with the studies of many other naturalists, I have tried in this book to convey to the reader a comprehensive account of a very remarkable family of birds.

To find an appropriate name for this family of ninety-four species of diverse birds has not been easy. In the past it was often known as the troupial family, but to name it for one of its least typical members hardly seems correct. "American blackbirds" is now more commonly used, although by no means all its species are black. A more appropriate designation is "American orioles," for the largest genus, *Icterus,* which has given the family its scientific name, the Icteridae. The adjective "American" is needed to distinguish these birds from the Old World orioles of a different family. Yellow-and-black songbirds that English-speaking colonists found in America reminded them of the Golden Oriole they knew in their homeland, and they transferred the name without regard to affinities. "Icterid" is probably the most precise and convenient designation for all members of the family Icteridae and will be used throughout this book.

When I wrote *The Life of the Hummingbird* and its successors, books on woodpeckers, tanagers, and pigeons, I treated each by topics—food, voice, nests, and so on. For this more heterogeneous although smaller family, this treatment did not seem the best. Accordingly, in this book I have, as far as possible, given a separate chapter, long or short, to each major group or genus—one for the orioles, one for the grackles, one for the cowbirds, and so forth—a treatment that emphasizes the outstanding characteristics of each.

The accounts of the Melodious Blackbird, Great-tailed Grackle, Giant Cowbird, Yellow-rumped Cacique, Yellow-billed Cacique, Montezuma Oropendola, and certain orioles are taken, with slight modifications, from the first volume of my *Life Histories of Central American Birds,* published in 1954 by the Cooper Ornithological Society and long out of print. That on the Oriole-Blackbird is from volume 10 of *El Hornero,* journal of the Asociación Ornitológica del Plata, and that on the Troupial from volume 81 of the *Wilson Bulletin,* published by the Wilson Ornithological Society. I have gathered other information from writings listed in the bibliography. To all these sources I am most grateful.

Orioles, Blackbirds, and Their Kin

1

The Bobolink

The delightful Bobolink, the only member of the genus *Dolichonyx,* is in several ways an unusual bird. The male differs from most birds that are not uniformly colored by being lighter on the back than below. Of all icterids, Bobolinks perform the longest migration, and the male undergoes the most striking seasonal changes in coloration—two facts that appear to be causally related. He sings on the wing with more verve than any other bird that I have heard, as though in ecstasy. In keeping with his ebullient character, he seeks long hours of daylight in the summers on both sides of the equator. Bobolinks are among the few long-distance migrants known to have helpers at their nests.

In breeding plumage, the six-inch (15 cm) male Bobolink is largely black, with golden-buff hindhead and nape; whitish patches on the upper back; whitish lower back, rump, upper tail coverts,

Bobolink, *Dolichonyx oryzivorous,* male

and shoulders; and often buffy edgings on the wings, flanks, abdomen, and under tail coverts. After nesting, he molts into a plumage similar to that of the female, which, like many grassland birds, is cryptically striped on the head and streaked on her remaining upperparts with dark brown and olive-buff. Her underparts are plain buffy. Bobolinks have short, conical, finchlike bills and stiff, sharply pointed tail feathers.

Bobolinks nest across southern Canada from Nova Scotia to interior British Columbia and widely through the northern half of

Bobolink, *Dolichonyx oryzivorous,* female

the contiguous United States. They appear to have been originally confined to the Northeast and to have followed advancing agriculture westward almost to the Pacific coast; while in their eastern homeland they have become rarer as abandoned farms have reverted to woodland or been covered by urban sprawl. Bobolinks winter in southern South America, mostly east of the Andes, from Peru, eastern Bolivia, and central Brazil to northern Argentina. Migrating in large flocks, mostly by night, at summer's end they converge on the southeastern United States. Some cross the Gulf of Mexico to Yucatán, but most fly from Florida to Cuba, thence to Jamaica, whence they launch forth over five hundred islandless miles (800 km) of the Caribbean Sea to the coasts of Colombia and Venezuela. From this point their route lies over the forests, savannas, and scrublands of the southern continent to their winter homes. Most Bobolinks that breed in the western United States and Canada retrace the route by which their ancestors apparently colonized this region to join eastern populations on their southward journey. By facing the hazards of migration for an extra two thousand miles (3,000 km) or so each way, they increase their mortality, suffering losses avoided by the minority that follows a more direct course across Arizona, New Mexico, and Texas to Central America (Lincoln 1950).

Avoiding marshes, Bobolinks nest in lush fields of pasture grass, alfalfa, clover, or grains. They forage largely on the ground for many insects and a variety of wild seeds, and are considered beneficial to agriculture. On their way southward in late summer and fall they continue to eat seeds of wild plants until they reach the rice plantations of the southern states, where they fall upon the ripening crop in such multitudes that, despite costly efforts to scare them away or otherwise combat them, they cause substantial losses. Given the abundant food, they lay up fuel for their long flight across the sea and become very fat, which—before they were given legal protection—caused them to be slaughtered in great numbers and sold in the markets. Reaching Jamaica in October, they scatter in multitudes over the lowland plains and hillsides to eat ripening seeds of coarse guinea grass. Known there as "butter birds," they continue to pay the penalty for being fat, and tasty when cooked (Gosse 1847). When they reach South America after flying across the sea, they have spent much of this fuel and are probably less sought for human food amid the rich avifauna of that continent.

Retracing their southward route in the spring, Bobolinks reach their breeding grounds mainly in the northern United States in April, but also in the far west in May, the males, now in full nuptial attire, about a week ahead of the females, as is usual in migratory passerines. Soon they spread over the flowery meadows and alfalfa fields, singing exuberantly the whole day long. They sing chiefly while they circle around on vibrating wings, low or high, but rarely much above the taller of the surrounding trees. They do not, like the Skylark, soar heavenward until they become specks in the blue, but they are perhaps the nearest approach to this celebrated singer among the birds of the eastern United States. Often, too, they chant from a perch high or low: a treetop, telephone wire, bush, the top of a weed stalk or a stake driven into the meadow. Here, with drooping wings and spread tail, they pour forth their notes as rapidly as when they sing in flight.

Their song is notable for vivacity rather than harmony as in the music of thrushes. The spirit of melody seems to bubble over in them too violently to be regimented into harmonious phrases. Their notes seem to stream out so rapidly that they fall over each other and become jumbled. At a little distance, the song sounds like a rapid tinkling. Often, especially at the conclusion of a flight song, a Bobolink raises his voice and exhorts, in a wholly different tone, *O, listen to me,* as though to make sure that a shy female lurking in the grass gives attention to his outbursts. Alighting on a twig or weed stalk, he often utters a peculiar nasal mewing, which contrasts strongly with the cheery phrases that preceded it. The most frequent call note sounds like *pink* or *ping.*

The streaked females stay well hidden in the tall grasses and weeds, where they are difficult to detect. The moment one reveals herself, perhaps inadvertently driven from cover by a person walking over the meadow, one or often two males dash in pursuit of her, twisting and turning as they follow every sudden shift that she makes to elude them, and not relaxing their chase until she dives into sheltering herbage, unmindful of an approaching human. While watching these birds as they began to nest, I noticed no territorial defense. One male might perch and sing near, or fly singing above, the very patch of meadow where another has been chanting, without being challenged or attacked. However, Bobolinks do not breed in colonies but scatter their nests over suitable terrain. A male who attracts a female courts her on the ground, spreading his tail and dragging it like a pigeon, erecting the buffy feathers of his nape, pointing his bill downward, and partly opening his wings, while murmuring a subdued version of his song. She is not always impressed.

In a field of alfalfa and grass where many birds were singing, I watched long but fruitlessly, hoping to see a female building a nest. Apparently, she works mainly under cover to avoid the too-ardent males. Bobolinks' nests are difficult to find. After searching in vain for a nest to which the parents were taking food, I set in the ground some small branches, on which the birds rested before continuing onward. In the direction they took, I set another branch, on which they also perched; and so, by establishing closer and closer points of reference, I finally located the slight structure amid daisies. James F. Wittenberger (1978) learned that the best way to find a Bobolink's nest was by watching a female return to it in the dusk from her day's last absence for foraging.

Bobolinks, like yellow-breasted meadowlarks, often set their nests in a shallow depression in the ground, scraped out by the bird herself or found already made, perhaps by the hoof of a horse. The slight shell of coarse grasses and weed stems is thinly lined with finer grasses. The female lays from four to seven eggs, most often five or six. The pearl gray to pale reddish brown or pale cinnamon-rufous ground is irregularly spotted and blotched with shades of brown, chocolate, heliotrope purple, and lavender (Bent 1958).

During the incubation period of eleven to thirteen days, the female alone covers the eggs. After they hatch, the male abandons his carefree, exuberant ways to become a good guardian of the nest and provider of food for the nestlings. Bigamy is frequent in Bobolinks, trigamy occasional. With two or three families on his land, a male's behavior can become rather complicated. The brood of his first, or primary, female has the strongest claim upon him; whether he attends the nests of his secondary or tertiary consorts depends upon circumstances.

I read with surprise Wittenberger's statement that in Oregon males brooded nestlings. Brooding by males, even for brief intervals, is extremely rare among icterids (it has been recorded of the Common Grackle), and I wondered whether the "brooding" of these Bobolinks was more than close guarding—a point difficult to determine in low nests screened by vegetation. However this may be, we have ample evidence that male Bobolinks do not shirk the labor of nourishing their families. In a four-year study, Wittenberger (1980, 1982) found that males delivered about 60 percent of the food brought to older primary nestlings when the weather was unfavorable and food scarce, about 50 percent when the weather was good and food scarce, and about 40 percent when food was abundant in fair weather. When in fair weather the nestlings needed less brooding, their mother could devote more time to foraging for them and herself and had less need of their father's contributions. He was then free to give more attention to the brood of his secondary consort. Caterpillars and other larvae are the nestlings' main food; grasshoppers come next, while small numbers of various other invertebrates complete their diet.

Polygyny arises in birds that are primarily monogamous because some females prefer to become secondary consorts of already mated males with highly favorable territories rather than sole partners of males with inferior territories. Or the qualities of the available males themselves may influence the females' choices; older, experienced males are usually more efficient fathers than inexperienced yearlings. Polygyny is frequent among passerines, such as northern meadowlarks, Red-winged Blackbirds, Tricolored Blackbirds, and Bobolinks that nest in grasslands and marshes, where the concentration of the energy of sunlight in a thin layer of vegetation—rather than its diffusion through the depths of a tall forest—makes food, especially insects, more concentrated and easier to find. Here a female can nourish a large brood with little or no help from a polygynous male.

About 30 percent of the males in Wittenberger's study area in Oregon had two consorts; a few had three. The polygynous males were two or more years old; yearlings had a single mate or remained bachelors. The attention that Bobolinks give to broods of their secondary females varies from population to population. In Wisconsin, where bigamy and trigamy were frequent and one male had four consorts, males tended to neglect secondary broods, but in Oregon they were more widely helpful. In four years, fourteen of eighteen polygynous males fed their secondary nestlings. In three of these years, all did; but in the fourth year, when food was scarce, only one of five males attended his secondary offspring. The first mate of a polygynous male always nested earlier than the subsequent consort and hatched her brood first. These young were the first to be fed by the male, but when they were seven to fifteen days old he began to divide his attention between them and his secondary nestlings, who were from a few days to a week younger. Whether these secondary nestlings received his contributions depended upon how well his first family was nourished. The smaller the primary brood, and the less food it needed, the more frequently he fed the secondary brood.

Arriving with a billful of insects, a polygynous father nearly always went first to his primary nest, perched above it, and cocked his head to peer down at its occupants, apparently listening to the thin squeaks given by hungry nestlings. If these revealed that they needed food, he gave what he had to them; otherwise, he took it to the less favored brood. Or, before going off to forage, he similarly inspected both families, later delivering what he gathered where the need appeared greater. Sometimes a careful father had difficulty deciding where to feed. On one occasion he flew back and forth between his two nests four times, alighting and cocking his head each time he reached a nest, before finally deciding to feed his secondary nestlings.

Except in a year of favorable weather and abundant insects, secondary females did not compensate for reduced male help by feeding their young more frequently than primary females did. Consequently, more nestlings starved in secondary than in primary nests; and although secondary females laid as many eggs as primary females did, they raised fewer fledglings. This was the price they paid for joining already mated males. They might have done better by seeking one of the bachelor males, even an inexperienced yearling.

One of the surprises of recent careful studies of Bobolinks is the presence of helpers at their nests. At three of fourteen nests in New York, Robert C. Beason and Leslie L. Trout (1984) found helpers bringing food; at each of two nests, a second female; at a third nest, one or two extra males. At another locality in the same state, Eric K. Bollinger and his companions (1986) watched two males feed nestlings. At none of these nests was the relationship of the helpers to the pair they assisted known. The tendency of Bobolinks to return to the locality where they had successfully nested in an earlier year, or where they were reared, suggests the possibility that the helpers were offspring of the pair they helped. More probably, they were unrelated adults who had lost their own young before the waning of their parental impulses, which they directed to the young of neighboring pairs who did not oppose their friendly intrusions. Helpers, usually progeny of the breeding pair but sometimes individuals less closely related, are widespread in the tropics and subtropics, where migration does not disrupt families, but rare among long-distance migrants like Bobolinks.

Whenever I approached a Bobolink nest with eggs, or nestlings only a few days old, the parents made neither cries nor demonstrations but withdrew to a discreet distance and watched me quietly, as though trusting in the adequacy of their nest's concealment. At most the male permitted a few tinkles of sound to escape his muffled bill. As the nestlings grew older, the parents, especially the father, became more demonstrative when I visited their nest. Instead of their almost silent withdrawal when their young were recently hatched, they now circled closely around me or clung to alfalfa stalks a short way off, repeating loud clucks, which the father interspersed with snatches of rapid, excited song. He reminded me how the male Black-cowled Oriole sang breezily when I looked into his nest slung beneath the huge leaf of a traveler's tree *(Ravenala madagascariensis)* in Honduras. These Bobolinks did not try to lure me from their nests by "feigning injury," but such distraction displays by Bobolinks have been reported by others.

2

The Marsh-nesting Blackbirds

Nine species of blackbirds of the genus *Agelaius* inhabit the Western Hemisphere from Alaska and northern Canada to Patagonia. Ranging in length from about seven to nine inches (18 to 23 cm), the males are largely black with touches of bright color. The Red-winged Blackbird displays on his lesser wing coverts a large bright reddish patch, bordered behind with buffy white. The corresponding wing patch of the Tricolored Blackbird is deeper red with a purer white border. The Yellow-hooded Blackbird has a bright yellow head and upper breast. The Chestnut-capped Blackbird's crown, throat, and breast are tawny rufous or chestnut. The Yellow-winged Blackbird has yellow shoulders and under-wing coverts. The Yellow-shouldered Blackbird also wears yellow patches on his wing coverts. On the Tawny-shouldered Blackbird, these patches are yellowish brown. The Pale-eyed Blackbird has only light orange eyes to relieve his blackness. The Unicolored Blackbird wears silky black plumage so glossy that he needs no additional adornment. The female Yellow-hooded Blackbird has a yellowish head and bright yellow eyebrows and breast. Female Yellow-shouldered and Tawny-shouldered blackbirds hardly differ from their largely black mates. Females of other species are mostly streaked with brown, black, olive, or a yellowish color and look very different from the males. Blackbirds' bills are short, conical, and sharply pointed.

The Red-winged, Tricolored, and Chestnut-capped blackbirds are partly migratory, withdrawing from higher latitudes as winter approaches. Most species of *Agelaius* breed in marshes (Orians 1961a), but they perch in surrounding trees and forage in uplands, meadows, and farmlands, where they walk over the ground, gathering weed seeds, fallen grain, and insects. In the evening they return to roost amid the grasses, sedges, rushes, and cattails of the marshes. In Cuba Tawny-winged Blackbirds prefer well-drained country, including savannas, parks, and gardens. Blackbirds are highly social, flying, foraging, roosting, nesting, and migrating in companies that may be multitudinous.

Marsh-nesting blackbirds are not brilliant singers; their simple songs have appealed to some hearers and been disparaged by others. In the north, the ringing *conqueree* of the Red-winged Blackbird is a welcome harbinger of spring; and far in the south W. H. Hudson (1920) loved the sweetness and purity of the Yellow-winged Blackbird's tones. The notes of other blackbirds have been stigmatized as harsh, rasping, wheezy, nasal, like the creaking of a rusty hinge or the braying of a donkey.

Red-winged, Tricolored, Yellow-hooded, and Chestnut-capped blackbirds are polygynous; Yellow-winged and Yellow-shouldered blackbirds, monogamous. For the other three species information is lacking. Blackbirds' colonies in marshes may be small or huge, with nests densely crowded or more dispersed. Yellow-shouldered and Tawny-shouldered blackbirds build in trees, and some individuals of marsh-nesting species breed in uplands. Polygynous species defend territories in which they gather harems, but monogamous Yellow-winged and Yellow-shouldered blackbirds are not territorial. Wherever they situate their nests, blackbirds make substantial open cups. In most species the female builds the nest alone, but the male Yellow-hooded Blackbird constructs an unlined shell, then invites a female to occupy it; she adds the lining.

The eggs of these blackbirds are white, pale blue, bluish green, or greenish, more or less heavily spotted, blotched, or scrawled, mainly on the thick end, with dark brown, blackish, or violet. In tropical and south-temperate populations, clutches range from two to four eggs, with three the most frequent number. In the North Temperate Zone, sets tend to be one egg greater—three to five. The wide-ranging Red-wing has larger clutches in the northern United States than at the southern extremity of its range in Costa Rica. Only female blackbirds incubate, but males vigorously guard the nests and rarely, as in the Yellow-shouldered, feed their incubating consorts. Incubation periods range from eleven to fifteen days and may be as short as ten days in the Yellow-hooded Blackbird. Males of the monogamous species appear regularly to feed their young both in the nest and after they leave it. Polygynous males are less dependable; they appear more often to feed fledglings than nestlings. Undisturbed young remain in the nest for ten to fifteen days, seldom longer.

The Red-winged Blackbird

After our rapid survey of the marsh-nesting blackbirds, let us look more closely at a few whose habits are well known, beginning with the Red-wing, abundant in marshes from Alaska and Canada to Costa Rica, and from the Pacific Ocean to the Atlantic. In the northern United States, where Red-wings have been intensively studied, adult males arrive from the south in flocks in March, before the ice has melted from streams and marshes. The first to appear are generally transients who soon continue onward to more northern breeding grounds, but before the month's end other males come to stay and breed. These summer residents settle promptly on stations in the marsh, often on the same ones they occupied the preceding year, for site fidelity is strong in Red-wings, as in many other territorial birds. Newcomers find such vacant areas as they can among these returning adults. Each night the resident males roost in their chosen places in the marsh, where they find part of their food. Since insects are still scarce there and seeds unavailable, around the middle of the morning the males fly to upland meadows and farmlands, to forage until they return to the marsh in the afternoon. As the season advances, they spend increasing hours there until, when the first resident females appear in April, they remain throughout the day.

While awaiting the arrival of potential mates, the males sing and display. To deliver the familiar *conqueree* or *oakálee,* a male thrusts his head forward, depresses and spreads his tail, fluffs out his plumage, opens his wings, and raises the brilliant epaulets that adorn them; together, sound and visual display form a "song-spread." At highest intensity, this display ends with the wings touching the corners of the spread tail, making the bird appear almost round in frontal view. The streaky females also perform the song-spread, posturing much as the males do, but showing at most a tinge of red on their erected wing coverts. The accompanying notes, quite different from the males' clarion call, have been rendered *check-check-a-skew-skew-skew* or *spit-a-chew-chew-chew.* A female performs a song-spread while facing another of her sex from a prominent position in her territory. Often most of the females in a marsh shrill and screech together in song-spreads directed toward a solitary female Red-wing circling above them. Males display in this fashion to other males at territorial boundaries and often also to the world at large. In confrontations with others, males tilt up their heads and point their bills skyward in a posture widespread among icterids (Nero 1963).

A territory-holding male often sings while he flies, especially as he leaves his domain or returns to it with a long, slow glide. The rapid, prolonged flow of notes sounds like *tseeee . . . tch-tch-tch-tch . . . chee-chee-chee-chee.* After successfully chasing a trespassing male from his territory, the male glides back to it with slowly beating wings and spread tail, voicing similar notes all the way. In another aerial display, males spiral upward, often a dozen or more simultaneously, to hover with wings barely aflutter for many min-

Red-winged Blackbird, *Agelaius phoeniceus,* female

utes before, with half-closed wings, they zigzag down to the cattails and sedges. Apparently, they seldom sing while on these prolonged ascents. Among other vocalizations of the Red-wings are short, dry notes and whistles. A flock passing overhead draws attention by the flight note, a *chuck* or *chack.* With a staccato *check* the blackbird expresses annoyance or scolds a human approaching its nest.

When, about the second week in April, the resident females arrive, the males become intensely excited. While a newcomer rests in a tree beside the marsh, males fly up to perch near her. With spread epaulets they hop along a branch toward her, then fly down to their territories with song-spreads. Sometimes the female flees as they approach; at other times she dives suddenly into the cattails. The quality of a male's territory, the sites it offers for her nest and its abundance of food for her nestlings, appears to influence her choice as strongly as his personal attributes. While she hops about amid the cattails assessing it, he descends to give a song-spread, then strut around her with spread wings and a soft whispering *ti-ti-ti-ti.*

Pairs are formed, often quite promptly, when a female remains in the territory that she has come to inspect. The territory-holding male already treats her as his consort; if she strays into a neighbor's domain, he may invade it to drive her roughly home with bites and bumps. At intervals, aroused by her arrival in his territory, her quarrel with another female, or for no apparent reason, he pursues her in a wild aerial chase, such as I have often watched in Great-tailed Grackles. To elude him, she may lead him far beyond his territory. Sometimes he overtakes her, hits her, or seizes the feathers of her rump and holds on for a fraction of a minute while she struggles to break away. At the conclusion of the pursuit, she returns to remain quietly in his territory. Such pursuits and seizures, reported also of certain ducks and finches, appear to be expressions of overwrought sexual excitement.

After the pair is formed, the male, as though impatient of his consort's delay to start building, tries to prompt her by a "symbolic nest-site selection" display. Crouching near her, he performs a song-spread, then flies to a cattail and clings to it, holding his wings above his back. If she follows, he climbs slowly downward through the marsh vegetation, still with his wings elevated. He bows with bill between his feet, tears off fragments of leaf, and manipulates them in the manner of a female gathering material for her nest. Often his partner follows his torturous descent through the cattails and quietly watches his display, which he may repeat several times. At a later stage, when a female has started to build a nest, her mate sits in it, plucks at its materials here and there, and with wings erect goes through the movements of shaping it. If, after watching for a minute or two, she moves away, he follows to lead her back and repeat his demonstration; but he does not himself build (Allen 1914; Bent 1958).

Male Red-wings do not acquire full adult plumage until the second year after they hatch, and few try to breed before they are about two years old. Females nest as yearlings. Consequently, a population contains many more breeding females than males, a situation that often leads to polygyny. The sex ratio varies with the environment, from two or three (rarely more) females to every breeding male in humid regions, chiefly in the northeastern and north-central United States, to three to six per male in drier environments of the West, where marshes are more productive of insect food. The number of females that a male attracts to his territory varies greatly, from one or two to an amazing thirty-three of a highly successful polygynist in the state of Washington, which was about three times the number recorded for any other male in the region (Orians 1980).

With such diverse numbers of consorts, it is not surprising that males' territories vary widely in size. In Wisconsin, Robert Nero (1956) found the average size of seventeen territories of well-established males whose females were nesting to be 3,550 square feet (330 m^2), or approximately one-twelfth of an acre. The smallest was 1,330 square feet (123 m^2), about one thirty-second of an acre, and the largest 6,280 square feet (583 m^2) about one-seventh of an acre. He reported a study by J. H. Linford in Utah, where Red-wings' territories averaged 31,603 square feet (2,936 m^2), with extremes of 17,292 and 45,903 square feet (1,606 and 4,264 m^2), slightly more than one acre. On uplands, territories tend to be larger than in extensive marshes. Adult males ordinarily avoid trespassing on the territories of well-known neighbors, but if a female enters a territory other than her mate's, he may invade it to drive her back. Yearling males sometimes give song-spreads and claim territories but appear to lack the drive to hold them long.

Although male Red-wings readily accept as many females as choose their territories for nesting, females often resist the intrusion of additional individuals of their sex. If fights arise between an established female and a newcomer, the male sides with the latter against his older consort. When two or more females occupy a male's territory, they tend to divide it between them, each trying to exclude the other(s) from her own subdivision. With continuing familiarity, they become more tolerant of one another, but they continue to resist the approach of their male's other consorts to their nests. Although male Red-wings welcome additions to their harems, they are selective; an already mated female who wanders into a neighbor's territory is often repulsed by him.

Red-wings' nests differ little from those of other marsh-dwelling blackbirds. Beginning from one to four days after the completion of her structure, the female lays, on consecutive mornings, three or four eggs, rarely five, and starts to incubate them early on the day when she completes her set. Her eggs hatch after an incubation period of usually twelve days, less often eleven, and rarely as much as thirteen or fourteen during cool weather of early spring. She warms her eggs alone. Usually all the eggs that hatch do so within an interval of twenty-four hours. The hatchlings have sparse whitish natal down and tightly closed eyes, which open when they are five or six days old. When they are six or seven days of age their wing feathers begin to expand from the tips of long sheaths, and thenceforth feathering of the body proceeds rapidly. If alarmed, nestlings jump from the nest when nine or ten days old; if undisturbed, they may scramble through the surrounding vegetation when only ten or eleven days of age, or they may delay in the nest for two weeks.

Gordon Orians (1973, 1983b) noticed interesting differences in the manner of feeding of northern and tropical Red-wings. In northern marshes aquatic insects—including mayflies, caddisflies, damselflies, and diptera—emerge from the water chiefly in the daytime, when they climb up the stalks to expand and harden their wings, and can be so readily gathered by Red-wings foraging for their progeny that they become the young birds' main food. The parent can fill her bill with a number of these creatures before she flies to the nest. In the tropical marshes of Costa Rica, aquatic insects emerge mainly at night, when they are unavailable to diurnal birds. Here the Red-wings gather other foods, chiefly caterpillars of butterflies and moths, grasshoppers, and other orthoptera that live amid the marsh vegetation above water and are available by day. Some of these insects are procured by forcefully opening the bill (gaping) amid the leaves and stems of plants, which a bird cannot well do while it already holds something in its bill. Accordingly, the tropical Red-wings bring their young only one item at a time, instead of several at once as northern birds do. Provisioned at a slower rate, the tropical nestlings develop more slowly.

The ornithological literature contains conflicting statements about male Red-wings' participation in feeding the young. The confusing situation has recently been clarified by Linda A. Whittingham (1989), who in an experimental study in Michigan shifted equal-aged nestlings between nests, increasing or decreasing the size of broods. She found that a male helps to feed nestlings only when their mother alone cannot provide enough food for them. The age and number of nestlings determines his participation; he assists only at nests with more than two young at least four days old, regardless of whether their mother is his first or a subsequent spouse and of her experience as a parent. In years of abundant food, unassisted females fledge more young than assisted females do in less productive years; but in these leaner years nests attended by both parents yield more fledglings than do nests attended only by mothers. To learn when the broods in his territory need him, and which of them needs him most, a polygynous male must pay close attention to the condition of the young that he fathers, much as polygynous Bobolinks do. Male Red-wings who feed nestlings come to them less frequently than their mothers do, and they remove fewer fecal sacs. More often than they attend nestlings, males feed fledglings while their consorts are occupied with second broods. The young are fed for about two weeks after they quit the nest and then become independent when about a month old.

Occasionally, the consorts of a polygynous male help one another. A female Red-wing who lost her own progeny two days after they fledged was seen by Ruth Strosnider (1960) to feed a neighbor's fledglings twelve times during the next ten days. Such helpfulness may be frequent but difficult to detect in marshes crowded with Red-wings.

Although male Red-wings often fail to feed the young, they are vigilant guardians of their families. Standing on the highest available perch, often repeating their *conqueree,* they survey their territories and hasten to confront any approaching creature that they regard as a menace. I could not walk across a field where several nests were hidden low amid tufts of alfalfa without an escort of protesting Red-wings. While I was still a hundred feet away, a male would fly up to hang with spread wings and dangling feet close above my head, and in this manner accompany me as I advanced toward his nest. When I reached it, his shrill protesting whistles redoubled as he darted toward me, without ever striking me. At one nest the female appeared rather unperturbed while the male menaced me, but at another she was the bolder of the two, clucking anxiously as she alighted on the alfalfa near my feet, as though to lure me away. Probably because the dense vegetation of marsh or upland meadow offers a poor stage for a broken-wing act, Red-wings have seldom been seen to perform it, except in Costa Rica. A bolder Red-wing repeatedly struck the head, back, and arm of a man who visited a nest, and attacked the camera that he set up to photograph it. Red-wings harass hawks, crows, and other large birds, pursuing them far beyond the limits of their territories.

In July, when the young have become self-supporting, Red-wings leave the marshes, where they have thrived upon insects, to frequent upland meadows and stubble fields where grain and weed seeds become their main support. Adult males forage in flocks separate from those of females and young, but all mingle with grackles and cowbirds seeking the same food. In the evening they return to the marshes to roost. In August the flocks of Red-wings disappear, as though they had already migrated southward. Now they are rarely seen because by day, as well as by night, they lurk obscurely in the dense growth of the marshes, extremely wary of an approaching human, while they molt and their flight is impeded—reminiscent of ducks that likewise seek seclusion while they molt. After the renewal of their plumage, Red-wings again visit the uplands by day, until in October their southward movement begins. Only a hardy minority remain through bleak winter months as far north as southern Canada.

The Tricolored Blackbird

In appearance the Tricolored Blackbird confusingly resembles the Red-winged Blackbird, from which the male differs chiefly in the deeper red of his epaulets and the purer white of their posterior borders; the female shows another difference, in her darker brown, less extensively streaked plumage. When breeding, however, the Tricolor differs greatly from the Red-wing, as from every other bird in North America. This singular bird inhabits the lowland valleys of California, west of the Sierra Nevada, whence its range extends into southern Oregon and northwestern Baja California. Undertaking no regular migrations, the Tricolor winters through all but the northernmost part of this range. The species probably descended from a population of Red-wings isolated in the far west of what is now the United States during the last Pleistocene glaciation.

When not breeding, Tricolored Blackbirds live in multitudinous nomadic flocks of their own kind, which spread over open fields and pastures eating weed seeds, fallen grain, ripening rice, and diverse insects. Outbreaks of nomadic grasshoppers attract them so strongly that they have been called "locust birds" and compared to the Wattled Starling of Africa and the Rose-colored Starling of Eurasia, which wander widely, searching for the outbreaks of grassland locusts that follow rain, and breed gregariously when they find such abundance. Occasionally Tricolors descend in swarms upon fields of ripening cereals, but the losses they inflict are, at least in part, compensated by the removal of many insects injurious to crops. At night the blackbirds roost in the marshes, often in vast numbers that congregate from many square miles of the surrounding country (Bent 1958).

Mainly in April and May, rarely in autumn, often following a rain that foments the multiplication of insects, the blackbirds, as though moved by a common impulse, descend, swarms of males and females together, upon an area that promises to favor their reproduction. Usually they invade a marsh covered with cattails, tules, or other emergent plants that offer support for their nests. No matter that breeding Red-wings already occupy the marsh; avoiding rather than resisting the Red-wings' attacks, they overwhelm by sheer weight of numbers the frantic defenders, who abandon the futile attempt to save their territories. Often the Red-wings remain on the margins of a marsh, leaving the preferred center to the Tricolors (Orians and Collier 1963). Even the bigger Yellow-headed Blackbirds fare no better when Tricolors invade their marsh. Groups of Tricolors will nest in streamside trees and bushes, along irrigation ditches, or in fields of grains, alfalfa, or other crops. These nesting colonies range in size from less than a hundred to two hundred thousand or more nests—by far the largest breeding aggregations of any North American land bird.

In any part of such a huge colony, nesting is highly synchronized. The first arrivals settle in thousands in the center of a wide marsh, where the males choose territories and are joined by females who promptly begin to build their nests. After an interval, another flock may settle in an adjoining part of the marsh and start nesting without delay. In every area, all the breeding females will be at the same stage. Separated by ten days or two weeks, up to four waves of simultaneous nest building may occur in the same locality (Lack and Emlen 1939; Emlen 1941; Orians 1960, 1961b; Orians and Christman 1968; Payne 1969a).

The displays of Tricolored Blackbirds closely resemble those of Red-wings. Although their song-spreads are similar, their notes are less clear and pleasing. They point their bills skyward just as Red-wings do, and males show nest sites to their consorts in much the same way. To territories that average only about thirty-five square feet (3.25 m^2), the polygynous males often attract several females, whose nests are necessarily crowded, generally with a density of one for every five square feet (0.46 m^2). Two occupied nests may rest side by side, or one close above another. The nests may be situated so low that they are flooded by a slight rise in the water level, or they may be more than head high. The female, building unaided, binds her nest to upright stems or leaves in the skillful manner of other blackbirds. Decaying vegetation, applied wet, with sometimes a little mud, plasters the inside of the cup, which when dry is lined with fine grasses. Often freshly plucked grass makes a bright green nest. After four days of building, the female lays three or four eggs, which she warms alone during an incubation period of eleven or twelve days. During this interval the males abandon their briefly held territories to forage on the uplands, returning in the evening to roost in the marsh, usually apart from their territories.

After the nestlings hatch, fathers help mothers to feed them. Since the tiny territories cannot supply enough food, the adults seek it on the grassy uplands, up to at least four miles (6.5 km) from the nests, to which they bring flightless young grasshoppers —a preferred food—crickets, beetles, spiders, and the like. Only a minor part of the nestlings' food—including dragonflies, damselflies, and midges emerging from the water, with a few tadpoles and snails—comes from the marsh itself. When eleven to fourteen days of age, the young leave their nests and gather in large groups where parents feed them for about two weeks more. Whether each parent confines its feeding to its own offspring, as in the crèches of penguins, flamingos, and other large birds, or spreads its contributions more widely through the crowd appears not to be known. From the start of nesting until the young can feed themselves, Tricolored Blackbirds take about ten days less than Red-wings do. Eight days after losing a nest, a female may start another. The Tricolored Blackbird's breeding schedule appears to have been condensed to take maximum advantage of a transitory abundance of insect food.

Unlike Red-wings, who menace with screams or boldly attack creatures that threaten their nests, densely massed Tricolors fail to defend their eggs and young from hawks, owls, and, above all, snakes of several kinds that devour them. While a Northern Harrier glides over the marsh, they fall silent and hide low amid the cattails until the raptor passes. The cries of nestlings seized by a hawk or serpent draw a small flock of blackbirds to hover over the site of the tragedy, but they lack the spirit to mob the predator or defend their progeny. Despite the water that surrounds the nests in an inundated marsh, many young are lost to predators, which may gorge to satiation yet pillage only a small proportion of many thousands of nests. Abbreviated synchronized nesting shortens the interval when any section of an immense colony is exposed to predation. However, too much disturbance, shortage of food, or reasons more obscure may cause Tricolored Blackbirds suddenly to desert a whole colony, perhaps leaving hundreds of thousands of nests with the remains of eggs broken by the Tricolors themselves or other agents. In any case, as soon as their young can follow them, these extraordinary birds abandon their marshes to roam in great flocks over the uplands, gathering in tules or cattails only to roost.

The Yellow-hooded Blackbird

The Yellow-hooded Blackbirds of South America and the neighboring island of Trinidad breed in contrasting environments, in two of which they were studied by R. Haven Wiley and Minna S. Wiley (1980). In Trinidad and northern Suriname, regions of moderate rainfall well distributed through the year, these blackbirds nested in abandoned or fallow rice fields, overgrown with grasses, sedges, cattails, and other plants standing above shallow water. In both countries laying spread over about half the year, although not evenly throughout the interval. Territories of the polygynous males ranged from 50 to 165 feet (15 to 50 m) in diameter, with active nests, usually in patches of tall, dense sedges, from 7 to 35 feet (2 to 10 m) apart. The birds foraged mainly beyond their territories, in newly plowed fields where they hunted food in the mud, and in the emergent vegetation of ditches and marshes, where they hopped from stem to stem just above the water, peering and probing in the sheathing leaves of grasses.

The widespread *llanos bajos,* or low plains of the interior of Venezuela, are a region of strongly contrasting seasons. In the nearly rainless four months from mid-December to mid-April, the ground is parched and plants wither. After heavy rains begin between May and July, water covers much of the land and marshes that have long been dry are flooded. After several months of rain they are covered with wild rice *(Oryza perennis)* and dense stands of *Thalia geniculata,* a tall, broad-leaved herb of the ginger family. The former yields food for the blackbirds; the latter supports their nests. By early October these plants and the grass *Leersia hexandra* have developed enough to promote nesting. Now the blackbirds, which arrived about two months earlier, begin a breeding season that lasts only two months and ends as the dry season approaches in late November. Here nesting is concentrated in space as well as

Yellow-hooded Blackbird, *Agelaius icterocephalus,* male

in time. In the densest part of a colony, amid *Thalia* plants 6 to 8 feet (2 to 2.5 m) tall, Wiley and Wiley found that fifteen males held territories 13 to 26 feet (4 to 8 m) in diameter, all in an area 50 by 82 feet (15 by 25 m). These small territories contained an average of 6.6 nests, 3.9 of which held eggs.

Male Yellow-hooded Blackbirds confront their neighbors at territorial boundaries with singing, song-spread, and agonistic displays much like those of Red-winged Blackbirds. They evict intruding males by approaching them, singing, or chasing. They are among the few male icterids known to participate in nest building and apparently the only ones that attract females by offering them nearly finished nests. By wrapping wet decaying grass around emergent stems about twenty inches (50 cm) above water, a male makes a nest and leads a receptive female to it by a special fluttering flight. If he fails in this endeavor, he may abandon his first nest and start another, or he may work on two nests alternately. He does not finish a nest but leaves this task to a female, who lines it with fine grasses, sometimes after bringing a little material for the shell. After a male has settled a consort in one of his nests and she begins to incubate, he starts another, and continues, successively, until he has from two to five females breeding in his little domain. Often, however, males, especially those at the edges of a colony, build nests that remain empty. Male Yellow-hoods, which do not acquire full adult plumage until the second year after they hatch, do not hold territories or breed when only one year old. Females may nest as yearlings.

After a female has accepted a nest, she becomes a regular visitor to it and spends long intervals resting and preening nearby. The brevity of the breeding season on the *llanos* permits no long delay between settling on a territory and starting to lay, such as do the Red-winged Blackbirds and other icterids with more time available. Only two to five days after joining a male, a female deposits the first egg of her clutch of three, laid at daily intervals before 9:00 A.M. On the day her first egg appears she begins incubation, which continues until her eggs hatch after a period of ten to eleven days.

Because of the early start of incubation, the nestlings tend to hatch on successive days and differ in size. At first they are fed only by their mother; she forages for them beyond their father's territory and usually brings only one large grasshopper or other orthopteron at a time, as do mothers in tropical populations of Red-wings. In less than 10 percent of her visits to the nest, she comes with two small items. After the nestlings are five to seven days old, most males begin to feed them, but only at one nest on a given day. Or a male might feed at one nest until the young leave it, and then, after an interval of several days, turn his attention to another nest. Males tend to feed less frequently than females and to bring smaller items, all found much nearer to the nests than where the females forage. They are less dependable, bringing food one day but not the next. When they start to attend a nest, they appear not to know its exact location; often they delay, singing, before they descend among the *Thalia* plants, perhaps at the wrong spot, to seek and feed the brood.

Male participation in feeding the young was found to be important, because food for nestlings was not abundant on the *llanos* and some starved. At nests attended by both parents, the nestlings gained weight more rapidly, and only at such nests were as many as three known to survive until eight days old. After the young fledged at the age of eleven days, males fed them only as long as they remained in the paternal territories. A male feeding his progeny had less time for singing and defending his territory, but none was known to lose it for this reason. On the Venezuelan *llanos,* parasitism by Shiny Cowbirds caused the early abandonment of small colonies. Large colonies, where vigilant males chased intruding female cowbirds, enjoyed a nesting success of 60 percent, which is exceptionally high for small tropical birds with open nests. The more widely spaced nests in Trinidad and Suriname were much less successful (Cruz, Manolis, and Andrews 1990). In these lands with a more equable climate, the blackbirds permanently resided in the vicinity of their nesting colonies (ffrench 1973). In the highly seasonal climate of the *llanos,* they were present for only four months of the year, of which the last two were occupied by breeding. As the *llanos* dried up, Yellow-hooded Blackbirds sought greener lands.

The Yellow-winged Blackbird

Widely spread and abundant over the more southerly parts of South America, from Bolivia and southern Brazil to Patagonia, clad in black with bright yellow shoulders, the Yellow-winged Blackbird is the south-temperate counterpart of the northern Redwing. Like the female Red-wing, the female Yellow-wing is dark and streaky, with pale eyebrows. Sociable birds, Yellow-wings flock through much of the year, with the males often in parties of thirty to forty individuals that remain apart from larger companies of females and young. Immature males resemble females and apparently do not breed as yearlings. Yellow-wings forage while they walk over wet pastures and the borders of marshes. In Chile they eat mainly insects, especially dragonflies, damselflies, and moths, with some seeds (Goodall, Johnson, and Philippi 1957).

Occasionally Yellow-wings alight on trees and sing in concert. Sometimes a male sits on a rush to sing alone, with notes whose sweetness and purity compensate for slight variety. After every note or two he pauses, as though to increase the effect of his small repertoire. Among his few notes is one unequaled for plaintive sweetness. Of all the icterids that Hudson (1920) knew, the Yellow-wing most endeared itself to him, by its grace and lovely black-and-yellow attire, its pretty social habits, and, above all, its unforgettable song, especially the one full, beautiful, passionate note on which it ends.

In an Argentina greatly altered from that which Hudson knew, Orians (1980) found Yellow-wings breeding in monogamous pairs, which apparently formed before they chose nesting areas. Together a male and female prospect for a nest site, the latter leading the way and dropping down into the vegetation for a closer look, while her companion perches above her and sings. When, satisfied by her inspection, she flies away, he follows. They choose sites for nests in cattails and clumps of pampas grass *(Cortaderia selloana)* in roadside ditches, and in patches of cattails or the sedge *Cyperus* in marshes dominated by another sedge, *Scirpus*. While females build typical blackbirds' open cups close to the water (De la Peña 1979), their mates often pick up materials only to drop them. Occasionally one carries a piece to a nest, then flies away without it. Dense vegetation makes it difficult to learn whether a male ever adds his piece to a nest. Females lay two to four eggs, most often three.

The Yellow-wings that Orians watched did not defend territories or protest the approach of other individuals to their nests. Even when a visiting male sang a few yards from a nest, neither the resident male nor the female tried to drive him away. The fights that Orians saw were over females, never over space. In four locations two or three active nests were situated within thirty-three feet (10 m) of each other, suggesting polygyny, but each nest was attended by a different pair. In nearly five hours of watching, Orians did not see a male visit a nest to incubate or to feed his incubating mate. Sometimes the male accompanied her when she went to forage, but usually the two individuals fed at different times and in different places. While the female incubated, her partner often sat nearby and sang. At two nests the incubation period was thirteen days.

The Yellow-shouldered Blackbird

Confined to Puerto Rico and nearby Mona Island, the Yellow-shouldered Blackbird neither nests nor roosts in marshes and differs from other species of *Agelaius* in other important ways; because of its close affinity to the marsh nesters, however, I include it in this chapter. Formerly widespread and abundant on Puerto Rico, its numbers and range have been shrinking alarmingly since the invasion of the island by Shiny Cowbirds in the mid-twentieth century, and apparently largely as a result of parasitism by that bird, whose preferred host it has become (Post and Wiley 1976, 1977a, 1977b). The sexes are alike in their black plumage with yellow wing coverts, but the female is smaller than the male.

On islands with smaller avifaunas than comparable areas of continents, birds tend to be less narrowly restricted by competition to modes of foraging in which they specialize; rather, they occupy a wider spectrum of ecological niches and forage in more diverse ways. Although, like other blackbirds, Yellow-shoulders walk over the ground searching for food, overturning small objects with their bills and occasionally scratching with one foot, they do most of their foraging in trees. Walking along branches, mainly in the subcanopy, or flying between clusters of foliage, they glean insects from leaves and twigs. Often they probe tangled clumps of the epiphytic bromeliad *Tillandsia recurvata*, standing upon them or hanging beneath them, inserting their bills up to the base to extract insects, and sharing this source of food with Black-cowled Orioles and Adelaide's Warblers. Clinging like woodpeckers to vertical trunks and the undersides of branches, they probe into crevices, pry off flakes of bark, and enlarge crannies. Inserting their bills, they gape inside fruits, buds, and cocoons. Yellow-shoulders dart into the air to seize insects that they put to flight while they forage, or sit on exposed perches to sally forth like flycatchers. They hover, manikin-like, to pluck edible objects from exposed surfaces.

Like orioles, honeycreepers, and many other birds, Yellow-shouldered Blackbirds enjoy nectar. Early in the year, when *Aloe vulgaris* blooms, they are strongly attracted to these shrubs, whose flowers they open by inserting their bills and gaping. Flowers of *Yucca, Erythrina,* and *Inga* trees also offer sweet drinks. For vegetable foods the birds visit places where animals are fed, including El Guayacán and La Cueva Islands with their large colonies of introduced rhesus monkeys, where they help themselves to biscuits provided for the primates.

William Post and Micou M. Browne (1982) watched about twenty Yellow-shouldered Blackbirds lying close together on bare ground and picking up ants that they applied mainly to their wings, but also to their tails, upper tail coverts, and breasts. Often their tails were bent under their bodies while they rubbed ants on their wing feathers. Five times individuals fell on their sides while for eight minutes they indulged in the puzzling activity called "anting." In the temperate zone of North America birds nearly always ant on the ground, but in the tropics they commonly do so on trees or shrubs. The Yellow-shouldered and the Oriole-Blackbird (chapter 6) are apparently the only birds for which terrestrial anting has been reported in tropical America.

Yellow-shouldered Blackbirds, unlike most of their relatives, roost not in marshes but in trees. William Post and Kathleen Post (1987), who studied their roosts, distinguished two kinds. Between spells of foraging, blackbirds, with Greater Antillean Grackles and Shiny Cowbirds, rested in diurnal or "secondary" roosts, which were shady trees near where they ate. These shelters were especially important during the midday heat, when intense insolation might be lethal to birds whose black plumage strongly absorbs the sun's rays. The Yellow-shoulders rested about five inches (12.5 cm) apart, but sometimes in contact, and occasionally they preened one another—behavior not often seen among icterids. The three species that rested together responded in the same way to potential predators, becoming suddenly silent and often diving together into denser vegetation when an American Kestrel or a Turkey Vulture flew over, and together mobbing flightless animals that disturbed their repose, mainly humans. When a Yellow-shoulder caught in a mist nest was handled and started to scream, grackles, cowbirds, and Yellow-shoulders flew close around its captors, threatening them.

The nocturnal or "primary" roosts were the safest places the birds could find: mangrove trees on tiny offshore islets or peninsulas; coconut palms in plantations where the trunks were encircled by metal rat guards; or the wires and beams of electric transformer stations, surrounded by high chain-link fences and floodlighted through the night. The use of these stations demonstrated once more that protection from nocturnal predators is more important to sleeping birds than darkness and tranquility. Throughout the year, several hundred to over a thousand of each of the three species—blackbirds, grackles, and cowbirds—slept together in these roosts, often with Gray Kingbirds, Mourning Doves, and other small birds. It is strange to find Yellow-shoulders and their enemies, the cowbirds, consorting so intimately together.

Several hundred yards of open water separated some of the mangrove cays where the blackbirds roosted from the mainland shore. In strong winds and high waves birds flying low risked being engulfed by the water. One evening in September, three blackbirds on their way to the roost fell into the water as they approached the cay. Probably their flight was impeded by the loss of wing feathers by molting. In slightly less than seven minutes, one of them swam the 130 feet (40 m) to the cay; the other two reached it from shorter distances away. All swam well, with only their heads, necks, and backs above water. Yellow-shoulders sleep in the communal roosts even during the nesting season, when only females incubating their eggs or brooding younger nestlings remain near their nests through the night.

Post (1981), who studied Yellow-shouldered Blackbirds for two years, found them monogamous and nonterritorial. The sexes seem to be present in approximately equal numbers. Females nest as yearlings but males do not until they are about two years old. Yellow-shoulders begin to breed when plants renew their growth with the spring rains of April and May and insects become more numerous. Pairs form from six to ten weeks before laying begins, in groups of birds that visit the nest sites of former years. In the presence of a female that a male is trying to win, he stands on the remains of an old nest, sings, pulls and jabs at its weathered materials, and crouches in it to make shaping movements with his breast. Sometimes he picks up fragments of material, carries them a short distance, and drops them when near the nest site. He follows the female around and defends her by singing, supplanting other males, and displaying with upward-pointing bill. The long interval of keeping company before breeding begins helps to bind the mates firmly together.

In southwestern Puerto Rico, Post found most nests built low in small mangrove trees recolonizing abandoned *salinas* (shallow enclosures for holding and evaporating seawater in the coastal mangrove zone during the dry season). Groups of four to six pairs nested on the main branches of red mangroves on tiny islets up to nearly a mile (1,460 m) offshore. In pastures beside mangrove swamps, Yellow-shoulders built their nests in large deciduous trees, especially the oxhorn bucida *(Bucida buceras).* On a university campus, massive rachises of royal palms supported the nests, and in plantations of coconut palms they were situated in the axils of the great fronds. Other blackbirds make their nests in hollows in trees, opening upward or sideward. The most surprising nest sites were in crevices or on ledges of the high, vertical cliffs surrounding Mona Island. The diversity of sites matches the diversity of foraging methods; probably no other blackbird is as versatile as the Yellow-shoulder. Nests were found from 8 inches to 65 feet (0.2 to 20 m) above dry ground, mud, or water, and higher on the cliffs. Although no large colonies were found in the declining population of the Yellow-shoulders, they tended to cluster their nests, sometimes placing them only 10 feet (3 m) apart. The average distance between nests in a colony was 52.5 feet (16 m).

Yellow-shouldered Blackbirds also demonstrate their adaptability by the diversity of materials they select for their open cups. At inland sites they prepare foundations of leaves, grasses, cotton, and occasionally paper, plastic bags, and string. The cups are composed of grasses and cotton and lined with fine grasses. Birds breeding on the mangrove cays build bulkier nests, largely of the seaweed *Sargassum,* mixed with such miscellaneous flotsam as pelican feathers, burlap, twine, and bamboo roots. Sometimes the blackbirds renovate old nests of the same or the preceding year.

Yellow-shoulders lay sets of three eggs, rarely two or four. Females start to incubate after laying their second egg. Thirty-seven sessions on the eggs by two females ranged from 2.5 to 99.5 minutes and averaged 23.5 minutes. The recesses of these two females varied from 1 to 36 minutes and averaged 9.6 minutes. One of them incubated with a constancy of 72 percent; the other spent 77 percent of her active day on her eggs. Males were never seen to incubate, but at least some of them fed their incubating partners, one of them eighteen times in twelve hours. They also mobbed predators and guarded the nest while the female was present or absent. At one nest the incubation period was twelve to thirteen days. One female incubated three of her own eggs plus six cowbird eggs, hatched all nine of them, and raised to fledgling two blackbirds and three cowbirds.

Males promptly began to feed hatchlings and continued throughout the nestling period to bring food as often as females did. Together they fed the young at an average rate of 12.7 times per hour, or 5.4 times per hour for each nestling. Because many nests were a mile or more from the nearest communal foraging ground, male participation in provisioning the young was important if not indispensable. In any case, feeding nestlings on offshore cays that provided little or no food was a strenuous activity. In an observation period of two and a half hours, two females made seventeen trips to the mainland, totaling twenty miles (32 km), while their mates flew twenty-eight miles (45 km) on twenty-four trips. Probably because they had to travel so far for their nestlings' food, they increased the amount they could bring by carrying some of it in their throats and regurgitating it, which few other blackbirds have been reported to do. Both parents removed fecal sacs, swallowing them or carrying them to a distance in their bills. Both preened the feathers of their nestlings and were apparently responsible for their freedom from the botflies *(Philornis)* that heavily parasitize the young of other birds in Puerto Rico, as in other parts of tropical America. Ten Yellow-shoulders remained in five nests for 13 to 16 days, for an average nestling period of 14.6 days. After fledging, the young are fed by their fathers for at least 24 days.

Although on offshore cays the difficulty of provisioning nestlings led to the starvation of many, this loss was compensated by reduced parasitism by cowbirds, with the result that the Yellow-shoulders nesting there were more successful in fledging young than were those on the mainland. On the cays, eleven of nineteen nests (58 percent) produced at least one fledgling, while on the neighboring mainland only fourteen of thirty-five nests (40 percent) yielded fledged young. As is generally true, nests in cavities were much more successful than those in the open, fledging about three times as many young per nest. Adults enjoyed an annual survival rate of 82.4 percent, much higher than is usual in small birds in the North Temperate Zone but equaled by a number of tropical birds. Juveniles had an exceptionally high annual survival rate of 65.5 percent.

Of twenty-five marked pairs, the twenty-four of which both partners survived remained intact throughout a breeding season. No cases of polygyny were noticed. Even the failure of a nest, which causes some birds to change mates, did not disrupt the pair bonds; sixteen pairs renested together in the same year. In several instances, partners whose nests had failed moved together to nest in a new locality, which might be nearly a mile from the previous site. Despite the firm pair bond during a breeding season, mates did not, as with many monogamous birds of the tropics, remain closely associated between nesting seasons. Nevertheless, attachment to earlier nest sites, if not to familiar partners, brought them together in the following year. Even two pairs whose last nestings failed in 1974 reunited in 1975.

Yellow-shouldered Blackbirds provide an instructive example of how a species on an island poor in marshes adapts to a situation very different from that of its closest relations on broad continents with no lack of marshes. Similarly, in Cuba the Tawny-shouldered

Blackbird has taken to light woods, farmlands, and other well-drained open country at low altitudes. It builds cup-shaped nests in palms and dicotyledonous trees and roosts with Greater Antillean Grackles and Cuban Blackbirds in towns, but little information on its habits is available.

3

The Jamaican Blackbird

Although the Red-winged Blackbird and most of its relatives in the genus *Agelaius* breed in marshes, two island-dwelling species, the Yellow-shouldered Blackbird in Puerto Rico and the Tawny-shouldered Blackbird in Cuba, build their nests in trees, including coastal mangroves as well as those that grow on well-drained land. The island-dwelling blackbirds' preference for trees is strongest in an inhabitant of Jamaica's montane forests. Closely allied to *Agelaius,* the Jamaican Blackbird has diverged far enough from the marsh nesters to be classified in a different genus, *Neopsar,* of which it is the only species, *nigerrimus.* True to their name, both sexes of the eight- to nine-inch (20 to 23 cm) Jamaican Blackbird are wholly black, with glossy upperparts. Their bills are of moderate length and sharply pointed. Restricted to the Caribbean island

Jamaican Blackbird, *Neopsar nigerrimus*

for which they are named, they rarely descend from their wet mountain forests to lowland woods.

For several seasons, R. Haven Wiley and Alexander Cruz (1980) and their helpers studied Jamaican Blackbirds in the vicinity of Hardwar Gap in the rugged Port Royal Mountains, at about 4,000 feet (1,200 m) above sea level. Here the birds lived in monogamous pairs in a closed forest of small trees on slopes so steep that many leaned downhill. The pairs foraged and nested in territories that measured about 500 to 1,200 feet (150 to 360 m) in greatest diameter and did not overlap. They found most of their food amid the dead leaves that draped around the bases of the bromeliad *Vriesea sintenisii* that grew abundantly in the low canopy of these humid woods, about six to eight to a tree. Clinging to the hanging leaves with body horizontal, less often upright or with head downward, they probed into the bromeliad to extract snails, millipedes, sowbugs, small frogs, and a diversity of insects including grasshoppers, beetles, ants, cockroaches, and the larvae of lepidoptera. Instead of thoroughly searching each bromeliad, they passed rapidly from one to another, spending about three-quarters of their foraging time in them. Among the mosses, ferns, and lichens that covered the boughs, and in the crowns of tree ferns, they found more food. Rarely did they hunt in the leaf litter on the ground or eat a berry. Instead of foraging together, as many paired birds of tropical woodlands do, the members of a pair tended to occupy different parts of their common territory.

Instead of the frequency-modulated songs that carry farthest through dense forests, Jamaican Blackbirds have buzzy songs similar to the those of blackbirds that live in more open places. Lacking songs appropriate for advertising their woodland territories, these blackbirds have developed two displays unusual for forest birds. A male and his mate often roosted as much as fifty to a hundred yards, or meters, apart. At daybreak he became active from fifteen to thirty minutes before she did. Unlike many birds that sing profusely at dawn in the breeding season, he began to fly back and forth with frequent abrupt turns, tracing an erratic course through the center of his territory. For about half an hour, on flights about ten to sixty feet (3 to 20 m) between changes of direction, he flew with rapidly whirring wings, repeating the *dzik* call that both the sexes frequently utter throughout the day. In one minute he made from two and a half to five of these flights. The sound of whirring wings and staccato calls proclaimed the male's presence as daylight waxed beneath the forest canopy. After the female became active and called *dzik* a few times as she flew, one or both birds sang for the first time, and the male's patrolling flights ceased, often abruptly.

Neither the blackbird's movements nor the sounds he made would attract attention at a distance in dense forest. He more effectively and widely advertised his territory by performing a song flight at irregular intervals throughout the day. Flying slowly with exaggerated wing-strokes, he rose above the canopy on a gently inclined course and began to sing as he approached his zenith high above the treetops. The moment he finished singing, he closed his wings and fell headfirst, like a rock, until he neared the foliage below him and spread his wings to retard his wild descent. Frequent among birds of open spaces where trees and shrubs for song posts are few, song flights above the canopy are rare among woodland birds. In mountain forests much taller than those inhabited by Jamaican Blackbirds, Resplendent Quetzals practice them, as do Ovenbirds in temperate-zone woodlands, at night.

In May and June, Wiley and Cruz found seven nests, situated from twenty to thirty-six feet (6 to 11 m) above the ground in small trees that grew on steep slopes and bowed downhill until the main trunk was almost horizontal in the lower canopy. Placed against and partly beneath the trunk, the nests were supported on the sides by twigs sprouting from it or branches of a liana, leaving the bottoms exposed. In form the open cups resembled those of marsh-nesting blackbirds, but they were made of very different materials, mainly small epiphytic orchids and thin dark roots of air plants. Only the female built them. In each nest she laid two spotted eggs and started to incubate after she deposited the second, so that both hatched at about the same time.

Only the female incubated. Twenty-two sessions of one female ranged from 9 to 41 minutes and averaged 26.5 minutes. Her twenty-six recesses ranged from 1.5 to 21 minutes and averaged 12.6 minutes. She covered her eggs with a constancy of 68 percent. Other females' sessions averaged 28.5 minutes; their absences, 16 minutes. While a female incubated, her partner perched for long intervals within about thirty feet (10 m) of the nest, resting inactive or preening. If an Orangequit, Bananaquit, Rufous-throated Solitaire, White-eyed Thrush, or White-chinned Thrush approached too near, he chased them away. When he sang, his mate answered with similar notes from the nest; when she sang first, he responded. At one nest the incubation period was fourteen days; at another, between thirteen and fifteen days.

The male started to feed the nestlings on the first day after they hatched, at the same time that their mother did. Throughout the nestling period at one nest, he delivered 46 percent of 148 meals. Compensating for his slightly fewer contributions, he tended to bring larger items than she did. With rare exceptions, a parent approached the nest with only a single object in its bill, either because food was difficult to find, or because their method of foraging precluded holding one item while procuring another. As the nestlings grew older, the articles delivered to them tended to become larger rather than more frequent. Overall, they were fed about eight times per hour, mainly with orthoptera, larval and adult lepidoptera, and adult beetles, with occasionally a spider, centipede, or small tree frog. Only the female brooded the nestlings, seldom for more than two minutes together after the first day. After the young had left the nest, their parents continued to feed them for at least forty days. For at least another month, the surviving juvenile of this pair remained with its parents.

The Jamaican Blackbird demonstrates changes in behavior that evolve when a bird evidently descended from a marsh-dwelling stock adapts to life in montane forests where its food, thinly scattered throughout the woodland's depth, is less readily gathered than in marshes, where it is concentrated in a thin horizontal expanse exposed to sunshine. Now polygyny, widespread among birds of marshes and grasslands, becomes impracticable because two parents are needed to nourish a brood. Large, exclusive territories are required to supply enough food. The rarity of predators (until brought by humans) in an insular forest permits modes of behavior that would be imprudent on continents where predators abound. Accordingly, the Jamaican Blackbird has become a monogamous holder of a large territory, where male and female take equal part in provisioning the young. The male begins his day with erratic flights near the nest, flights that on predator-infested continents might draw hostile attention to it. He retains, and reveals his ancestry by, a mode of singing more appropriate for birds in crowded colonies in marshes than for forest dwellers, and he compensates for its inadequacy by flying above the vegetation to sing, in a manner much more frequent among birds of open places than among inhabitants of woodlands.

4

The Meadowlarks

Five species of meadowlarks of the genus *Sturnella* inhabit the Western Hemisphere from Canada to Tierra del Fuego and the Falkland Islands. The two northern species have yellow breasts; the three southern species, red breasts. All, as their name implies, inhabit meadows, grassy savannas, dry or moist fields, and similar open habitats, over which they walk, picking up food with long, sharp bills.

The Northern Meadowlarks

Most widely distributed and most studied is the Eastern, or Common, Meadowlark, a short-tailed bird that in length ranges from eight to ten inches (20 to 25 cm). Both sexes have bright yellow underparts marked on the chest with a bold black V that makes

Eastern Meadowlark, *Sturnella magna*, sexes alike

them unmistakable. The head is broadly striped with dark brown or black and buffy white. The upperparts are streaked with blackish brown and buff; the wings and central tail feathers are barred with the same colors. The outer tail feathers are largely white. This color pattern permits the meadowlark to make itself conspicuous or highly cryptic, according to circumstances. When it creeps through low herbage with its breast held low, it is hard to detect. When it stands upright on a perch with breast outward, singing, it catches the eye from afar. When alarmed, it fans out its white tail feathers in a warning signal that is widely visible. Eastern Meadowlarks fly with bursts of rapid wing-beats alternating with short glides on wings held stiffly downward.

From southeastern Canada through eastern and central United States westward to Nebraska, Texas, and Arizona, the Eastern Meadowlark ranges widely through Mexico, Central America, and northern South America to Amazonia. In the West Indies it inhabits Cuba. Tolerant of climatic extremes, it thrives from lowlands up to timberline, reaching an altitude of 11,500 feet (3,500 m) in the Andes of Colombia. Wherever it appears, it prefers open places well covered with grass or other herbaceous vegetation and avoids closed woodland. Although some individuals remain through the winter about as far in the north as they nest, many migrate in autumn to the more southerly parts of the North American breeding range, traveling by day as well as by night. Southern populations are sedentary. Like other icterids of open country, Eastern Meadowlarks are spreading more widely as forest barriers are destroyed. Here in southern Pacific Costa Rica, I first saw one twelve years ago.

About three-quarters of the meadowlark's diet consists of insects and other invertebrates that it gathers on or beneath the ground. Pushing its pointed bill into the soil, it gapes to widen the hole and reveal whatever grub or worm may be hiding there. Similarly, it employs the widespread icterid powered gape to spread apart grass tussocks for lurking insects. Although flycatchers, tanagers, and many other passerines can add item after item to the collection in their bills without dropping any of them, a meadowlark cannot gape while holding something. When gathering food for its young, it lays down the first article that it finds while it probes the ground for another, continuing until it has collected a full load. Then it gathers up the harvest and carries it off in its bill or mouth to the nest. Because they eat so many cutworms, grasshoppers, caterpillars, and beetles injurious to crops, meadowlarks are highly beneficial to farmers.

When insects become dormant or scarce in winter and early spring, northern populations of meadowlarks turn to seeds, including fallen grain in stubble fields and those of various wild plants such as smartweed and ragweed. They scratch to uncover shallowly buried seeds, and pull down stalks to hold beneath a foot while they pick off the seeds. Fruits do not attract meadowlarks greatly, but occasionally they take wild berries and such cultivated fruits as cherries, blackberries, or strawberries. In the southern United States they occasionally attack sprouting maize, digging down to remove the softened grain, leaving the young shoots to wilt, but this injurious habit appears to be exceptional (Bent 1958).

In northern lands, the songs of newly arrived meadowlarks, delivered from the top of a fence post, a service wire, a branch at no great height, or some other exposed station, are welcomed as harbingers of spring. The sweetly plaintive song typically consists of four whistled notes, which to me fit the words "Spring o' the year," although others hear them differently, sometimes as *la-zéeekill-dee.* Colombians have given the meadowlark the onomatopoeic name *chirlobirlo,* which might also be applied to northern races and reminds us that the Eastern Meadowlark's voice, like its appearance, changes little throughout its great range. Occasionally a meadowlark sings in flight, rising swiftly upward and vibrating his wings while he rapidly pours forth a volley of sharp, chattering notes, quite different from those of his usual song. After circling around in the air, he slowly descends to the ground. When suspicious or alarmed, meadowlarks voice an unpleasant, rasping *zzzrt zzzrt* while they rapidly spread and close their tail feathers, sending forth twin flashes of white.

In the spring of northern lands where Eastern Meadowlarks are migratory, males arrive before females and establish in meadows, fields of alfalfa, and similar places, territories that range in area from three to fifteen acres and average about seven acres (roughly three hectares). Although disputes over territorial boundaries are settled mostly by singing and displaying, they may escalate into stubborn fighting, with the contestants clinging together, rolling over the ground, and jabbing each other with sharp beaks.

The appearance of the first female excites fresh rivalry among males. When a female first enters a male's territory, he reacts to her as to an intruding male, from whom he fails to distinguish her. Confronted by the resident male, an invading male would either retreat or resist the attempt to expel him by fighting, but a female in search of a mate with a territory does neither; she passively endures his threats, thereby revealing her sex and perhaps winning acceptance as his partner.

Finally, he courts her, standing before her with body elevated, neck stretched upward, and bill pointing skyward. Spreading his tail to expose the white outer feathers, he flicks it up and down, at the same time waving his wings above his back. The central feature of this animated display is his golden breast, turned squarely toward the female and fluffed out to form a golden shield boldly marked with a black V. In this impressive posture he may spring into the air. The female responds to this allurement by stretching up her body and neck, pointing up her bill, flashing her wings and tail, and chattering or calling *zzzrt*.

Pair formation is followed by nest building, which only the female does. She always chooses a site on the ground, amid meadow grasses or other screening herbage. The nest may rest upon the level surface, but often it is set in a little depression, perhaps made by the foot of a grazing quadruped. If this is not ample, she may enlarge it by loosening the soil with pecks and throwing out particles with her bill, behavior not unexpected in a bird that finds much of her food by probing in the ground. The nest may be so deeply sunken that the tops of the eggs are level with the surrounding surface. After preparing the site, the meadowlark begins to build by arching surrounding grass or other herbs above it, interlacing them or binding them together with lengths of grass that she brings. Then she proceeds to build up the walls and cover the bottom with grasses, which become finer in the lining.

Especially early in the season, the female may start several nests before concentrating on one of them. At this time, when to complete a receptacle for her eggs is not urgent, she may spread building over as many as eighteen days. Later, at the height of the breeding season, three days may suffice to finish a nest. The roughly globular domed structure with a side entrance is six or seven inches (15 or 18 cm) in diameter and encloses a chamber four or five inches (10 or 12.5 cm) wide. The ample doorway measures about three inches wide by four inches high (7.5 by 10 cm). Some nests, probably hurriedly built, have incomplete roofs or are mere open cups; they suffer heavy predation and are less successful than well-roofed structures. Sometimes a covered passageway leads to the entrance. These nests, well concealed amid herbage, would be less often found if incubating or brooding females sat more tightly when a person passes near them.

Often before the female has brought the last few wisps of grass to her nest she starts to lay. In the north, the set consists of three to seven eggs, with five the most frequent number for first broods, four for second broods. At lower latitudes, clutches tend to be smaller; but despite the wide distribution of a not uncommon bird, I have found no significant information on how many eggs Eastern Meadowlarks lay in the tropics. Usually pure white, sometimes pale pinkish or greenish, the eggs are speckled or blotched with shades of brown, rust color, purple, or lavender over the whole surface, but most densely on the thicker end (Roseberry and Klimstra 1970).

A female with a five-egg clutch may start to incubate after laying the third, fourth, or fifth egg, but most often she begins when she has four. A female watched for twelve hours and forty minutes by George B. Saunders (in Bent 1958) spent nine hours and forty minutes, or 76 percent of the time, on her eggs, not continuously but with brief recesses for foraging. Apparently, her mate did not feed her. While he sang, she "hummed," and when she heard his flight song, she voiced low, sweet, chuckling notes, unlike any uttered by meadowlarks at other times. These subdued notes were audible only by one standing close to a nest. A rather restless sitter, the female preens, catches insects passing close to her doorway, rearranges the materials of her nest, and frequently turns her eggs. The incubation period is usually fourteen days, exceptionally thirteen or fifteen.

Hatchling meadowlarks have fairly abundant pearl gray down on orange-red skins. Their mother promptly carries the empty shells to a distance but permits unhatched eggs to remain in the nest. Only the mother broods the nestlings, and she feeds them much more frequently than their father does. Although she brings insects in her bill and delivers them directly to her young, if by restless behavior one reveals its hunger while she broods it, she rises up and regurgitates to it.

The nestlings' eyelids start to separate on the third or fourth day of their lives, and by the fifth day their eyes are fully open. By the eighth day they are alert and attentive to sounds from beyond their nest. By frequent preening they help their sprouting plumage to escape from its horny sheaths. They stretch their legs and necks and beat their wings so vigorously that they sometimes wreck and unroof their nursery, leaving themselves exposed to sunshine, which makes them pant violently to reduce their temperature. Sometimes they crawl spontaneously beyond their doorway, to return after receiving a meal. If disturbed, they may abandon their nest prematurely by the eighth day, but if unmolested they remain attached to the nest until they are eleven or twelve days of age, when they can fly in case of need. Their parents continue to feed them for two weeks or more after they leave the nest, or until they are about four weeks old. Their reiterated *tseup tseup* guides the attendant parents to them. Their mother may start a second nest only two or three days after her first brood abandons its nest. While building and laying she continues to feed these fledglings, but after she starts to incubate their father becomes their chief attendant. After they have become self-supporting, he apparently chases them from his territory.

Near Ithaca, New York, in the summer of 1931, I found two female meadowlarks simultaneously feeding nestlings in nests only thirty-five feet (11 m) apart. I noticed not the slightest antagonism between them. The male who was apparently the bigamous father of these broods of four and five nestlings perched on service wires nearby, whistling at intervals his melodious *la-zéeekill-dee* and occasionally pursuing one of the nestlings' mothers. I did not see him take food to either of the nests. In contrast to Red-winged Blackbirds nesting nearby in the same field of alfalfa, these parent meadowlarks showed little concern when I approached their nests. When they saw me coming, they flew off with a warning *zzzrt* and remained at a distance as long as I was in view. But when a mother heard the squawks of a feathered nestling that I lifted up to examine, she became more excited and flew so directly toward me that she almost struck me on the back. Then she fluttered to the ground and tried to lure me away with a broken-wing act that I could not see clearly because she was so well hidden in the alfalfa. Others have watched meadowlarks give distraction displays, which are widespread among ground-nesting birds.

In the same population of meadowlarks where I found two females nesting close together, Saunders, in a much more prolonged and thorough study, learned that about half the males were bigamous, and one had three simultaneously nesting consorts.

One of the factors favoring polygyny in this species is the absence of hostility among females of the same male, who forage and accompany him together.

Saunders followed the development of four meadowlarks that he took from the nest when eight days old and reared, giving them long hours of freedom in the fields. When fifteen days old, all dust-bathed, fluffing out and shaking their plumage in the manner of adults, after which they drank freely. On the following day they started to probe the soil with their bills, as though searching for hidden insects. At the age of seventeen days, they began to hold their bodies erect on upstretched legs whenever they heard a startling sound. A twenty-day-old fledgling bathed thoroughly in water, then spent minutes arranging his plumage. When twenty-two days old, two of the four meadowlarks started to feed themselves; up to this time, Saunders had been giving each young bird about 175 grasshopper nymphs daily.

In addition to the predators and insect parasites that plague many other nesting birds, meadowlarks suffer appalling mortality from the mowing machines that slash through their nests while harvesting the clover, alfalfa, or timothy that conceals them so well. However, by rearing two broods each year, parents who escape this unintentional slaughter produce enough flying young to keep meadowlarks abundant. As autumn approaches, the survivors gather in flocks, which may continue to increase until they contain several hundred individuals, foraging together in weedy fields and the stubble of harvested crops or in salt marshes along the Atlantic coast, where many sleep on winter nights. Rarely, meadowlarks roost in trees with Common Grackles. Some remain through the coldest nights as far north as the coast of Maine. Severe snowy weather may drive meadowlarks that fail to migrate far southward into barnyards, to seek relief from hunger amid domestic animals, or even into the streets and vacant lots of towns, where they forage with House Sparrows and European Starlings. In spring these large aggregations break up as the meadowlarks disperse to find territories and begin another annual cycle.

In southwestern Canada and the western and central United

States, the Eastern Meadowlark is replaced by the Western, which ranges southward to the highlands of central Mexico. Not highly migratory, it remains through the winter in all except the northern fringe of its breeding range, while some individuals wander eastward as far as Alabama, northwestern Florida, and western Kentucky.

Although the two sibling species differ slightly in bodily proportions and markings—the black V on the Western's yellow breast tends to be narrower—they are so similar in appearance that in the field they are best distinguished by their voices. The Western's song, lower in pitch, richer in tone, sweeter, and more musical, has won the highest encomiums from all who have written their impressions of it. A more subtle difference between these two species is the Western's preference, over most of its range, for regions drier than those inhabited by the Eastern.

Like the Eastern Meadowlark, the Western is often polygamous. The nests of the two are similar, but those of the Western often appear to be set farther into the ground, occasionally in a hollow as deep as eight inches (20 cm). The covered runway leading to the nest's doorway may be two to five feet (0.6 to 1.5 m) long. The eggs, three to seven in number but usually five, closely resemble those of the eastern species.

These two meadowlarks almost certainly evolved in populations of the same ancestral species that later became isolated from each other, but today they meet in a long, broad zone that stretches from the Mexican state of Jalisco to the Great Lakes in the north. Ornithologists have long wondered how such similar species remain distinct. Although hybrids are rare in the wild, Wesley E. Lanyon demonstrated that individuals of the two species, held together in the same enclosure and given no choice of mates, would unite as a breeding pair. Whether the male was an Eastern and the female a Western, or vice versa, the eggs of these mixed matings were no less fertile than those from normal pairings. However, when the hybrids that hatched from them were mated with a male or a female of either of the parent species, the resulting eggs were almost invariably sterile. This study demonstrates that interbreeding in the zone of contact is not likely to blunt the distinctness of the two species. But it does not tell us how the meadowlarks avoid mismatings that lead to sterility in the hybrid offspring.

Male meadowlarks treat males of the other species much as they treat males of their own species. Most territorial passerine birds repel from their domains all individuals of their own species except mates and young, but they do not exclude members of other species, so that the territories of several species may broadly overlap or be more or less conterminous, as when a tanager and a finch share a territory. The failure of male meadowlarks to distinguish between their own and the related species, at least in their territorial behavior, leads to interspecific territoriality, which is much rarer among birds than intraspecific territoriality. The territories of these two species might form a mosaic, like those of either species alone, but this arrangement would not prevent hybridization.

Evidently not the males but the females keep their species pure. They do this by choosing mates of their own kind, mainly by paying attention to call notes: the *chupp* of the Western is quite different from the *zzzrt* of the Eastern. These calls are either innate or learned by nestlings. The birds' primary song is learned later, by listening to the renditions of older individuals. Juvenile males with no opportunity to hear adults of their own species imitate the songs of other species with whom they are associated during their first summer and autumn. A captive Eastern Meadowlark learned to repeat the songs of both his own and the western species. Lanyon concluded that the two sibling meadowlarks are kept distinct in the region of overlap primarily by behavior, namely, the females' discriminating choice of mates. Even if they make frequent mistakes, specific distinction will not be lost because hybrids are seldom fertile (Lanyon 1956a, 1956b, 1962, 1966, 1979).

The Southern Meadowlarks

Of the three currently recognized species of Southern Hemisphere meadowlarks, the largest is the ten-inch (25 cm) Long-tailed. The male's upperparts, wings, and tail are black, variegated by grayish brown feather edges. A narrow streak above each eye is red in front of the orbit and white behind it. His chin, throat, foreneck, breast, and abdomen are vivid red, as is the bend of each wing. His underwing coverts are white. The paler female has a white eyebrow and throat. Her chest is brownish and black, changing to pale reddish on her posterior underparts. The Long-tailed Meadowlark occupies the narrowing tip of South America from southeastern Buenos Aires province in Argentina to Tierra del Fuego, with a northward extension to Atacama in Chile. A different race inhabits northwestern Argentina, and a third is established on the Falkland Islands. From the coast of Chile and adjacent islands the Long-tailed Meadowlark ranges upward to about 8,200 feet (2,500 m) in the Andes. At the approach of winter, southern populations migrate as far as northern Argentina (Goodall, Johnson, and Philippi 1957).

In Chile, where it is a widespread and well-known permanent resident, the Long-tailed Meadowlark prefers low, moist ground; and across the Andes in Patagonia, its presence in valleys is said to indicate the proximity of water. Although a strong flier, it walks much over meadows and cultivated fields, gathering a variety of fruits and seeds, mature and larval insects, and small crustaceans. In winter it eats many bulbs of certain wild plants that it digs up with its long, sharp bill. Its song, often delivered in flight, is vigorous and cheerful but of little variety. One can often hear it on winter's coldest days. The Long-tailed Meadowlark's nest, a bulky structure of dry grasses, well bound together and lined with finer pieces, is built upon the ground, usually well hidden amid tall grasses, and sometimes domed. It is exceedingly difficult to find because the female never flies directly to it but approaches by creeping through the surrounding grass in a crouching posture. She leaves in the same cautious manner. In Chile the clutch usually consists of three eggs, occasionally four, but in Argentina, Hudson (1920), whose egg numbers sometimes seem too high, reported five. Among the most beautiful eggs of Chilean birds, they are pale buff or very pale mouse gray, profusely sprinkled with spots, blotches, and scrawls of reddish brown, ocherous or dusky. Incubation, by the female only, lasts about fourteen days, and the young, fed by both parents, remain in the nest about eight days. Males are sometimes polygynous, like the Eastern Meadowlark.

From southwestern Ecuador through coastal Peru to northern Chile resides the Peruvian Red-breasted Meadowlark, which differs from the preceding species in being smaller (8 inches, 20 cm) and more deeply colored, with a darker back and more intensely red breast. On this arid coast it frequents the moister ground along streams, irrigated pastures, fields of cotton, and other plantations. Its nests, otherwise similar to those of the Long-tailed in site and construction, are usually half roofed over. It lays three eggs, distinguishable from those of the Long-tailed only by their smaller size.

The Lesser Red-breasted Meadowlark resembles the preceding species but is distinguished by its black underwing coverts and thighs. From east-central Argentina it ranges northward through Uruguay to the state of Paraná in southern Brazil. In the nineteenth century, Hudson found these birds abundant on moist, grassy pampas, where they were tame and flew reluctantly; when approached they crouched down, hiding their red breasts, and remained motionless to escape detection. He never saw one perching in a tree, but they frequently alighted on the roof of a rural dwelling or on some other elevation that offered a broad support. They fly up a few yards in the air when they sing. They are pacific birds, and so sociable that when one becomes separated from its fellows it often seeks the company of birds as different as plovers or flycatchers. He remarked how on the great monotonous plains, where most of the small birds are gray or brown and in winter no flowers gladden the eye, it was delightful to meet a flock of these meadowlarks, whose crimson bosoms glowed with a strange splendor on the somber green of the earth, a sight exhilarating to the mind. He called these meadowlarks "military starlings" because of the way they migrated slowly northward at the approach of winter, not flying but walking over the ground. A flock of four or five hundred to a thousand or more individuals would spread out to present a very broad front, and at intervals the hindmost would fly forward to alight just ahead of the foremost rank. The broad front, the precision of their movements, and the scarlet breasts all turned the same way suggested to Hudson a disciplined army on the march.

Lesser Red-breasted Meadowlarks nest in loose colonies and are polygynous. In a nest of grasses and rootlets hidden amid pasture grass, the female lays four white eggs marked on the thick end with spots and fine streaks of dark chestnut (De la Peña 1979). Shiny Cowbirds rarely parasitize these nests, probably because they are usually distant from elevated perches from which the parasites can watch revealing movements of their builders (Gochfeld 1975; Short 1968).

5

The Yellow-headed Blackbird

The handsome Yellow-headed Blackbird is the only species in the genus *Xanthocephalus.* The ten-inch (25 cm) adult male is largely black, with a bright yellow or orange head, neck, and chest. A small patch of black extends from around each eye to the base of the bill. A white patch on the coverts of each wing becomes conspicuous when they are spread. The eight-inch (20 cm) female in summer is mainly dusky grayish brown or sooty, with dull whitish or yellowish cheeks, chin, and throat. Her chest is light yellow; her lower breast broadly streaked with white and grayish brown. In winter her chest becomes deeper yellow, the white streaks on her breast less distinct. Both sexes have brown eyes and black bills and feet. Yellow-headed Blackbirds breed in western and central North America, from west-central Canada south to northeastern Baja California and northern Texas, and from central Washington and

Yellow-headed Blackbird, *Xanthocephalus xanthocephalus,* male

Oregon east to southern Ontario, northwestern Ohio, and northwestern Arkansas. They winter from central California and central Texas south through Mexico to the Isthmus of Tehuantepec.

Arriving from the south in March or April, male Yellow-heads settle amid cattails, tules, or bulrushes growing in fairly deep standing water in marshes or at the margins of open ponds and lakes. Here they claim territories and wait for females, who tend to winter farther south than the males, have a longer journey, and arrive a week or two later. Yellow-heads forage widely, not only on their watery territories but beyond them on exposed mud in higher parts of the marsh and farther afield on meadows and cultivated land. They eat many kinds of insects including dragonflies, damselflies, caterpillars, beetles, grasshoppers, crickets, and weevils. They catch many in the air. With other blackbirds, they follow the farmer's plow, picking up the grubs and worms that it exposes. Although the animal component of the Yellow-heads' diet is varied, it is exceeded in bulk by seeds of wild plants and cultivated grains, of which oats appears to be the favorite and maize the second choice. When foraging they walk sedately, seldom hopping (Orians 1980, 1985).

Yellow-headed Blackbirds are not the most melodious of feathered creatures. Few naturalists have written a good word for their utterances; some have roundly disparaged them. Among the former was Arthur Cleveland Bent (1958) himself, who, seated with two congenial companions in a comfortable buckboard drawn by a lively pair of unshod broncos, had driven for many miles across the vast rolling prairies of North Dakota on a bright warm day in June at the beginning of this century. At last the travelers reached a depression ten feet (3 m) below the level of the prairie, from the edge of which they surveyed a great slough, alive with ducks of many kinds, coots, gulls, terns, shorebirds, and great flocks of blackbirds. By far the most abundant birds were Yellow-heads, the most characteristic inhabitants of every North Dakota marsh, who swarmed everywhere, making a constant din. The song Bent most often heard suggested the syllables *oka weé wee,* the first a guttural croak, the last two notes loud, clear whistles in descending pitch. Fifty years later, the rhythmic swing of the many-voiced chorus seemed to ring in the ears of the indefatigable biographer of North American birds whenever he thought of a North Dakota slough and its Yellow-headed Blackbirds. Probably Bent was so favorably impressed by the Yellow-heads' notes because of the pleasant circumstances in which he first heard them, for others have found them disagreeable.

As previously noted, icterids of several genera assume a special posture when they sing, making a song-spread. Although Yellow-heads' songs have many variations, diversely heard by different human ears, two main types, each delivered with a distinctive pose, have been recognized. The buzzing or growling song, which differs little from male to male, begins with several short, slightly descending, rather low-pitched melodious notes, followed by a loud, harsh, prolonged, wavering buzz or wail, suddenly increasing in volume that is sustained to the end. The bird utters the introductory notes of this performance with his bill closed. While delivering this song, he contorts his body in the strangest manner.

As it begins, he turns his head to the left until his bill points at a right angle to the axis of his body. When he comes to the buzz or wail, he decreases the angle that his bill makes with his body and holds it in this position until the song ends. His neck is stretched out, his bill pointed upward, his tail spread widely, and his wings slightly opened; he appears to sing over his left wing. His body vibrates slightly while he seems to eject his notes with a painful effort (Nero 1963).

The second type of song, called the accenting, differs greatly from the buzzing song, as does the accompanying display. It consists of clearly defined, slightly separated notes that are harsh and unmelodious but more liquid than the buzzing song. The pose that accompanies the accenting song is nearly or quite symmetrical rather than highly asymmetrical. The bird spreads his wings, fully exposing the white patches; he fans out and slightly depresses his tail; and he directs his head forward and upward at an angle of thirty to forty-five degrees. As he utters the final note, he may snap his head backward until the bill points to the sky. Yellow-heads commonly use both of these displays in boundary disputes, and often when no exciting cause is apparent. Female Yellow-heads sing less often than males, with similar but less pronounced postures. Usually they direct their song-spreads toward other females flying overhead or invading their territories while they are building their nests or starting to incubate. After displaying, they may attack the intruder, assisted by their mates (Orians and Christman 1968).

Among the many utterances of Yellow-headed Blackbirds is the hawk alarm call. This harsh, raucous rattle is sometimes sounded by a perched male, but usually he flies up to chase the raptor, most often a Northern Harrier, also known as Marsh Hawk, the diurnal aerial predator that attacks Yellow-heads most frequently. On hearing the alarm cry of one male, others repeat it and join him in mobbing the enemy.

Male Yellow-headed Blackbirds do not acquire their adult plumage and begin to breed until they are about two years old; as yearlings they mostly remain aloof from the nesting areas. Females breed when a year old. Yellow-heads breed colonially, or perhaps more properly, in grouped territories, apparently always in marshes inundated with water 2 to 4 feet (0.6 to 1.2 m) deep, and rarely much deeper, which protects the nests from some terrestrial predators. So important is the water to the birds that if it becomes too shallow or dries up while they are building, they abandon their sites and seek better ones. In productive marshes, their territories average about 2,300 square feet (215 m^2), but in less favorable habitats they may be much larger. Densities as high as twenty-five or thirty nests in an area of 225 square feet (21 m^2) have been reported. A territory may contain up to eight consorts of a polygynous male, but seldom more than four or five. Some males have only one; others acquire territories without attracting a female. A colony may contain from a few dozen to hundreds of nesting females.

A male Yellow-head who has attracted a female to his territory shows her a nest site. He starts a nest-site demonstration, after the song-spread his most conspicuous display, by elevating his wings high above his back in the form of a V. Then, flying at stalling speed, he leads the female down into the marsh vegetation where, still with elevated wings, he bows and crawls through the close-set stems until he reaches a promising nest site. He pecks at it, or picks up a fragment of vegetation and goes through the motions of building. When another female joins him, he repeats the sequence to her. His consorts do not always build their nests in the sites he has pointed out to them.

Without help, a female constructs an open bowl, attaching it to upright stems or leaves of cattails *(Typha)*, reeds *(Scirpus)*, other tall aquatic plants, or shrubby willows *(Salix)* on land flooded by swollen streams. Most nests are situated from one to three feet (30 to 90 cm) above the water. The builder gathers from its surface dead leaves, rootlets, and similar wet and pliable materials and wraps them around four or five supporting stems, to which, when they dry and shrink in wind and sunshine, they cling firmly, preventing the nest from slipping down. She stretches the earliest strands across the space between the supports, forming an open network, which she proceeds to fill by bringing more pieces and winding them around more stalks, until as many as twenty or thirty of these uprights may be bound securely into the cup's walls. When the shell has become thick enough, the builder simply drops her contributions into it, then enters and compacts the sodden materials by stamping her feet, alternately and rapidly, on the bottom and sides, with sounds audible several yards away.

After the nest dries, the female Yellow-head lines it with narrow strips of grass blades or broad dry reed leaves. Some nests are garnished around the rim with plumelike inflorescences of reeds. A skillful, industrious female Yellow-head can complete in two to four days a beautifully woven, smoothly finished, sturdy, basketlike nest 5 or 6 inches wide and 4 inches high, with an inside diameter of about 3 inches and depth of 2½ inches (12.5 or 15 by 10 cm and 7.5 by 6.3 cm). In one colony nearly half the nests started were abandoned unfinished. One builder, probably a novice at such work, made four defective nests before completing one that would hold her eggs. Unfortunately, the unequal basal growth of the monocotyledonous stems that support many well-made nests soon tilts them so far that the eggs roll out.

Newly completed nests remain empty for one to five days before the female starts to lay, on consecutive days, her set of usually four eggs, frequently three, rarely five. The shell's grayish white or pale greenish white ground is everywhere profusely blotched and speckled with shades of brown, cinnamon-rufous, and pearl gray. Incubation often begins with the laying of the second egg. Only the female covers the eggs, but her consort sometimes feeds her. During eighty-three hours of observation by Reed W. Fautin (1941b) in Utah, sessions on the nest ranged from 1 to 41 minutes and averaged 9.1 minutes. They were longest during the middle of the day, when the bird shielded her eggs from intense sunshine beating down through upright stems and leaves. Recesses ranged from 1 to 18 minutes, averaged 5.4 minutes, and tended to be longest in the early morning and evening, when Yellow-heads foraged most actively. Individual females were on their nests for 53.1 to 69.0 percent of the observation periods; their constancy averaged 63.9 percent.

After an incubation period of twelve or, less often, thirteen days, the nestlings hatch thinly covered on head and back with buffy down. They are fed mainly by their mothers, during their first two days with items that are too tiny to be detected in the parents' bills and are apparently regurgitated. After this interval, their parents bring food visible in the bill, at first spiders and small insects. Older nestlings receive damselflies, dragonflies, grasshoppers, and other fairly large insects. Most males who are not too busy attracting additional females to their harems bring some food to their offspring, usually giving most attention to the young of their first, or primary, consorts.

However, if the number of nestlings in the primary nests drops to two, which the mother can adequately nourish alone, a father may shift his care to a secondary nest with a larger brood. By experimentally reducing the size of secondary broods, males may be led to bring food to still later ones. Although male Yellowheads feed their nestlings less than females do, they guard them more zealously, sometimes forcefully striking the head of a person banding their young. The alarm call of a parent whose nest is menaced by a predatory bird draws other colony members to help repel the enemy.

When, at the age of nine to twelve days, the young abandon their nests, they are still unable to fly but adept at scrambling through the marsh plants and hopping from stalk to stalk. They do not return to their nests but remain low, sometimes resting on floating dead vegetation. After four or five days they can fly a few feet. When three weeks old, their flight range has increased to about twenty-five yards (23 m), and soon they pursue their parents with noisy pleas for food (Fautin 1941a).

Despite Yellow-headed Blackbirds' concerted efforts to drive their enemies from the marshes where they nest, and the protection from certain predators afforded by standing water, their reproductive success is much lower than that of many other north-temperate birds. Their broods are destroyed not only by a diversity of predators but also by rainstorms and gales that sweep over the open marshes, shaking eggs or nestlings from nests supported by slender stalks that bow before the wind. Rising water sometimes inundates the nests. In Fautin's colonies 443 eggs yielded 314 nestlings, of which 215 were lost from their nests. The 99 that survived until they could scramble out represented only 22.4 percent of the eggs laid. A colony studied by George S. Ammann (in Bent 1958) in Iowa also suffered heavy losses. There, 504 nests contained 1,565 eggs, or an average of 3.1 eggs per nest. Of these eggs 53.6 percent hatched, and 27.5 percent yielded young who lived to hop from their nests. Small colonies are sometimes totally destroyed.

Not only do Yellow-heads mob or drive away large predatory birds that fly over their marshes, they also exclude the much smaller Marsh Wrens. Strangely, interspecific territoriality, which most frequently arises between closely related or quite similar birds, like the two species of yellow-breasted meadowlarks, has developed between birds as dissimilar as Yellow-headed Blackbirds and Marsh Wrens. Although these wrens might compete with the blackbirds for certain kinds of food, such competition rarely causes interspecific territoriality; many birds whose diets coincide broadly occupy overlapping territories. Apparently, Yellow-heads drive away Marsh Wrens because the latter puncture and destroy eggs and kill nestlings of marsh-nesting passerines. In a marsh in Manitoba, Marty L. Leonard and Jaroslav Picman (1986) found Yellow-heads nesting above water in the center and wrens around the drier margins. Yellow-heads arrived in the marshes earlier than the wrens and departed sooner. After they left, the wrens expanded their territories into areas abandoned by the larger birds.

Yellow-heads are not known to raise second broods, and they seldom lay again after losing a first brood. In July, when the latest generation is on the wing and adults and young molt, they abandon the nesting area and gather in large companies amid the densest stands of cattails and bulrushes in the marshes, often with adult males and females in different parts and young birds with their mothers. Here they remain hidden through much of the day, emerging to forage in the early morning and in the evening. In August, when they have about finished molting, Yellow-heads become more active and roam widely over the western plains, in pastures and cultivated fields, along roadsides, in barnyards, and even on the streets of towns. Often these wandering hordes are composed of old and young of several species of blackbirds and grackles; at other times the handsome adult male Yellow-heads travel by themselves, while females and immatures forage in larger flocks. As night approaches, all seek the marshes, flying in from all sides in successive waves, with much protesting chatter and whirring of wings as each later company dislodges from their perches birds already settled. This uproar continues until the latest arrivals have dropped down out of sight into the broad green expanse of foliage stirred only by wandering breezes. The Yellow-heads' southward migration may begin as early as July. By September the marshes where multitudes nested are no longer embellished by bright yellow heads.

Oriole-Blackbird, *Gymnomystax mexicanus,* sexes alike

They forage mainly on the ground but perch on the topmost boughs of tall trees, where they might escape notice if they did not call attention to themselves by their scratchy notes. After resting there a while, they suddenly take wing and continue to fly high in the air until they vanish in the distance. Like other blackbirds and grackles, they roost in large companies, sometimes in swampy woods.

Foraging Oriole-Blackbirds walk over open ground with alternately advancing feet and hop over obstructions. In the wet season, when the herbage is fresh and lush, they vanish beneath it and emerge with their golden plumage wet and bedraggled. They eat winged insects, caterpillars, earthworms, small frogs, and such other diminutive creatures as they find on the ground or amid the grass. They vary their diet with fruits. One June, when in the hills south of Valencia in Venezuela abundant tall *Cordia* trees ripened multitudes of small, yellow, sweetish astringent berries,

6

The Oriole-Blackbird

A unique bird, the only species in the genus *Gymnomystax,* the handsome black-and-gold Oriole-Blackbird looks like a large oriole but behaves like a grackle. Nearly a foot (30 cm) long, both sexes are bright yellow on the head, neck, shoulders, and all underparts and deep black on the back, rump, tail, and wings, which have a patch of yellow on the lesser coverts. Their brown eyes are set amid black bare skin. The species ranges from eastern Colombia, Venezuela, Guyana, and French Guiana to Ecuador, eastern Peru, and northern Brazil, and in Venezuela from low altitudes up to about 3,300 feet (1,000 m) in the vicinity of water. Avoiding closed forests, Oriole-Blackbirds frequent gardens, pastures, cultivated lands, savannas, marshes, and light woodland, especially along streams.

Oriole-Blackbirds fly in pairs or loose flocks with a broad front.

I saw Oriole-Blackbirds gather many of them. Perching beside a cluster of these fruits, the birds plucked them one by one and pressed the pulp from the skin, which they dropped. The name *maicero,* or *tordo maicero,* given to them in Venezuela, refers to the habit, widespread among blackbirds, of tearing open ears of maize to reach soft, milky, maturing grains. Since I left Venezuela before the maize formed ears, I did not witness this.

One has only to hear the screechy, scratchy notes of the Oriole-Blackbird to suspect that it is more closely related to the dusky grackles than to the melodious orioles, despite its resemblance to the latter. Its long-drawn, nasal screech, ascending sharply toward the end, sounds much like the squeaking of a gate with rusty hinges and seems unworthy of a bird so splendidly attired. A frequent utterance, the nearest to a song of all that I heard, sounds like *chaa chaa chrick chaaa,* with the *chrick* clearer in tone than the nasal *chaa* and sharply rising in pitch. This sequence of notes is usually delivered from a treetop. When taking wing, the blackbird often calls *chrick chaa.* While building, a female frequently voiced a single sharp *cluck* or a longer *tuc titit,* as she flew down to the ground for more material. Parents fearful for the safety of their nestlings repeated a sharp *chip* and harsh, rasping notes as they circled around a climber. I (Skutch 1967) noticed little difference in the voices of the two sexes, except that the male's songs tended to be longer than those of his mate.

A male displaying to another Oriole-Blackbird, apparently his mate, who perched near him in a tree, raised his head, puffed out the plumage of his nape and back, spread his tail fanwise, and slightly opened and drooped his wings while he emitted the usual screech. Sometimes, while they call, Oriole-Blackbirds point their bills almost straight upward in a posture often assumed by grackles and related birds.

In April I often saw trios of Oriole-Blackbirds resting high in trees beside or in pastures, but as May advanced pairs became more frequent. Again and again I watched hopefully for them to build; but after an interval of motionless perching they abruptly left, flying so far over the barren hills that I did not attempt to follow. Finally, in mid-morning of June 5, after the rains had started,

I at last found a bird building in a yagua palm standing in a wide pasture with scattered trees. The site of the nest was twenty-three feet (7 m) up, in the dense crown of the massive palm tree, between the bases of the younger feathery fronds, where it was almost invisible from the ground.

Next morning I watched the birds build. While the male *maicero* perched conspicuously high in trees, repeating at intervals his nasal screech, his mate walked over the pasture, where the new grass was sprouting, gathering pieces of dead herbage for her nest. Often she picked up, tested, and dropped bits before she found satisfactory material. Then, her black bill laden, she flew up to a frond of the palm and promptly vanished into the heart of the dense crown, where I could not see what she did. Usually she stayed out of sight for a good while before she emerged and flew down, often gliding on set wings, to gather more material. From time to time her partner descended from his high lookout to pick up pieces of vegetation from the pasture and carry them to the nest. He did this both while the female was present there and while she was on the ground, when he probably placed in the nest what he took to it. Once, however, he emerged from amid the palm fronds still bearing his material. Soon after nine o'clock the pair flew away and remained absent so long that I left. In little more than two hours, the female had taken thirty billfuls of material to the nest, the male only five.

The next morning the female was gathering what appeared to be cow dung mixed with fragments of plants. In a quarter of an hour she carried five billfuls to the nest. Once the male descended from a treetop and tugged at wiry stems that he could not tear loose. Then, apparently having found an ants' nest, he went through the movements of anting—the only time that I ever saw a bird engage in this strange behavior on the ground in the tropics, where it is nearly always done in trees or shrubs. After eating something, he gathered a billful of dead grass blades and carried them up to the palm tree, only to fly away still holding them. Again he took his billful to the crown of the palm but failed to deposit it in the nest. Instead, he bore it to a neighboring *Cordia* tree, where he and his mate went at intervals for berries. In the middle of another morning the female gathered a generous billful of fine material from the ground to line her nest, then started to pant in the bright sunshine, dropped her carefully selected load, and went to perch in the shade; she seemed most intolerant of heat.

Building continued at a leisurely pace for at least ten days. The female did most of the work, with her partner making occasional gestures of helping. The completed nest was a broad, shallow bowl constructed of materials loosely arranged, without weaving or interlacing. The foundation was of long coarse straws mixed with smaller fragments of plants. Black, curving rachises, up to six inches (15 cm) long, from the compound leaves of an acacia-like tree that grew nearby, composed the bulk of the nest. Although one morning I thought I saw the female gather mud or cow dung, I found no such material in the nest after the young had gone; if only a little had been brought, it might have been washed away by the heavy rains.

A nest found by George K. Cherrie (1916) on May 6, 1907, at Caicara, Venezuela, rested about twenty feet (6 m) above the ground, amid the thickly tangled branches of a parasitic plant in the top of a chaparro tree. The somewhat thick-walled open cup was made of weed and grass stems, loosely but neatly woven together, and lined with moderately coarse rootlets. This nest measured 6¾ inches in diameter by 4¼ inches high; the cavity, 3¼ inches in diameter by 2⅛ inches deep (17 by 11 cm and 8.5 by 5.5 cm). Neither Cherrie nor I noticed other nests near the single one that each of us found; apparently, Oriole-Blackbirds do not breed in colonies. Each of these nests contained three pale blue eggs, marked with spots, blotches, and speckles of brown or black and pale lilac, chiefly on the thick end. Cherrie's nest also held a single egg of the Shiny Cowbird.

Late in the morning of June 16, I found the female on the nest in the palm tree for the first time, suggesting that incubation had begun. I watched this nest all of one afternoon and all of the following morning. Only the female incubated. On June 21 she retired for the night as a shower began at 6:34 in the evening, and she did not leave her eggs until 6:47 the next morning, so that her nocturnal session lasted 12 hours and 13 minutes. On the two days

I timed ten sessions on the nest, ranging from 17 to 113 minutes in length and averaging 43.3 minutes. Ten recesses, or absences from the eggs, varied from 8 to 35 minutes and averaged 23 minutes. She spent 65 percent of her active day incubating. During her longest session in the late afternoon, a rainstorm with much wind continued for three-quarters of an hour. This session was followed by her shortest absence, returning from which the female settled on her nest for the night.

Watching the only Oriole-Blackbird's nest that I could find was frustrating, for except occasional glimpses through the palm fronds of her golden head I could see nothing of the incubating female. Sometimes she left the nest in response to her partner's calls, or joined him to forage in the *Cordia* tree or on the ground, after which he accompanied her back to the palm. She had no fixed route for returning to her nest; sometimes she alighted far out on a palm frond and walked to it; at other times she flew right into the center of the crown and promptly returned to her eggs. She rarely used the same approach twice in succession. Her partner was most attentive. At intervals he visited the nest, while she was present or absent, but I could not see just what he did there. He carried no visible food and apparently did not feed her. Once, peering up between his legs, I glimpsed movements that might have been preening, or gently pecking, her head but, exasperatingly, I could not see enough to be sure what he did.

The nestlings hatched between July 3 and 5. Their yellowish flesh-color skins bore long but sparse gray natal down of the usual passerine type. The interior of their mouths was purplish red, and at the corners were prominent white flanges, about one-twelfth of an inch (two millimeters) broad. An athletic boy climbed a long pole to the crown of the palm tree, and with a cord lowered a nestling in my binocular case. While he was at the nest, the two parents circled so close above him, repeating a sharp *chip* and angry, rasping sounds, that he became alarmed and shook the palm fronds to keep them away. I could not persuade him to remain still, as he was in no great peril from birds so small, and see what the *maiceros* would do.

When the nestlings were nine or ten days old, I spent the morning watching the parents attend them. At 6:13 their mother, who had brooded through the night, left the palm and called with sharp *chips* from a neighboring tree. Two minutes later, her mate joined her in the dim light of dawn. In the first five hours of the day, the nestlings received thirty-five meals. Both parents fed them, sometimes arriving together but more often singly. After delivering a meal, however, one would frequently wait in a nearby tree until the other had fed the nestlings, then they would fly off together. From the first, food was brought visibly in the parents' bills rather than regurgitated. Most items were unrecognizable, but I distinguished three small frogs, two spiders, caterpillars, earthworms, and a black cricket. Fairly large articles were brought singly, but several small caterpillars were often carried together. The parents removed the nestlings' droppings in their bills and released them where they fell free of the palm tree's crown.

This nest was situated on the flyway between the Oriole-Blackbirds' undiscovered roost to the south and the foraging or breeding areas of a number of them to the north. Morning and evening they passed over the pasture, sometimes, chiefly in the morning, interrupting their journey to rest briefly in trees near the palm where the nest was hidden. Usually the resident pair ignored these transients; occasionally they appeared to be mildly annoyed, but not enough to chase them. They did not defend a territory. While building, incubating, and feeding nestlings, the parents often flew afar, as though they knew no territorial boundaries.

By July 18, the nestlings had vanished. At most two weeks old, they seemed too young to have flown far. From the hilltop where the nest was situated, I looked long and eagerly over the wide expanse of surrounding pasture for some sign of them or their parents, but in vain. While the blackbirds nested, a pair of American Kestrels, or Sparrow Hawks, spent much time in surrounding trees, but the blackbirds paid little attention to them. When one of these raptors swooped down at the female while she gathered material on the ground, she avoided it by jumping into a bush conveniently nearby and then calmly resumed her occupation. While the male blackbird perched on a dead twig at the very top of a leafy tree, a kestrel flew at him, making him duck his head. When the raptor again dived at him, he moved to another branch a few yards away, and the kestrel alighted on the branch he had just left. For a while the Oriole-Blackbird and the American Kestrel rested in the same open treetop. Although the raptor appeared not to endanger the adult blackbirds, he might have taken the young. Other possible culprits were a pair of Yellow-headed Caracaras nesting nearby and one of the larger hawks that from time to time appeared at this open spot. Or, as I hoped, the young blackbirds may have been led to sheltering vegetation beyond my view (Skutch 1967).

7

The Melodious Blackbirds

Of the diverse birds that in different lands are called "blackbirds," none has a better claim to the name than the two species of *Dives*, for they are among the blackest of birds. As though to compensate for the plainness of their attire, they have exceptionally fine voices. The continental Melodious Blackbird ranges from northeastern Mexico to northwestern Costa Rica, and upward in the highlands of Guatemala to about 6,500 feet (2,000 m). In the early 1930s, when I (Skutch 1954) traveled rather widely in Guatemala, I found it only on the Caribbean side of the country. When, after an absence of four decades, I returned to Guatemala in 1973, I was surprised to find a large flock of Melodious Blackbirds roosting in a tree beside Lake Atitlán, on the Pacific slope. This blackbird has been extending its range not only westward but southward, reaching Costa Rica in the late 1980s. After a wide gap in their

Melodious Blackbird, *Dives dives,* sexes similar

distribution, Melodious Blackbirds reappear on the Pacific coast of South America from southwestern Ecuador to southern Peru. Opinions differ as to whether these South American birds are a different race of the Melodious Blackbird or a different species, called the Scrub Blackbird. I found these blackbirds abundant in rice-growing country around the Guayas Estuary of Ecuador, where I heard them called *tilingas,* an onomatopoeic name. Another species, the Cuban Blackbird, is confined to that island.

Male blackbirds are ten to eleven inches (25 to 28 cm) long; females about one inch (2.5 cm) shorter. The plumage, strong conical bills, legs, and feet of both sexes are black; their eyes, as seen in the field, appear almost equally dark. The male is glossier, more strongly tinged with blue or purple, than the female, but the sexes are often difficult to distinguish. Avoiding closed forests, they inhabit open and semiopen country of many types: fields, pastures, marshes, the shores of streams, light woods, and the vicinity of human habitations. In Caribbean Guatemala and Honduras they live in regions of high rainfall, such as the Alta Verapaz, where I found them abundant, but their western extension brings them into drier country. On the arid west coast of South America they are mostly confined to irrigated fields and the borders of streams.

The habits of these attractive birds, both in foraging and reproduction, suggest affinity to the orioles on one hand and to the grackles on the other. They hunt among trees and shrubs, picking insects and their larvae from foliage in the manner of orioles but with less agility, and they also forage much on the ground, like grackles and cowbirds. They investigate curled leaves by inserting the bill and gaping. They drink nectar from the staminate flowers of bananas, like orioles, oropendolas, and Brown Jays. Often they pluck a white banana flower and hold it beneath a foot while they probe it with the bill. At other times they hang head downward beside the huge, dull red flower bud and drink from blossoms still attached. The large white flowers of balsa trees also offer them nectar. They eat a variety of fruits, including berries of the bountiful melastome family, discarding larger seeds but swallowing small-seeded berries whole. They climb over the backs, faces, and ears of cattle, apparently relieving them of ticks, thereby benefiting both quadrupeds and their owners; but they distress farmers by tearing open ripening ears of maize for the milky grains. At finding food Melodious Blackbirds are hardly less versatile than grackles.

They frequently forage on the ground, preferably in damp and open places, where they walk instead of hopping. In the Motagua Valley of Guatemala they joined the stone-turning parties that toward evening I watched on the shingly floodplain of the Río Morjá, a tributary stream. Here they moved small stones in the manner of the Great-tailed Grackles, Giant Cowbirds, and Bronzed Cowbirds that accompanied them. Inserting the upper mandible beneath the near edge of a stone, they shoved forward with the bill slightly open until they overturned it, or pushed it aside if it was large, then swallowed anything edible that they disclosed. In other places these blackbirds employ the same procedure to turn over or shove aside fallen leaves, sticks, dry cow dung, and other small objects.

After their evening meal, the Melodious Blackbirds along the Río Morjá retired for the night into a dense stand of giant canes that had sprung up on newly formed land beside the barren flats. Many Bronzed Cowbirds joined them, while the larger Giant Cowbirds flew off noisily to roost downstream. I knew no more delightful way to spend the evening than to sit on the sand beside the canebrake and listen to the clear, soothing notes that issued from its depths. As the assembled blackbirds grew drowsy, the notes came lower and more widely spaced, until in the deepening dusk the birds fell asleep. Then I climbed up to the plantation house high on a spur of the Sierra de Merendón where a pair of these birds nested.

The loud, clear, mellow whistles of the Melodious Blackbird earn for it the first part of its name. Usually of two or more syllables, these notes are almost always full and round. *Whit-wheer; whit whit whit wheer; twit twit twit twit tuwait tuwait; chic weer chic weer; whar rit whar rit* are some of their songs; but these few attempts to paraphrase them hardly do justice to their rich variety. The flight call is a clear, silvery *tink tink tink*. Whistling on a perch, the blackbirds puff out their plumage, stretch up their legs, and prolong the last syllable until it becomes high and sibilant. The female's notes are only slightly less full and mellow than those of the male. Often she sings on the nest or duets with her mate.

Like orioles, Melodious Blackbirds breed in monogamous pairs. Both sexes defend a nesting territory, beyond which they find much of their food in areas shared with others of their kind, as on the exposed shore of the Río Morjá. The three nests that I found in the foothills above this stream sat ten to twenty feet (3 to 6 m) above the ground, well screened by the dark, glossy foliage of orange and lemon trees near buildings. In the pine woods of Belize, Gordon Orians (1983a) watched a nest about twenty-three feet (7 m) up, close to the trunk of a pine tree. In central Peru he found two nests of the southern race inaccessibly high in large eucalyptus trees, where they hung beside moderately thick vertical branches. The Cuban Blackbird places its nest at the base of a palm frond or in a cavity in a tree or building.

The nests I found in Guatemala were bulky open cups, woven of coarse fibrous strips from the leaf sheaths of banana plants, shreds of epidermis, and lengths of vines. They were securely attached by looping strands at the rim around slender, upright supporting branches. The bottom of the concavity was heavily plastered with cow dung and mud, which extended up the sides to within an inch or two of the rim. Within this was a lining of thin rootlets and fine fibers. The inside of the bowl measured 4¾ inches in diameter by 2¾ inches in depth (12 by 7 cm). These nests closely resembled those of the Great-tailed Grackles in the coconut palms nearby. The nest Orians saw in the pine trees was quite different. Made almost wholly of dry pine needles, with a thin lining of fine grasses, it appeared flimsy, which the Guatemalan nests certainly were not. Two of these nests held three eggs, that in Belize had four. The Cuban Blackbird also lays three or four eggs. Melodious Blackbirds' eggs are clear light blue, marked with scattered, large and small, black or dark brown blotches and dots.

On April 6 I found a pair of Melodious Blackbirds starting a nest in an orange tree on a hilltop above the Río Morjá. Male and female shared the task of building. They worked in a very desultory fashion, bringing long fibers once or twice in succession, then flying beyond view for half an hour to an hour before returning to resume operations. The first bird to arrive in the nest tree would call with loud, clear notes and be answered from the distance by its partner, who would soon appear, sometimes bringing a contribution for the nest and sometimes with empty bill. Once the male entered the nest and proceeded to weave with materials already present. On other occasions he brought long strands. Weaving the cup occupied the pair for about nine days. Less than two days sufficed to plaster the inside with mud and fresh cow dung. Adding the lining occupied two more days, making a total of about thirteen days for the entire construction of the nest.

Like many tropical birds, these blackbirds nested at a leisurely pace. A week elapsed between the completion of the nest that I watched them build and the appearance of the first egg on

April 25. Next day the nest was empty. The replacement nest of this pair, about fifty feet (15 m) from the first, was so well hidden in another orange tree that I did not find it until May 19, when the set of three eggs was complete. Since the sexes were so similar that I could distinguish them only in the most favorable light, I tried to mark one of them by an expedient I had found successful with flycatchers—that of sticking into the nest above the eggs a slender twig with a tuft of paint-soaked cotton on its end. Twice a parent pulled the cotton from the twig without acquiring a distinguishing mark. Stretching an inch above the nest a string coated with white paint gave no better result with birds so careful to keep their plumage spotlessly black. Fearing to provoke desertion, I desisted from the attempt to mark a bird before I watched the nest from a blind. In seven hours I timed four full sessions of incubation ranging from 50 to 81 minutes in length and averaging 62.5 minutes. Six recesses varied from 8 to 23 minutes and averaged 17.7 minutes. The eggs were covered for 78 percent of the seven hours, always by the female, as far as I could learn.

Although the male did not incubate, he was attentive to his partner. Before sunrise he arrived near the nest, probably from the canebrake far below, and greeted her with deep-toned double whistles, *whar rit whar rit*. He repeated them over and over, while she answered from the nest with notes higher pitched than his. Then together they flew to forage down in the valley, both whistling *tink tink tink tink*, clear and silvery, as they went. This flight call was to my ear the most pleasing of the birds' many notes. In twenty-three minutes they returned together and the female entered the nest, whither he brought her food, an attention with which he favored her only once or twice in a morning. Much of the forenoon he stayed near the nest, frequently repeating songs that his partner answered while sitting in the deep shade of the orange tree. Sometimes she called first from the nest, to be answered by him if he remained within hearing. On leaving the nest the female always flew down into the low valley lands for food. Often the male accompanied her, but frequently he stayed to guard the nest.

More than twenty-four hours after the eggs in this nest were pipped, all three hatched. The nestlings bore a few tufts of blackish down and had tightly closed eyes. When they were two days old, I spent three and a quarter hours watching and saw their mother bring food fifteen times, the father thrice. Sometimes it was difficult to distinguish the sexes, but I believe my identifications were correct; twice the two came almost at the same time with food. Only the mother brooded, for intervals of 5 to 14 minutes. Before the nestlings were a week old, two had vanished, but I watched the parents attend the lone survivor. In three hours, the seven-day-old nestling was fed eleven times by its mother, twice by its father, and once more by a parent of undetermined sex. The male's attentiveness had not increased during the intervening five days. This nestling also vanished before it could fly.

Almost everything these blackbirds did was accompanied by melody. While they stood on the nest's rim delivering food they voiced low, liquid monosyllables. Despite their apparent tenderness, they were unspirited parents who neither made remonstrance nor showed concern when I handled their nestlings in their presence. The same was true at a late nest, possibly a second brood, that I found with nestlings in mid-July. Whether Melodious Blackbirds try to protect their young from creatures smaller than man, I had no opportunity to learn (Skutch 1954).

8

Some South American Icterids

This chapter is devoted to a number of species that await more prolonged or thorough studies.

The Brown-and-Yellow Marshbird

Two species of marshbirds of the genus *Pseudoleistes* inhabit southern South America. The Yellow-rumped Marshbird, which ranges from eastern Brazil to Paraguay and eastern Argentina, is about nine and a half inches (24 cm) long. Both sexes are dark brown or blackish, with bright yellow rump, upper- and underwing coverts, and underparts. The slightly smaller Brown-and-Yellow Marshbird is found from extreme southern Brazil to northwestern and central Argentina. Mostly glossy olive-brown, its shoulders and breast are bright yellow. In Argentina, where it is an abundant and well-known permanent resident, it is called *pecho amarillo,* or yellow breast.

W. H. Hudson (1920), who knew these Brown-and-Yellow Marshbirds well, told how they associate in flocks of twenty to thirty birds that forage on the ground in open places. While his companions eat, one stands guard on a stalk or thistle top; when he flies down, another takes his place. At the approach of a human, the sentinel sounds the alarm, and the whole party flies up in close formation, filling the air with loud, ringing notes. "After feeding, they repair to the trees, where they join their robust voices in a spirited concert, without any set form of melody such as other song-birds possess, but all together, flinging out their notes at random, as if mad with joy. In this delightful hubbub there are some soft or silvery sounds."

Even in the breeding season these peaceful birds do not claim nesting or feeding territories and rarely quarrel with other birds of their own or different kinds. In September or October they start to nest in pairs, choosing sites in pampas grass, cattails, and sedges or, in drier places, in large cardoon thistles, thick bushes, and low trees. Occupied nests may be separated by as little as four yards, and several are often placed in a small cluster of pampas-grass tussocks. Other nests are widely separated. The female builds without help, but her mate often accompanies her on trips to gather materials. The thrushlike nest is a deep cup or bowl, compactly made of dry grasses and slender sticks, plastered inside with cow dung or mud, and lined with hair or soft, dry grasses. The interior is about four inches in diameter by three inches deep (10 by 7.5 cm). Four is the most frequent clutch size, but sets of three or five eggs are not rare. The elongated white eggs are spotted with reddish chestnut, most heavily on the blunt end.

Orians and his family (1977) studied Brown-and-Yellow Marshbirds in eastern Argentina. As far as they could learn, only the mated female incubated, but while at the nest she was fed by two to four others, including her mate. After the eggs hatched, up to eight birds visited the nests, and at least some of them fed the nestlings. The only brood that could be followed after they left the nest continued to be fed by the same helpers. The sexes and ages of the helpers and their relationship to the parents were unknown, but apparently none had nests of their own in the vicinity.

The Saffron-cowled Blackbird

Closely related to the marsh-nesting blackbirds of the genus *Agelaius*, but sufficiently different to be classified in a genus all its own, is the eight-inch (20 cm) Saffron-cowled Blackbird, *Xanthopsar flavus.* The male is splendid in bright golden yellow, with glossy black lores, back, wings, and tail. The female is yellow or dull orange-yellow on her eyebrows, shoulders, and underparts. Elsewhere she is brown, slightly streaked above. Both sexes have black bills and feet. These birds range from the far south of Brazil to northern Buenos Aires province in Argentina, where in Hudson's day they were well known to the country people, who called them *naranjos* (orange-colored) in allusion to their orange tints, brilliant in sunshine. Resident throughout the year, Saffron-cowled Blackbirds live in flocks of twenty to thirty individuals that frequent agricultural lands, following the plow to pick up worms and grubs, and settling in orchard trees to sing in concert with powerful notes lacking in variety.

The social habits of these blackbirds persist through the breeding season, when they often build their nests close together in the rushes and reeds of marshes or in cardoon thistles two or three feet (60 or 90 cm) above the ground. The open cup of dry grasses and interlaced fibers receives four white or bluish eggs sprinkled with reddish brown spots and blotches that are sometimes confluent on the thicker end. These beautiful birds deserve to be better known.

The Chopi Blackbird

The all-black Chopi is the sole representative of the genus *Gnorimopsar.* The ten-inch (25 cm) male is distinguished from other black icterids by the narrow, pointed feathers of his crown and nape and the ridge and groove at the base of the lower mandible of his long, sharp black bill. The female is slightly smaller and duller black than the glossy male. From northeastern and central Brazil to eastern Bolivia, Paraguay, Uruguay, and northwestern and central Argentina, the Chopi inhabits marshy pastures, reed beds, palm groves, cultivated lands, and the vicinity of human habitations. A sociable bird, it walks elegantly over the ground seeking adult and larval insects.

The most vivid account of the Chopi's habits I have read was written long ago by Felix de Azara, who knew the bird well in Paraguay, where it frequently visited the courtyards and verandas of houses. Azara's charming account has been made accessible to English readers by Hudson, in his *Birds of La Plata.* With strong and easy flight, the Chopi readily attacks any large bird passing near, following it persistently through the air and pouncing down to cling to its back. If a Crested Caracara alights to shake off its persecutor, the Chopi perches a few feet away, alert to renew his insults at the earliest opportunity. In similar fashion, the redoubtable blackbird holds aloof all unwelcome intruders. From afar he recognizes an enemy, by its form or even its shadow, and with a loud whistle warns all neighboring birds of the approaching peril. While they rush into shelter, he bravely goes forth to confront it. After routing the enemy, he celebrates his victory with a loud song that begins with the repetition of his own name.

The Chopi's song has been highly praised for its variety, vigor, and melody. He welcomes the dawn from the eaves or tiled roofs of the houses where he sleeps. Perching erect, he vibrates his wings while he pours forth his clear notes. The effect is heightened when several sing together. Considered one of the best singers among Brazilian birds, the Chopi is commonly caught and kept in a cage, with the result that his kind has become rare in regions where many people live (Sick 1984). The female also sings, but more often she utters a sharp *zwirr.*

Few other icterids so frequently build their nests in enclosed spaces, such as a hollow in a tree or an earthen bank, an old woodpecker's hole, a spot beneath the eaves of a house, the abandoned clay nest of an hornero, or the old nest of some other bird. Chopis also place their nests in crowns of palm trees or amid the sticks of a Jabiru's bulky structure. If no cavity is readily available, the Chopi chooses a high site amid close-set twigs, well out from the trunk of an orange or some other tree with dense foliage. The cup-shaped nest is made of sticks, straws, and fibrous materials, artlessly put together, and lined with a few feathers. The four eggs are white or pale blue and spotted with brown (De la Peña 1979).

It is a pity that no thorough study of this interesting bird is available.

The Bolivian Blackbird

The Bolivian Blackbird, single member of the genus *Oreopsar,* is known only in the highlands of the country for which it is named. About nine inches (23 cm) long, both sexes are dull brownish black, with dark, dull brown wings. The bill is short and black. In flocks of about four to eight, these birds frequent canyons with steep cliffs in dry intermontane valleys. One April, Orians and his companions watched Bolivian Blackbirds eating the seeds of grasses they were able to reach by fluttering or hopping upward, seizing an inflorescence in the bill, holding it while they bore it downward, and then grasping the stem with one or both feet while they plucked off the seeds. Bunch grasses *(Stipa)* on steep slopes where little else grew were especially attractive to them. They ate fruits of a large arborescent cactus, gleaned insects from leaves and stems of small shrubs, and gaped into bromeliads growing on high branches of *Prosopis* trees. They also foraged for insects on the ground, frequently gaping in the widespread icterid manner to extract them from crevices and the bases of small shrubs.

The Bolivian Blackbirds' song is a series of sharp *chips* mixed with *chu-pits.* In flight they voice a loud, clear, far-carrying whistle —*chu pee*—and churr harshly. While perching, they repeat a clear whistle. A short, sharp *chip* expresses alarm or concern for their nest. A courting male elevates his tail and points his bill almost straight upward in the posture widespread among blackbirds and grackles.

These blackbirds nest in deep crevices in rocky cliffs and quarries. Orians and his party were able to extract only a single nest from its narrow cranny—a shallow bowl composed largely of tough fibers, mostly fine roots apparently gathered on the sides of the cliffs. It was lined with fine, dry grass stems and two black feathers. The three pale greenish gray eggs were marked with spots and streaks of gray and larger blotches of black, scattered over the whole surface and concentrated on the blunt end. While a female incubated, another bird brought food to the nest. On leaving her eggs, she whistled loudly and clearly while she flew to join her flock. With open bill pointing upward and fluttering wings, she begged from another bird who gave her a large insect. On another occasion, a single blackbird was fed five times in one minute by four others. A bird who found food was never seen to withhold it from a flock member. Bolivian Blackbirds give every indication of breeding cooperatively, but the extent of this cooperation remains to be studied (Orians, Erckmann, and Shultz 1977).

The Austral Blackbird

The genus *Curaeus* consists of two South American species, of which the larger and better known is the eleven-inch (28 cm) Austral Blackbird. The glossy black male wears on his crown, nape, and sides of his head narrow, pointed feathers with stiff, shiny black shafts. The female is browner and duller. The sharply pointed, flat-topped bill is black, as are the legs and feet. In Chile, where it ranges from Coquimbo in the north to the Cape Horn archipelago, and from lowlands up to 5,000 feet (1,500 m), it is so widespread and familiar that it is known as the *tordo común*, or common blackbird. In this country its preferred habitats are dense shrubbery along streams flowing between cultivated fields and thickets on hillsides above them. In the foothills of the Cordillera it lives in well-watered valleys and flies over open ridges. Except when nesting, it gathers in such large flocks that it has been called the most gregarious bird in the country. In the far south, where it crosses the Andes into Patagonia, it is the only icterid in the forests of southern beech *(Notofagus)*. Its pleasant song is admired by rural people, who domesticate it and try to teach it to talk (Goodall, Johnson, and Philippi 1957).

The Austral Blackbird eats a wide range of vegetable and animal foods. It compensates for the losses it causes farmers when it pulls up seedlings of wheat by ridding the fields of great numbers of injurious larvae and other insects. It has the bad habit of devouring the eggs and nestlings of other birds.

From October to December the Austral Blackbird breeds, usually hiding its nests in such dense thickets that they are difficult to find. The bulky open cup is made of twigs and straws bound together with clay. The usual clutch is four or five eggs, three or six are occasionally laid. The pale blue shells may be plain or spotted and scrawled with black; most often they are marked with only a few fine black dots.

Orians and his family (1977) watched Austral Blackbirds at three localities in southern Argentina in early December, after the young had fledged. When they caught one of a group of at least four fledglings, the captive's screams incited six adults to mob the captors. When a fledgling attempted to fly across a road, six grown birds accompanied it to the trees on the other side. More than two adults carried food in their bills, but the watchers did not see any of them feed a youngster. Elsewhere, four adults with at least one fledgling protested the observers' presence. However, at another locality, a single fledgling was fed by at least three of the four grown birds accompanying it. These and other brief observations suggest cooperative breeding, but they leave open the possibility that after the young fledge, two or more pairs that have attended separate nests unassisted cooperate in caring for the fledglings.

Forbes' Blackbird

This second species of *Curaeus* is smaller than the Austral Blackbird and equally dark. The sexes are alike and the shafts of their head feathers are glossier and more striking. A diminishing species, it is found in two widely separated regions of eastern Brazil: the states of Alagoas and adjacent Pernambuco in the north and southeastern Minas Gerais. The northern population was studied from 1981 to 1986 by Anita Studer and Jacques Vielliard (1988), who found about 150 individuals confined to the edge of a shrinking forest, where they nested, and adjoining wet meadows and marshes, where they foraged. Their song is a buzzy *beechchch* twice repeated; their alarm call a series of rattling *clicks.* With spread wings and mouth widely open to display the vivid red of its throat, a courting bird throws its head upward and rearward until it touches its back. The individual so addressed responds with a similar performance. Three or four of these cooperatively breeding birds were seen displaying together, but their sexes could not be determined because they looked so similar.

In any month, the advent of the irregular rains stimulates nesting. The blackbirds build in the densely foliaged crowns of trees —mainly mangos in the northern population, occasionally in a thorny tree, rarely in an epiphytic bromeliad, a banana plant, or a liana—at heights of ten to forty feet (3 to 12 m) with an average of twenty-three feet (7 m). As a site for its deep, open cup, the blackbird chooses three or four diverging twigs at the end of a branch, from two of which it suspends the structure, using the other(s) to support the sides or bottom. The nest is composed of herbaceous stems and dry twigs, up to thirty-one inches (80 cm) long, gathered in the marshes and interlaced in a thick wall. The bottom of some nests is smoothly plastered with cow dung and mud; others have only a scattering of this plaster or none. The clutch consists of one to four (average 2.84) light blue eggs spotted and streaked with black, mainly on the thick end of the pointed ovals. The incubation period is usually thirteen days, rarely one day more or less.

Most nesting pairs of Forbes' Blackbird are assisted by two to four helpers of the same species; Studer and Vielliard could not determine the ages and relation of the helpers to the breeding adults they observed. Although in many cooperative breeders the helpers do not assist at the nest until the nestlings hatch, in this species they participate in building. Usually one member of the group sits in the growing nest, arranging contributions that the others bring from as far away as 660 feet (200 m). With so much help, the bulky nest is built in three or four days but then remains empty for from four to ten days before the first egg is laid. During the incubation period, the adults are silent. Often a parent arrives with a helper or two, and all perch for many minutes at the summit of the nest tree. At intervals of two or three hours one bird replaces another on the eggs, but whether the helpers share incubation has not been determined. Although adults eat both insects and fruits, for the nestlings parents and helpers bring only insects, mainly grasshoppers, caterpillars, and butterflies from the meadows and marshes, delivered entire to the young.

Forbes' Blackbird is heavily parasitized by the Shiny Cowbird, which foisted its eggs on 64 percent of all the nests found in this study, reaching 100 percent in the last year, when none of the host's young fledged. Formerly abundant in the Alagoas-Pernambuco region, where the cowbird was unknown late in the nineteenth century, these blackbirds have declined in number alarmingly since the arrival of the parasite.

The Scarlet-headed Blackbird

Another blackbird with a genus all its own is the Scarlet-headed, *Amblyramphus holosericeus.* Nine and a half inches (24 cm) long, both sexes are glossy black with brilliant scarlet head, neck, upper breast, and thighs. The tip of the long, straight bill is broad and flat, chisel-like. From northern Bolivia and southern Brazil through Paraguay and Uruguay to east-central Argentina, these birds reside throughout the year in marshes and reed beds, living in flocks of from half a dozen to about thirty birds when not nesting, and in territorial pairs when breeding. Hudson (1920) found "chisel-bills" abundant along the swampy shores of the Río de la Plata, where they perched at the summits of tall rushes. Seen in the distance, the flame-colored heads shone "with a strange glory above the sere, sombre vegetation of the marshes." Their only song is a prolonged, variable whistle that often "sounds as mellow and sweet as the whistle of the European Blackbird." Scarlet-heads forage by inserting their flat-tipped bills into the soft stems of cattails and sedges, then forcefully gaping to split them and expose whatever insects lurk within. Somehow they know just where to open a stem to extract the food they like, for Orians (1980) split many of them without finding the kinds of insects the birds were bringing to their nestlings.

In early spring in Argentina, Scarlet-headed Blackbirds apparently form pairs in the winter flocks. Then the partners go together to claim in the marshes territories that may be as large as a hundred acres (40 hectares). The sexes join in territorial defense, calling noisily, while in the air they display with dangling scarlet thighs, made more conspicuous by fluffing out the feathers. While the male sings from an exposed perch, his consort moves obscurely through the dense vegetation, probably seeking food and a nest site. With a little help from her partner, who may bring material only to drop it before he reaches the nest, the female builds an open cup. Securely fastened to several upright stalks, three to five feet (0.9 to 1.5 m) above the water, it is composed of strips of cattail leaves, which for the lining are finely shredded. In central Argentina she lays three or four eggs, which are light blue and sparsely spotted with black. Sometimes unequal basal growth of the stalks that support a nest tilts it until the eggs roll out. Only the female incubates. During her brief absences for food her partner does not accompany her but perches conspicuously near the nest, guarding. Apparently, he does not feed her. After the nestlings hatch, only the mother broods them, but the parents take equal shares feeding them, at the average rate of one visit every five minutes. They alternate their visits so the nest is seldom left unguarded. They bring some food from cattail marshes as far as one kilometer (three-fifths of a mile) from the nest.

At first wholly black, young Scarlet-heads become pale terracotta red on head and neck, which gradually acquire the vivid scarlet of the adult plumage. Apparently, they do not breed until they are about two years old, for yearlings flock during the nesting season.

The Red-breasted Blackbird

Smallest and most widespread of the grassland icterids with ruddy breasts, the Red-breasted Blackbird is sometimes classified with the meadowlarks in the genus *Sturnella;* but it appears sufficiently different, especially in its shorter, stouter, more finchlike bill, to be retained apart in the genus *Leistes,* where it has long been placed. The seven-inch (18 cm) male is black with a bright red breast and

Red-breasted Blackbird, *Leistes militaris,* male

throat, and at the bend of the wing is a patch of brilliant red conspicuous only in flight. The smaller, very different female is nearly everywhere streaked with dark brown and buff, with only a tinge of red on her buffy breast to reveal her identity. From northern South America to Trinidad, Panama, and southern Costa Rica, males lack the long, buffy white streak behind the eye that distinguishes birds from northeastern Brazil to central Argentina, which are sometimes considered a separate species, the White-browed Blackbird. Females of the two forms differ chiefly in the shorter bill of the southern race.

In central Argentina white-browed blackbirds are migratory, arriving in Buenos Aires province in early October. When returning northward in the austral autumn, males travel singly or in parties of six to a dozen birds, flying laboriously in straight courses, conspicuous well above the ground. Females migrate separately, singly or two or three together, flying low with intermittent wing-beats and gliding, occasionally darting sideward, as though frightened. More northern Red-breasted Blackbirds appear not to undertake regular migrations, but after the breeding season they wander in flocks that are sometimes large and may include Bobolinks and Eastern Meadowlarks. Like a number of other icterids, they have been expanding their range. In Brazil they have in recent decades appeared in many localities where they were previously unknown; and they arrived in southern Pacific Costa Rica as recently as 1974.

Red-breasted Blackbirds inhabit borders of marshes, wet pastures and savannas, rice fields, and low, wet scrub, where they forage on or near the ground for grasshoppers, crickets, beetles, caterpillars, and seeds. They gather berries from low shrubs. Males spend much time perching on tall herbaceous stalks, low bushes, or fence posts, their glowing red breasts catching the eye from a distance like brilliant flowers, while females lurk inconspicuously amid the herbage. A male whom I watched in eastern Peru repeated a metallic buzz while perching. I did not see the aerial display described by Hudson (1920), who told how at intervals of two or three minutes, a male rises vertically upward to a height of twenty to twenty-five yards or meters, to utter a single, long, powerful, and rather musical note, ending with an attempt at a flourish while he flutters and turns about in the air; "then, as if discouraged by his failure, he drops down, emitting harsh guttural chirps, to resume his stand." When, evidently attracted by the male's colorful and songful display, a female emerges from the grasses, she starts up with a wild zigzag flight, "like a snipe flushed from its marsh," darts from side to side, and after her brief appearance, drops again into the herbage that hides her so well. The moment she reveals herself, the male pursues her until they vanish from view.

The nest, on the ground amid dense concealing vegetation, is a deep open cup of dry grass, lined with finer grass and often seed down. Sometimes it is approached by a tunnel through the herbage, which screens the female and makes her nest difficult to detect. Occasionally these blackbirds breed in loose colonies. The two to four eggs, occasionally five in Argentina, are greenish white or deep cream and densely blotched over most of the surface with chestnut or pale reddish brown (De la Peña 1979). Apparently, nobody has carefully watched Red-breasted Blackbirds at their nests.

The Velvet-fronted Grackle

Single member of the genus *Lampropsar,* the Velvet-fronted is a small grackle, the male eight and a half inches (22 cm) long, the female about an inch shorter. Both are wholly black, glossed with blue above. The male's forehead is densely covered with short, plushlike, jet black feathers. Both sexes have short, conical, sharp-pointed bills and dark eyes. Velvet-fronted Grackles range through the lowlands, up to about 1,300 feet (400 m), from northern Guyana, Venezuela, and southeastern Colombia through western Amazonia to northern Bolivia. In active parties of about six to twelve birds, they forage along the shores of streams and lakes, seeking insects amid the foliage of shrubs and trees at all heights. Sometimes they venture into marshes but rarely far from forest. Their calls include a *chack* or crackling note, and a half-musical *cheziit,* uttered in flight or as they alight. After gathering to roost in a clearing in low, wet forest, the grackles join their voices in a chorus of rapid gurgling notes that sounds almost musical in the dusk. In Guyana a nest was found above a small creek (Meyer de Schauensee and Phelps 1978; Hilty and Brown 1986).

The Red-bellied Grackle

Among the least known of South American icterids are the forest dwellers. One is the Red-bellied Grackle, only species of the genus *Hypopyrrhus,* appropriately named *pyrohypogaster*—firebelly. The twelve-inch (30 cm) male and the ten-and-a-half-inch (27 cm) female are alike in plumage, both black except the bright red belly and under tail coverts. The feathers of the whole head, nape, and throat are narrow, thick, and pointed, with shiny black shafts. The conical, pointed bill and the legs are black; the eyes pale yellow. These grackles are confined to Colombia, where they inhabit forests and moist scrub in the subtropical zone, mostly from 4,000 to 8,860 feet (1,200 to 2,700 m) above sea level. In noisy parties of about six to eight individuals, they forage in forest trees, often at the woodland's edge, hopping and climbing through the outer foli-

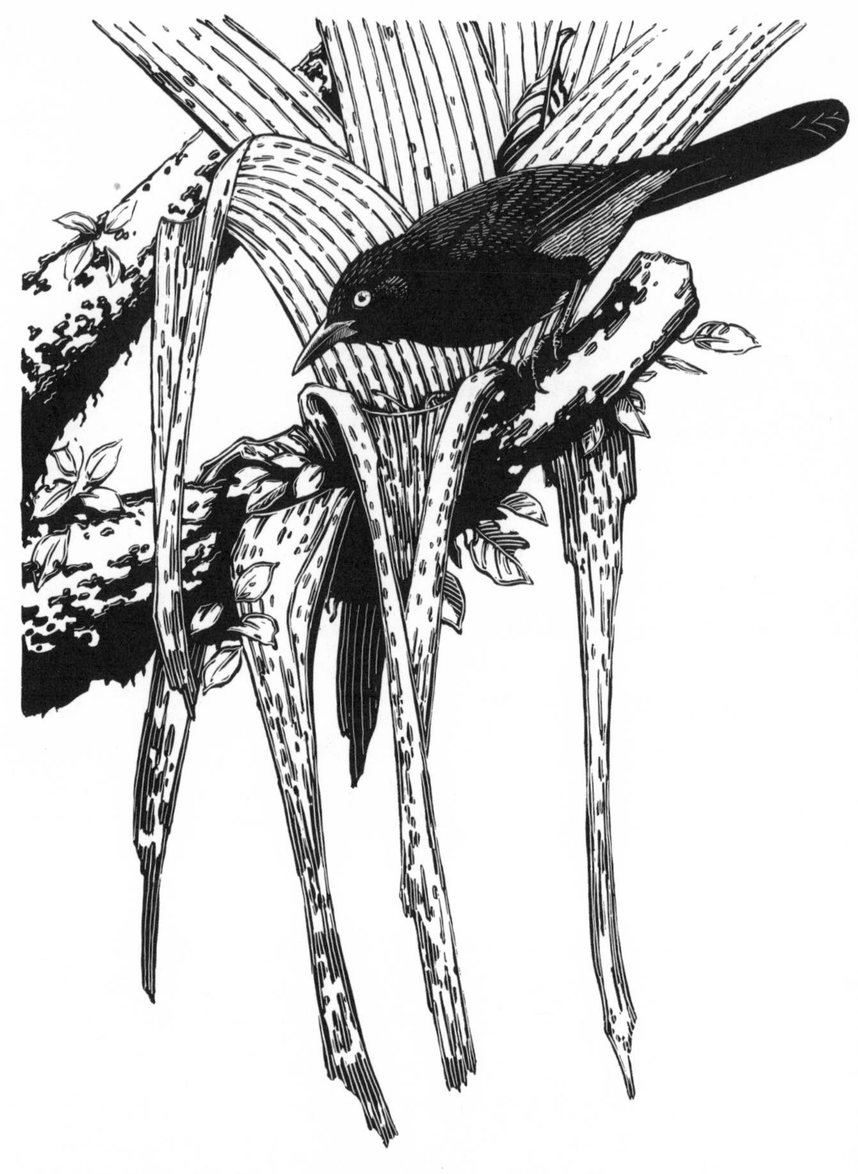

Red-bellied Grackle, *Hypopyrrhus pyrohypogaster,* sexes alike

age, sometimes clinging upside down. Often they associate with oropendolas or join mixed flocks of forest birds. They breed in solitary pairs, in cup-shaped nests of large dead leaves and sticks, loosely set in forks of small trees. The eggs, of unstated number, are greenish gray, spotted and streaked with dark brown and lilac (Hilty and Brown 1986).

The Mountain and Golden-tufted Grackles

Another little-known genus of icterids is *Macroagelaius*, with two species in northern South America. The male Mountain Grackle measures twelve inches (30 cm) in length, including his long tail; the female is about one inch (2 cm) shorter. Both are wholly dull bluish black, except the chestnut beneath their wings that is rarely visible. The conical bill and the legs are black, the eyes dark. These dusky birds are endemic to the mountain forests of Colombia, mostly between 6,400 and 10,200 feet (1,950 and 3,100 m), and their nests appear never to have been found (Hilty and Brown 1986).

The slightly smaller Golden-tufted Grackle is also a long-tailed, black-billed bird, whose glossy blue-black plumage is relieved by metallic feather edges and golden yellow tufts on the sides, where they are mostly covered by the wings. Confined to Venezuela south of the Orinoco, Guyana, and adjacent regions of Brazil, at altitudes from 1,640 to 6,600 feet (500 to 2,000 m), it flies in noisy flocks through trees along streams, in forest clearings, and on the rain-forested slopes of the tepuis—great, steep-sided, flat-topped mesas that soar into the clouds above the savannas of southern Venezuela. Its notes are reported to be squeaky and rasping. Nothing more appears to be known about it (Meyer de Schauensee and Phelps 1978).

9

The Rusty and Brewer's Blackbirds

Two species of North American gracklelike birds make up the genus *Euphagus*. Although their ranges overlap to some extent, the Rusty Blackbird has a more northerly distribution than Brewer's Blackbird.

The Rusty Blackbird

About nine inches (23 cm) long, the breeding male Rusty Blackbird is wholly dull black, with a faint greenish tinge on his head and upperparts. After breeding he molts, becoming rusty brown on his crown, back, and chest, with a buffy streak above each eye and a buffy throat, and dark gray cheeks, rump, and posterior underparts. In winter the male and female differ little. By spring the buffy tips and edges of the feathers have worn away, leaving

Rusty Blackbird, *Euphagus carolinus,* winter plumage

the male black and the female almost as dark as he or dull slate-gray. Adults of both sexes have yellow eyes; those of juveniles are often dark. Rusty Blackbirds breed in Alaska and across the whole of Canada to the limit of trees beyond the Arctic Circle, much farther north than any other icterid. Southward, their breeding range extends only to the central parts of Canada's western provinces, but in the east it crosses the international border to northern New York and New England. In winter these birds are found from about the southern limits of their summer range southward, east of the Rockies, to the Gulf coast from Texas to northern Florida, most abundantly in the southern states (Bent 1958).

Rusty Blackbirds frequent the vicinity of water. While foraging they walk or run over preferably wet ground or wade up to their thighs in shallow water. As do other birds that hunt over the ground in large parties, individuals continually fly from the rear to the front of the flock in a sort of rolling movement. They eat grasshoppers, beetles, and many other insects, supplemented by an occasional crustacean, snail, salamander, or small fish. Their vegetable food includes corn, wheat, and oats, gathered largely in stubble fields; seeds of weeds and trees such as the ash; and berries of trees and shrubs. When snow deeply covers the land and they are starving, they may become fiercely predacious, attacking and killing other birds trying to keep alive in the same small patch of open ground, such as Common Snipes and American Robins, then pecking out their brains and leaving the remainder of their corpses.

The song of the Rusty Blackbird is a continuous rhythmical alternation of a musical phrase and a squeaky single note higher in pitch: *tolalee—eek—tolalee—eek* . . . The pauses between the phrases, twice as long as the phrases themselves, prevent a prolonged recital from becoming monotonous. A less rhythmical song is the rapid repetition, two or three times, of a phrase that sounds like *kawicklee.* The blackbird's calls include a short *kick* and a rattling *turururo.*

In February Rusty Blackbirds that have wintered in the southern United States begin to move northward by day in multitudinous noisy flocks that pass overhead in somber waves. At intervals they pause in leafless trees to sing in many-voiced choirs; or they alight on the ground and follow the plowman for the worms and grubs he turns up; or they blacken stubble fields while they pick up spilled grain. At nightfall they rest in cattail marshes or bushy swamps, to resume their northward journey on the morrow. They often travel with Red-winged Blackbirds, Common Grackles, Brown-headed Cowbirds, or Starlings. They migrate to mosquito-infested northern woods of spruce, balsam fir, and other conifers, where they arrive in March or, in the far north, not until April or early May. When they reach their wilderness, the flocks disintegrate into pairs who seek widely separated nest sites, in spruces or other evergreens; in clumps of button bushes, sweet gale, or other deciduous shrubs; and sometimes in dead trees.

Rusty Blackbirds usually make their nests two to eight feet (0.6 to 2.4 m) above water or the bank of a stream; rarely nests can be found as high as twenty feet (6 m) up. On a thick foundation of *Usnea* lichen, the blackbird—doubtless the female, although observations are lacking—builds a shell of twigs, lichens, and occasionally a few dry grasses. Within this frame she shapes a bowl, not of mud, which may be difficult to find in these woods, but of duff, the partly decayed vegetation that covers the ground, which after drying becomes as hard and stiff as papier-mâché. Pressed into the framework of twigs, this plaster binds them firmly together. After the builder has carefully rounded and smoothed the bowl, she lines it with long, fine green leaves of swamp grasses, fibers, and a few thin, flexible twigs. The ample nest is about 7 inches in overall diameter by 5½ inches high, the inside 3½ inches wide by 2½ inches deep (18 by 14 cm and 9 by 6.5 cm). Four or five eggs, rarely six, are deposited at daily intervals. On a light bluish green ground they are blotched and spotted more or less profusely with chocolate, chestnut, beige, and pale gray.

In the cool early spring of these boreal woods, the female usually starts to incubate after she has laid her first egg. Her mate does not share this task with her but he often brings her food, sometimes calling her from the nest to take it from him with fluttering wings and low *chuck*s. After an incubation period of fourteen days, the young hatch covered with long, thin dusky down. Fed by both parents, they develop rapidly and by their tenth day are well clothed in feathers. At the age of eleven days the young abandoned the only nest that Frederic H. Kennard (in Bent 1958) was able to watch, and for the next two days they climbed and hopped through surrounding bushes, closely attended by both parents. When thirteen days old, they could fly across the stream beside which their nest was situated and, followed by their mother, disappear into the swamp beyond it.

As soon as the young can fly well and feed themselves, which occurs about mid-July in northern New England, the solitary breeding pairs begin to join others of their kind in small flocks that wander from lake to lake and forage along the shores, preparing for migration. This preparation begins in September and gathers impetus in October, when the small flocks have amalgamated in immense hordes that sometimes stretch across the sky as far as the eye can see, all flying southward on a broad front. On their leisurely journey they stop to forage on upland fields and pastures as well as in marshes, often with American Robins or Blue Jays. By mid-October the majority have settled for the winter in the southern United States, but a few remain as far north as Canada, enduring temperatures many degrees below the freezing point and finding food where cattle are fed.

Brewer's Blackbird

About the size of the more northerly Rusty, the adult male Brewer's Blackbird is likewise wholly black but in favorable light may be distinguished by the purple iridescence of his head and the more greenish sheen of his back. His eyes are pale yellow or creamy white. The adult female's upperparts, wings, and tail are dusky brown, sometimes faintly glossed with green. Her underparts are lighter gray, her eyebrows and throat still paler. Her eyes are dark brown. Juveniles resemble adult females. Brewer's Blackbirds breed in southwestern and south-central Canada, over much of the western United States, in northwestern Baja California, and eastward to the Great Lakes region. Recently they have been extending their nesting range eastward. They winter from southern Canada to southern Mexico and southeastern United States as far as the Gulf coast and western North Carolina (Bent 1958).

Birds of open country, Brewer's Blackbirds have profited by human activities in western North America, where they frequent farms, villages, and towns, as well as marshes and other wilder regions, at elevations up to 10,000 feet (3,050 m) above sea level. They forage in flocks, walking with bobbing heads and running over the ground in pastures, dry or irrigated fields, freshly plowed land, lawns, golf courses, the more open parts of marshes, sandy beaches, mudflats, and even the sidewalks of towns. A foraging flock advances with the same rolling movement as Rusty Blackbirds practice. Bird feeders attract them. With gaping bills, they overturn or push forward dry cow pats, fragments of wood, stones, or other small objects lying on the ground, to expose edible creatures hiding beneath them, but they do this less frequently than grackles with stouter bills. They wade up to their abdomens in shallow water, and on deeper ponds they walk, hop, or flutter

Brewer's Blackbird, *Euphagus cyanocephalus*, male

over a supporting mat of aquatic plants floating just beneath the surface. They hover above open water to snatch up floating creatures. Mainly insectivorous, with a taste for damselflies and grasshoppers, they diversify their diet with seeds of many wild plants and cultivated grains. They welcome tidbits of human food such as popcorn and scraps of bread, which if hard they soften by dipping in water. Although they also eat some fruits, these are evidently not their favorite food. A flock feasting in an orchard of ripe cherries abandoned it when a farmer started to plow a nearby field, to which they promptly descended, eager for the worms and grubs his plow turned up.

Brewer's Blackbirds roost in large companies, often with Red-winged and Tricolored Blackbirds, in cattails, tules, and other marsh plants, or in thick tufts of foliage in evergreen trees. Often they gather on overhead service wires before they fly, one by one, to congregate at the roost site. They retire earlier in the evening, and become active later in the morning, than associated birds, including the Red-wings.

A creaking *ksheeik* appears to be Brewer's Blackbird's nearest approach to a song. Among the dozen other utterances recorded for this bird, a metallic *check* appears to be the most characteristic.

Brewer's Blackbirds are not territorial. While flocking, they court and pair early in the year. A male and female attracted to each other walk, foraging, over the ground a few feet apart, somewhat aloof from other flock members. They perch on wires, separated by about eighteen inches (46 cm), or on the tips of pine boughs, and display to each other by fluffing out their plumage, spreading and depressing their tails, and relaxing their wings. Holding their bills horizontally or tilted slightly upward, they call *squeee* and *schl-r-r-r-up*. They may continue to display for several minutes before they fly away, usually the female first and the male following, but sometimes almost simultaneously.

In a quite different display, usually given by two males confronting each other, they stretch up their bodies and point their bills skyward, in a pose widespread among blackbirds. Rarely a female displays in this fashion toward an intruder at her nest site. As in other colonial icterids, males frequently supplant one another on favored perches, but they seldom fight. Females, quarreling over a nest site, often clash and spiral downward clutched together.

Brewer's Blackbirds breed in colonies of three or four to over a hundred nests, exceptionally as solitary pairs. They may build high in tall trees, lower in streamside willows or sagebrush, amid sedges, in a depression in the ground like meadowlarks and Bobolinks, and occasionally on ledges of cliffs. In trees nests are usually situated amid clusters of leaves near the ends of branches. The female builds the open cup without help from her partner, who accompanies her closely when she goes to gather materials and while she arranges them perches nearby, frequently displaying with fluffed out plumage and unmelodious calls, as already described. The completed structure is a typical blackbird's open bowl, tidily made with interlaced twigs and grasses, lined with coiled rootlets and horsehairs. Some nests are strengthened with a layer of dried mud or cow dung, which others lack. They are about 6 inches in overall diameter, and inside they measure 3 inches in width by 1½ inches in depth (15 and 7.5 by 3.7 cm).

Brewer's Blackbirds lay three to seven eggs, most often five or six. The pale gray or greenish gray shell is spotted and blotched with grayish brown or sepia, on some eggs so densely that the ground color is obscured. Males do not incubate but spend much time guarding their sitting mates from nearby perches. A bigamous male may alternately guard at nests of two simultaneously incubating consorts. The incubation period is twelve or thirteen days. Eggs in some nests hatch over an interval of two or three days because the female started to incubate before she had finished laying.

For six breeding seasons Laidlaw Williams (1952) studied a colony of color-banded Brewer's Blackbirds in Monterey pines growing beside a marsh at the mouth of the Carmel River on the central coast of California. The males in this colony were sometimes monogamous and sometimes polygynous; the same individual might have one female one year and more than one the following year, the number of his consorts changing with the sex ratio of the population. When the colony contained thirteen males and fourteen females, only one of the former had two of the latter. In a different year with eighteen males and thirty-six females, twelve of the males were polygynous: seven had two consorts, four had three, and one had four. In this colony, covering nine acres (3.75 hectares), the greatest density of nests was each year in the central acre, which in different years contained from fourteen to twenty-three. Several nests might be simultaneously occupied in the same pine tree; but no two were closer together than about nine feet, nor farther apart than about thirty-seven feet (2.7 and 11.3 m). The several nests of a polygynous male were sometimes widely separated, with nests of other males intervening—an arrangement one would not expect to find if these blackbirds were territorial.

Brewer's Blackbirds hatch with nearly black down. In the Carmel River colony, males assisted females feeding young both in and out of the nest. At 72 of 99 monogamous nestings and 76 of 109 polygynous nestings they brought food to their offspring, reminding us of the paternal diligence of Bobolinks. The discrepancy between the number of nests and the number of attendant males resulted largely from the disappearance of males, the loss of nests, or the lack of enough observations. A male with four consorts brought food to the nests of all of them, but not all on the same day. Although the mother usually brought food more frequently than the father, occasionally the reverse was true, especially while the female brooded younger nestlings. When a male brought food to two nests during the same hour, the number of visits to both nests did not exceed the maximum rate of feeding at a single nest, which was fourteen times per hour. The male might either feed the nestlings directly or pass his contribution to their brooding mother for delivery to them.

In the state of Washington, Orians (1980) learned that dragonflies and damselflies made a third of nestling Brewer's Blackbirds' food, butterflies and moths a quarter, flies a fifth, other insects and spiders the remainder. He made the interesting observation that when parents foraged within about fifty yards (50 m) of their nests, they returned promptly with two to four items in their bills; when they went more than twice as far, they brought fifteen to twenty or more items from each excursion, thereby saving wingbeats, as in similar circumstances we would try to save steps.

At the Carmel River colony, the young blackbirds remained in their nests for about thirteen days. Their first flights took them from three to seven feet (1 to 2 m). Both parents fed them for about twenty-five days, or until they were about thirty-eight days old. Females frequently rear second broods, sometimes carrying material for a second nest while they are still feeding fledglings. Those whose first nests fail may try once or twice more to raise a brood. A much greater proportion of nestlings (40 percent) died in nests of polygynous males than did in nests of monogamous males (7 percent). The causes of this difference are not clear. Possibly it occurred because a male could not guard two or three nests as well as he could guard one. Possibly, also, secondary females were younger, less experienced and efficient parents than primary females. Of 99 nests of monogamous males, young were reared to fledging in at least 64 (64.6 percent), whereas of 109 nests of polygynous males only 53 (48.8 percent) were successful. However, each monogamous male reared, on average, only 0.81 broods, while the average for polygynous males was 1.54 broods. Although males with more consorts fathered more fledglings, those who concentrated their attention upon a single nest increased its probability of success.

After the nesting season, Brewer's Blackbirds become more gregarious than while breeding, gathering in large flocks, often with Red-winged Blackbirds, Tricolored Blackbirds, and cowbirds. In the northern parts of their range, the flocks consist largely of male Brewer's, the females having migrated southward. In the fall and winter flocks, pairs are not evident; established mates separate and each goes its own way. Nevertheless, as the next breeding season approaches, they reunite. Of forty-five cases when both members of a primary pair returned to the Carmel colony from the preceding year, forty-two pairs were reunited, only three couples divorced. One pair remained intact for each of the five years that both were known to survive and nest.

In contrast to this constancy, no secondary female was known to breed with the same male in successive years. Most failed to return to the colony; the few that did reappear joined different males, sometimes improving their status by becoming primary consorts. A male and female already familiar with each other from having earlier nested together readily renewed their bond to raise a new brood. In other kinds of birds ornithologists have found that old established pairs cooperate more efficiently to rear the latest brood than do a male and female nesting together for the first time. This may be one of the reasons why monogamous pairs, a larger proportion of which had previously bred together, nested with greater success than did polygynous associations.

10

The Grackles

Noisy and aggressive, gregarious and abundant, often frequenting towns and villages, grackles call attention to themselves wherever they occur. With six extant species, grackles of the genus *Quiscalus* spread widely over the Western Hemisphere from northern South America and the Antilles to Canada. The Common Grackle—including the forms formerly known as the Purple Grackle, the Bronzed Grackle, and the Florida Grackle—inhabit the United States east of the Rockies and Canada as far west as northeastern British Colombia. In autumn northern birds migrate, mostly by day in great flocks, to the southerly parts of the species' range; but when supported at feeding stations some may remain through the winter as far north as Canada and Minnesota. The Boat-tailed Grackle lives, mostly in marshes, along the Atlantic and Gulf coasts of the United States from Long Island to Texas, and throughout peninsular Florida. It winters mostly from Virginia southward. In eastern Texas it mingles but rarely hybridizes with the rather similar Great-tailed Grackle, which ranges to Kansas and southwestern California, and through Mexico and Central America to northwestern Venezuela and northwestern Peru (Selander and Giller 1961).

The small Nicaraguan Grackle is confined to the southwestern quarter of that country and adjacent parts of Costa Rica. The Greater Antillean Grackle inhabits Cuba, Hispaniola, Puerto Rico, Jamaica, and neighboring small islands. The Carib Grackle ranges through the Lesser Antilles from Montserrat southward to Trinidad, Venezuela, the Guianas, and the northeastern corner of Brazil. A seventh species, the Slender-billed Grackle, lived in marshes in the vicinity of Mexico City until its habitat was destroyed and it vanished. Four little-known South American species called grackles belong to different genera and are apparently not closely related to *Quiscalus*.

A General Survey

In size, male grackles range from the eighteen-inch (46 cm) Great-tailed to the eleven-inch (28 cm) Carib. Females are noticeably smaller, from fourteen to nine inches (35.5 to 23 cm) long. Clad wholly in black, adult male grackles are slender, handsome birds, their sleek plumage glossed with iridescent shades of purple, violet, blue, green, or metallic bronze. Their long tails are keeled—folded upward in the form of a V. Their moderately long, sharp bills are black, as are their legs and feet. Except in the brown-eyed Florida race of the Boat-tailed Grackle, they have yellow eyes, which contrast with their dark attire. Adult females are duller black or olive brown, with paler underparts. Their eyes are less intensely yellow than the males'. Juveniles of both sexes resemble their mother, but males often look like adults of their sex before they are a year old.

Grackles inhabit open or sparsely wooded country, from coastal marshes to farmlands, suburban areas, and small towns, in both humid and arid regions and from lowlands to high plateaus. They

Common Grackle, *Quiscalus quiscula*, male

avoid the interior of tropical rain forests. Sociable birds, they forage in loose flocks, roost in large companies, and often nest in colonies. They find much of their food on the ground, over which they walk instead of hopping. Pushing their sharp bills into the soil, they forcibly open their mandibles in the manner of starlings, loosening the earth and enlarging the hole (as Philip Henry Gosse wrote of the Greater Antillean Grackle in Jamaica in 1847) and extracting grubs, earthworms, or whatever small creatures they find there. Pushing forward with their bills, they raise small objects lying on the ground to expose hidden food.

No single source of nourishment, animal or vegetable, satisfies grackles' omnivorous appetites. To a diversity of invertebrates, including spiders, amphipods, sowbugs, crayfish, and snails as well as insects, they add snakes, frogs, toads, salamanders, lizards, and mice. To catch small fishes, they plunge into the water from an overhanging branch and wade up to their breasts in shallows (Follet 1957). They frequently pillage birds' nests with eggs or young, and at least the larger of them catch and devour older birds, such as House Sparrows and wood warblers (Lamb 1944). To fruits of many kinds they add grains, distressing the farmer by tearing husks from ears of maize to peck out the milky unripe seeds, leaving the greater part of them exposed, to dry or decay. In harvested fields they glean spilt maize, wheat, oats, rice, or other grains. Scraps from human tables attract them. If a grackle finds a piece of bread too dry and hard to be swallowed, it softens the morsel in the nearest water. The year-long diet of Common Grackles contains more than twice as much vegetable as animal food (Bent 1958).

The utterances of grackles are hardly what one expects from relatives of the melodious orioles. Interspersed with a medley of screeches, squeals, clucks, and rattles, one often hears clear, ringing notes that increase one's respect for grackles' vocal powers. The Greater Antillean Grackle begins his song with two or three finely modulated, bell-like notes that degenerate into harsh, grating noises like those of ungreased cart wheels. The opening notes earn for him the name "tinkling grackle" in Jamaica. Grackles demonstrate their versatility by their utterances as well as by their ways of nourishing themselves.

Among mainly monogamous Common Grackles, pair formation begins in flocks in early spring. A male and female show their preferences by flying together, usually the male following the female but sometimes the reverse, and singing responsively with notes distinctive of individuals. As a male and female become more intimate, they leave the flock to fly and sing together. After cementing the pair bond, the female chooses a nest site, sometimes several weeks before she builds actively there (Petersen and Young 1950; Bent 1958; Wiley 1976a, 1976b).

Of the courtship of a promiscuous grackle, I shall tell beyond. In this family, diverse breeding systems correlate with sexual differences in size. Males of the mostly monogamous Common Grackle measure about 12 percent larger than females. Also mainly monogamous are the Carib and Greater Antillean grackles, with differences, respectively, of about 11 and 13.5 percent. Male Great-tailed Grackles of different races are 20.5 to 22.5 percent larger than females; male Boat-tailed Grackles, 21 percent larger; both are promiscuous or polygynous. These differences are based upon measurements of wings, tails, bills, and tarsi (Orians 1985). Male Great-tailed Grackles in Costa Rica weigh 8.1 ounces; females 4.4 ounces (230 and 125 g) (Stiles and Skutch 1989). Adult Boat-tailed Grackles in Louisiana average 8.5 ounces; adult females, 4.12 ounces (241 and 120 g) (McIlhenny 1937).

Nevertheless, a male Boat-tailed Grackle's success in winning females depends more upon age than upon size. In a cattail marsh in coastal South Carolina, where on average 4.7 males were present for each breeding female, William Post (1992) learned that the former were organized in an absolute dominance hierarchy, which remained stable over four years, with the older males at the top. Before the spring arrival of the females, males displayed at many potential colony sites, but after the former came, the males concentrated at sites that were occupied by the other sex, performing most freely at spots where more fertilizable females gathered. Most matings occurred at the nests, which were close together. In the absence of males of the highest rank, others almost as high in the hierarchy visited the nests and were able to mate. Middle-ranking males hovered about the edges of the grouped nests, moving from one to another, apparently seeking receptive females. Males of lowest rank rarely associated with colonies but gathered in separate areas, where they tried unsuccessfully to copulate with foraging females.

Grackles breed in colonies that may contain over a hundred active nests, but even in the more gregarious species some females prefer to rear their families in solitude. Boat-tailed Grackles breed mainly in freshwater or brackish marshes, placing their nests at no great height in grasses, cattails, and other plants. Occasionally they build in shrubs or trees, up to forty or fifty feet (12 or 15 m) above dry land. Other grackles nest in trees or shrubs, often in evergreen conifers or at low latitudes in palms. Again, these birds reveal, by the diversity of their nest sites, the versatility that has made them so successful. Common Grackles have been found nesting in holes carved by the larger woodpeckers and other cavities in trees, on the rafters and under the eaves of barns and of other man-made constructions, and, often a number together, in crannies amid the sticks of bulky Ospreys' nests. Nests of the several species range in height from a foot or two above ground or water to a hundred feet in the air (0.3 to 30 m).

Occasionally Common Grackles nest in marshes with Red-winged Blackbirds, building their open cups in clumps of cattails above water, much as Red-wings do. The latter resent the grackles' intrusion into the center of their territories but permit them to nest around the edges, in a situation that might be called partial interspecific territoriality, as described by John A. Wiens (1965).

Without help, female grackles of all species build bulky open bowls. However, when a female who had lost her nest delayed to build a replacement, an impatient male in a colony watched by R. Haven Wiley built for three days. The outer shell of grasses, weed stems, and coarsely fibrous stuffs is usually plastered inside, to within an inch or so of the rim, with mud or cow dung, which by binding the materials together makes the nest strong and durable. When present, this hard plaster is lined with fine grasses, vegetable fibers, and other soft materials, often including bits of string, rags, or paper. Grackles' nests measure four to five inches in internal diameter and three to four inches in depth (10 to 12.5 cm and 7.5 to 10 cm), being slightly more ample in the larger than the smaller species.

As in many other families of birds with wide latitudinal ranges, the number of eggs in a set increases with distance from the equator. In Trinidad at ten degrees north, Carib Grackles lay two to four eggs (ffrench 1973). In Guatemala, at fifteen degrees, Great-tailed Grackles lay sets of three twice as frequently as sets of two (Skutch 1954), while in the southwestern United States another race of this species lays three or four eggs, occasionally five. Boat-tailed Grackles usually lay three eggs in Louisiana and Florida, but farther north they produce sets of three to five. Common Grackles, which range to higher latitudes than any other species, lay four to six eggs, sometimes seven. Grackles' glossy eggs are blue, greenish blue, pale bluish gray, or pale greenish white, sometimes suffused with brown or pale lilac. They are variously scrawled, dotted, blotched, or streaked with brown and black. Incubated by the female alone, those of the smaller species, the Common and the Carib, hatch in twelve or thirteen, rarely fourteen days, whereas for those of the larger Boat-tailed and Great-tailed Grackles the incubation period is thirteen or fourteen days.

Hatchling grackles have pinkish skin shaded by sparse gray or brownish down on the head, back, wings, and thighs. Their eyes are tightly closed. Their mouths, gaping widely for food, reveal a red interior. As far as known, only in the Common Grackle does the male regularly help his mate to nourish their nestlings. At some nests he brings food slightly more frequently than the female does; at others he feeds much less often but may compensate for delivering fewer meals by making them bigger. Consisting mainly of insects and other invertebrates, with a little grain, food for the nestlings is carried in the parents' crops as well as in their bills (Hamilton 1951). In the botanical garden of the University of Michigan, Henry F. Howe (1979) watched a marked male feed nestlings in two nests at the same time, suggesting bigamy. Another male Common Grackle, who abandoned his first mate after she laid her eggs and acquired another, fed only the nestlings of this second mate. Later, he consistently followed another female, who might have been his third consort. Similar nuptial inconstancy by male Common Grackles was detected in a colony on the grounds of the New York Zoological Society by R. Haven Wiley (1976a). As a rule, only females brood the nestlings, but George R. Maxwell and Loren S. Putnam (1972) watched a male Common Grackle cover nestlings on five occasions totaling thirty minutes. Male Common Grackles also remove fecal sacs.

Common Grackles sometimes serve as helpers. Howe (1979) watched a male who had lost his brood to a predator feed young in a neighboring nest. His mate did not join him in this service. To another nest two males brought food throughout the nestling period. When they met at the nest, neither was antagonistic toward the other. The mother was equally tolerant of both, one of whom was evidently her mate. In this colony, as many males were helpers as were polygynous. In Quebec, Canada, a female Common Grackle fed and guarded three nestling Chipping Sparrows—unexpected behavior of a bird that sometimes preys on other birds and their young, but revealing once more how the strong appeal of hungry babies can overrule the predacious impulses of both birds and mammals (Bent 1958).

Common and Carib Grackles leave their nests when thirteen or fourteen days old, but for big Great-tailed Grackles the full nestling period is twenty to twenty-three days. Fledgling grackles resemble their mothers but have brown instead of yellow eyes. Promiscuous or polygamous males who fail to feed their nestlings nearly always neglect their fledged young. However, exceptions occur; in late July in Texas, Robert K. Selander (1970) watched a male Great-tailed Grackle walk across a lawn, persistently followed by two begging juveniles, whom he fed.

Young grackles acquire plumage essentially adult by postjuvenal molt, and before the following breeding season they differ little in appearance from adults of their sex. Females commonly nest as yearlings, but male Great-tailed and Boat-tailed Grackles seldom breed before they are about two years old. Yearling males tend to remain aloof from breeding colonies, and older ones may be present less constantly than females attending nests. This situation leads to counts of two, three, or more females to every male, and to the conjecture that this strongly unbalanced sex ratio goes back to the eggs. However, the sexes hatch in approximately equal numbers, with sometimes a small preponderance of males. Although in the same clutch grackles' eggs may differ in size, this does not affect the sex of the embryos they contain. Male and female grackles also tend to fledge in equal numbers, except when food is scarce and males, who grow faster and need more of it, may starve more often than their sisters. Similarly, in winters when food becomes scarce, big male Great-tailed Grackles at the northern extremity of their range may succumb more frequently than the smaller females, who can survive with less nourishment (Selander 1958, 1960, 1961; Howe 1976, 1977, 1978; Bancroft 1984, 1986).

The Great-tailed Grackle

In northern Central America, the Great-tailed Grackle has long been one of the most abundant and familiar of birds. In Costa Rica

Great-tailed Grackle, *Quiscalus mexicanus,* male

it was confined to the mangrove swamps of the Pacific coast until the 1960s, when it began to spread widely over the country. Wherever these big grackles occur, they nest in palm trees in the central plazas of towns and in the evening stream in noisy flocks from the surrounding fields where they forage to roost in village shade trees. This close association with humans has earned for male and female grackles different names in countries where a single vernacular name serves for a whole family of birds, as *carpintero* for all the woodpeckers, and *chorcha* for a diversity of orioles. The big, ostentatious male grackles are known by everybody as *clarineros* (trumpeters), the more modest females as *sanates.*

I have never willingly remained long in towns, but in 1932 I dwelt on a plantation in the midst of Great-tailed Grackles (Skutch 1954). The house at "Alsacia" stood on the upturned end of a ridge projecting from the Sierra de Merendón, on the border between Guatemala and Honduras, into the broad valley of the Río Motagua and its tributary, the Río Morjá. From this elevation we looked over great expanses of banana plants and much resting land covered with various stages of second growth, several hundred feet below. Around the house grew many tall coconut palms in which about a hundred grackles roosted and nested.

From February to July, I awoke every morning to the sounds of these close neighbors. In the earliest dawn, the clarineros repeated, over and over, in a calm, subdued voice, a prolonged note between a screech and a whistle, sounding pleasant and contented, and reminding me of a musician running up the whole scale of a stringed instrument with one deft stroke. How different from the shrill notes these birds would utter later in the day, at the height of their nuptial zeal!

As daylight grew stronger, with much commotion and clucking by the sanates and calling by the clarineros, the grackles emerged from amid the coconut fronds where they slept and flew down to seek breakfast. Many went directly to the *Conostegia,* a shrub of the melastome family with small pink flowers that grew abundantly on the steep, grassy slope below the house, to eat the little, sweetish black berries. Others settled in the cow pen, on the road, or on the lawn, where they walked about gathering small creeping things from the bare ground or the grass. One morning I watched four sanates plucking parasites from a gaunt old cow. One stood on her back, picking vermin from her hair and, after slight resistance, permitted another to alight beside her and share the feast. Two more sanates moved about on the ground at the cow's feet, at intervals jumping up to snatch something from her belly or flanks. They clambered over her legs and tail, performing the same cleansing service, while the cow stood patiently still.

On leaving the roost, many of the grackles flew directly down to the valley. As the sun rose higher, the rest followed, singly or in small groups, to the shores of the Río Morjá, which wound through a banana plantation half a mile away. Here they foraged along the moist shore or in the shallows, or searched among the piles of driftwood and washed-out banana plants stranded in the shoals. The clarineros walked sedately over the shingly beach, turning over small stones by inserting their bills beneath the near edge and pushing forward, to uncover small invertebrates lurking beneath them. The less powerful sanates engaged less in this activity. Frequently the grackles waded in shallow water to capture tadpoles, small fishes, and other aquatic creatures. I did not see them fish by plunge-diving or by flying low over the surface and seizing prey with a deft snap, hardly wetting their feathers, as others have reported.

On hot afternoons the grackles delighted to bathe at the stream's margin, shaking wings and tail so vigorously that they sent up a spray of crystal drops to sparkle in the sunshine. One afternoon I saw a sanate stand as close as she could to a bathing clarinero, profiting by the shower that he raised, much as suburban grackles bathe in the spray from a lawn sprinkler. Finally, all the bathers flew up to the boughs of the willow and cecropia trees along the banks, vigorously shook the water from their plumage, and carefully preened their feathers with slender bills.

As the sun sank low and the air cooled, these grackles flew up the hill in small flocks, often calling like a company of Common Grackles, to congregate again in the coconut palms. On the way many would settle again in the *Conostegia* shrubs for a dessert of berries before going to roost. From their arrival until dark, our hilltop presented a lively spectacle. The varied calls and squeals of the clarineros mingled with the constant chatter of the more numerous females. Many would alight on the fronds of a single palm, only to fly to another, back and forth a dozen times before settling down for the night. The fresh breeze that generally blew up from the valley at sunset, tossing the great fronds of the coconut palms, made it more difficult for the birds to roost. The clarineros' long tails flagged back and forth as they perched on the leaves, obviously annoying them. I delighted to watch their graceful maneuvers while they hovered, soared, and posed with dangling legs above the treetops, reminding me of gulls playing above a windy seashore.

On some exceptionally windy evenings the clarineros engaged in spectacular encounters, meeting face to face and rising far above the crests of the palms, until the breeze took hold of them and twisted them around, forcing them to turn away from their opponents and devote their attention to the maintenance of their own equilibrium. These confrontations appeared to be playful rather than hostile, a way of increasing their enjoyment of the wind by vigorous exercise in it. It appears that male grackles of all species seldom fight. In the months that I lived amid this colony, I saw not a single serious conflict among them; but later, while traveling by train through southern Mexico, I watched two male Great-tails clinched together and tumbling about on the ground as long as I could keep them in view from the car window.

The sun hung well above the western hills when the grackles began to congregate in the coconut palms; the last red glow was fading from the sky when finally all had ensconced themselves out of sight amid the inner fronds, and their last drowsy notes gave way to the awakening calls of the Common Pauraque. But, especially at the outset of the breeding season, the clarineros slept lightly and repeatedly awoke during the night to shatter with shrill calls the monotonous humming of insects.

At first I was happy to have such active, spirited birds as close neighbors, but before long I began to wish them elsewhere, for like other grackles they plundered the nests of small birds. A number of thrushes, tanagers, flycatchers, and other birds that I desired to study built their nests on the hilltop, but few succeeded in rearing their young. The grackles so well kept all other large birds at a distance that I did not doubt that they themselves were responsible for most of the losses, especially after I surprised a clarinero in the act of pillaging a Yellow-throated Euphonia's nest.

The big male Great-tailed Grackle glides downward with spread wings set, the tips of his primary feathers distinctly curved upward by the weight of his body, his long tail folded upward until the feathers lie in a vertical plane, like a rudder, and vibrating from side to side in the air flow. Usually he flies upward with resonant wing-beats, like those of a male Montezuma Oropendola, but at times he flies silently. The female's flight is almost noiseless, but when she labors upward with long strands for her nest streaming from her bill, her wing-beats may be sonorous like the male's, although not as loud. Occasionally she folds her tail upward in the manner of the male, but less completely. In sustained flights on a horizontal or slightly inclined course, both sexes fly with regular and rapid wing-strokes, neither folding their wings intermittently nor spreading them for gliding.

I never heard sustained singing by the clarineros, but I admired the range and flexibility of their voices. At one extreme was a rapid sequence of pleasant little tinkling notes; at the other, calls so loud that they were best heard at a distance. A characteristic utterance was a prolonged slur, between a squawk and a whistle, rising slowly through the tonal scale. Flying or at rest, they repeated a resonant *tlick tlick tlick;* and I often heard a spirited, rollicking *tlick a lick tlick a lick*. A rolling or yodeling call, very vigorous, contrasted with the lazy, screeching note that reminded me of a gate swinging slowly on rusty hinges. Occasionally a perching clarinero puffed himself up, swelling out all his feathers, and with half-open bill slowly expelled air with a low, undulatory sound, like whistling through one's teeth.

As musicians, the clarineros did not lack originality. From time to time they invented a verse I had not previously heard, and if it pleased them, they repeated it over and over. One bird discovered a pretty phrase—*wheet-tóck*—which he voiced many times for a week or more, until, like a popular song, it grew stale and was neglected. Another clarinero delighted in a buglelike call that sounded like *ta-dee ta-dee ta-dee,* very martial and stirring, especially when heard at a little distance, for it carried far. After calling, the clarineros usually stood with their long, sharp bills pointing straight upward, a pose that displayed well the sleek glossiness of their purplish necks. They assumed this posture on other occasions. Sometimes two resplendent clarineros, perching side by side on a coconut frond, pointed upward simultaneously and held this pose for a good fraction of a minute, looking very self-conscious. Widespread in icterids, this display is held to be aggressive, but I never saw it lead to a clash or a chase. Females, too, sometimes assume this pose. It appears to be a way that icterids fend off an overly close approach by other individuals, of preserving their "individual distance," which swallows and other birds also insist upon maintaining, rather than an expression of hostility.

Sanates are not only smaller but quieter than clarineros. Their most characteristic utterance is a rapid, clicking *tlick tlick tlick,* sharper and less sonorous than the corresponding note of the males. I heard this while they flew, built their nests, and quarreled with their neighbors. Both sexes utter single throaty clucks. Sometimes a sanate attempts to reproduce the clarineros' prolonged slur, but hers is a weak, squeaky imitation.

When, in the last week of February, the sanates started to build nests in the coconut palms, the noise and excitement in the colony reached their highest pitch. The clarineros did not appear to claim territories or to form more than random and temporary alliances with the females, who outnumbered them two or three to one. This situation, fraught with possibilities for conflict, did not lead to struggles among the males. Sometimes, after standing side by side on a perch, calling peacefully, for several minutes, one would suddenly rush at the other and drive him away; but the bird thus threatened never turned to fight, and the other did not pursue him.

The clarineros courted ardently. Often one alighted beside a sanate foraging or gathering nest material on the ground, to address her with half-raised, quivering wings, tail level with his body, head depressed, plumage all fluffed out to make him appear bigger than he was. With half-opened bill he uttered pleading calls, shrill and insistent, or as soft and appealing as the peeps of a lost chick. No matter what tone he chose, the insistent suitor was sure to be ignored by the busy sanate, resolutely continuing her occupation. At other times a clarinero alighted beside a sanate in a tree and wooed her in much the same way, with no more favorable result. When the sanate failed to respond, the clarinero's interest died almost as suddenly as it had flared up.

The courtship flights of the grackles were spectacular. They started when a sanate flew away from a clarinero who addressed her on the ground or on a coconut frond, or when he tried to overtake her as she went about her usual business. As she fled, he voiced shrill nuptial calls and increased his speed to reach her. Doubling and twisting and dodging, she used every stratagem to escape him. Far out over the valley they sped, until they were high above the tallest silk-cotton trees. Closely as he pursued her, she always managed to avoid him; I never saw one of these chases lead to a capture. The wild flight over, the birds doubled back to the hilltop separately or together, or continued their flight to the river.

Although the clarineros were so spirited in courtship, the sanates decided when they desired this attention, which was often about midafternoon. Then a sanate vibrated her wings and called with pleading peeps, much weaker than the male's. Sometimes she might continue for a considerable interval without attracting one of the several clarineros in sight. When a clarinero responded, he flew with shrill, ear-splitting cries and quivering wings to mount her. After a union that lasted only seconds, they separated, perched not far apart with wings strongly vibrating, and continued to call with voices weaker than at first and soon fading away.

The preferred site for the sanates' nests was at the very center of a palm tree's crown, between the two youngest of the expanded fronds, which stood upright, face to face, offering between their broad green expanses a secluded nook where the structures could be hung. The wind sent ripples over the pleated surfaces of these young leaves and tossed the older fronds below them; the sun at high noon poured its rays between the vertical new leaves. A more attractive site for a bird's nest is hard to imagine. The first sanate to build in a coconut palm always selected this location, and sometimes she permitted another sanate to hang her nest between the same fronds, but on the opposite side of their rachises. If any mishap befell one of these nests and left the site vacant, it was most likely to be occupied again, as long as the sanates continued to build.

Here the nests were suspended, vireolike, by strands looped around the ribbonlike divisions of the frond and woven back into the wall, much as grackles build amid upright stems or the branches of shrubs and trees. This was not only the most sequestered site but the cleanest that a coconut palm offered. However, it had one great defect. As the two fronds that supported a nest continued to grow from the base at unequal rates, they tilted the heavy structure, inadequately upheld by slender green leaflets. Even if the whole nest did not fall, the eggs rolled out. We tried to save these nests by tying them up with string, but despite this help nearly all met disaster.

Lower in the palm tree, the broad, concave bases of the stalks of mature fronds offered more stable sites for nests, which usually rested there without being attached, unless nearby inflorescences with flowers or immature fruits offered branches to which the nests could be bound. In the bases of the lowest and oldest fronds fallen flowers, blasted fruits, and other debris accumulated and turned to mold, in which graceful pendent ferns, grasses, and other plants rooted to form little aerial gardens. Among these lowest fronds stinging ants established their colonies, spiders spun their webs, and other small creatures lurked. Few grackles nested here.

At first I hesitated to climb amid the coconut fronds, for these great growths are leaves, and people who grow up in a temperate zone are not accustomed to stand upon leaves. But after watching the boy half my weight whom I sent up to report climb among the fronds with perfect safety, I overcame my qualms and ascended myself to examine the nests and their contents. Avoiding the oldest fronds about to fall, I rested confidently on the bases of the others.

Five or six sanates often nested at one time in the same coconut palm. Each gathered from the ground weed stalks, small uprooted grasses, and miscellaneous vegetable materials and piled them on the base of a frond for a foundation. Above this she wove a bulky open cup with coarsely fibrous stuffs, chiefly uncleaned strips from decaying outer leaf-sheaths of banana plants, mixed with straws, bits of rags and string, and fibrous herbaceous stems made flexible by decay. Next she plastered the shell with cow dung or mud, then lined the cup with fine fibers.

In contrast to the peaceful coexistence of the idle clarineros, the busy sanates were quarrelsome. Conflicts arose when two desired the same nest site, or when two building close together interfered with each other. With open bills and high-pitched, irritated cries, they menaced each other, until they clashed face to face and fluttered earthward. After these flurries, they separated to gather more materials or to fly off to forage. These tiffs, which apparently never resulted in injuries, caused the sanates to scatter their nests instead of crowding them excessively.

Like colonial birds of many kinds, the sanates often tried to steal nest materials. Frequently a bird would seize the end of a long strand that dangled from a neighbor's bill. Side by side on a coconut frond, the two would tug at the coveted piece until one tore it from the other's grasp and carried it to her own nest. Sometimes one or two sanates pursued a third who had found an exceptionally desirable piece. The sanate whose fibers were rudely snatched from her bill never tried to retaliate but soon went off to search for more. The clarineros viewed with indifference these feminine quarrels, from whatever cause they arose.

The sanates who worked hardest finished their nests in five days, but others, probably less experienced, took twice as long. Four or five days after a nest was finished, it received eggs, the earliest in the colony on March 3. Usually an egg was deposited each day until the set of three, or less often two, was complete; rarely two days elapsed between the appearance of the first and second eggs in a set of two. The bright blue or pale bluish gray eggs were dotted, blotched, and intricately scrawled in patterns so diverse that, if all the eggs in the colony had been mixed together, each sanate might conceivably have recognized her own.

The incubating sanates were too well hidden by the palm fronds to be watched, but soon it was evident that all did not go well with them. Not only from the nests precariously hung between the youngest fronds, but even from those firmly set on the bases of mature fronds, eggs began to disappear in alarming numbers. The chief cause of loss seemed to be the grackles' unfortunate habit of roosting in the trees where they nested. At any one time, roosting birds were more numerous than those incubating or brooding nestlings. As this majority settled in the palms each evening, the excitement and disorder were so great that I wondered how sanates managed to remain on their nests. The angry cries that at this time emanated from birds unseen amid the coconut fronds seemed to come from sanates trying to protect their eggs. Not only was the safety of the nests jeopardized by the disorder so prevalent at nightfall, but the nests and their surroundings were soiled by droppings from the roosting birds. As the nesting season advanced, increasing numbers of grackles roosted in nearby orange and grapefruit trees, improving the situation in the palms. Colonial oropendolas, caciques, and doubtless other groups of grackles arrange things better. At night all members of the colony not incubating or brooding inside their nests roost at a distance.

After an incubation period of thirteen to fourteen days, the surviving eggs hatched. Only rarely in nests with two eggs did both hatch on the same day. More often, one egg hatched each day, so that in sets of three hatching spread over as many days; or two might hatch on one day and the third on another day. In a few nests where I marked the eggs as they were laid, they hatched in the order of laying.

The sanates hunted most of their nestlings' food on the ground and routinely bore it a long distance, sometimes from the shore of the Río Morjá half a mile away and several hundred feet below. They brought grubs extracted from the grass roots, green caterpillars, and occasionally small lizards. Although the clarineros failed to attend the nests, they defended them valiantly. Whenever they spied a human approaching the coconut trees, their sharp *tlick lick, tlick lick* warned the mothers to flee from their nests, with the result that I could rarely glimpse them while they incubated or brooded. When I or my young helper climbed into the crown of a palm that sheltered nestlings, the noise and excitement were overwhelming. Clarineros and sanates, including those whose nests were safe from intrusion in neighboring trees, circled around and filled the air with excited clucks.

No hawk or other big bird could fly with impunity close to our hilltop. Clarineros joined sanates in harrying Turkey Vultures and Black Vultures that approached the palms or attempted to alight on them, pursuing the scavengers far down the hillside, striking them repeatedly on the back until they retired to a satisfactory distance. The grackles also attacked a Great Curassow, possibly the first they had ever seen, who with labored wing-strokes flew over the hilltop. As the heavy bird approached the summit, two clarineros and several sanates flew out from the palms to buffet him. Probably already fatigued by this unusual journey, the curassow wavered in his course and lost altitude as his assailants beat down upon him, but he managed to remain airborne until he rounded the brow of the hill and passed from view.

In the marshes of Louisiana, colonies of Boat-tailed Grackles containing 100 to 250 crowded nests were found to lack male guardians, but groups of 6 to 20 nests were commonly protected by a single male (McIlhenny 1937). Similarly, although I found no evidence of territoriality among the Great-tails in the palm trees around the house, an isolated palm at a distance appeared to be the domain of a clarinero who guarded the single sanate nesting there. While I rested in the crown of this palm, examining the nestlings, he ventured closer to me than any of the others dared to approach. Often he alighted on the end of the frond against which I leaned, which bent perceptibly under his weight and made me instinctively clutch another support. He interrupted his clucks with rapidly repeated tinkling notes and indescribably harsh, agonized shouts, which set sanates circling around the tree into faster movement and louder calling. The mother of the nestlings in the palm relieved her feelings by angrily pecking the frond where she rested.

At the age of two weeks or a little more, the young grackles were feathered. When sixteen to nineteen days old, they would try to crawl from their nests when I visited them. The rasping cry of distress that nestlings of this age voiced when touched drove the adults to a frenzy. Nineteen-day-olds could not yet fly but remained climbing over the broad bases of the coconut fronds for two or three days longer. The full nestling period was twenty to twenty-three days.

Despite many failures, by persistent efforts the sanates succeeded in rearing a goodly number of offspring. Soon after leaving the nest, the young grackles began to follow their foraging mother, before long going as far as the river, where they perched on banana leaves arching over the bank while awaiting her return from searching for food along the shore; or else they pursued her along the sandy margin of the stream, begging with quivering wings. The youngsters' half-pleading, half-imperious *witit witit* mingled with the clucks and whistles of older birds. Young clarineros continued to solicit food from mothers smaller than themselves. I watched two fledglings, a clarinero and a sanate, alternately beg and receive food from their mother and help themselves to the ripe banana she was eating. While young birds waited to be fed in the hibiscus hedge beneath the coconut palms, they plucked off leaves and bright red flowers, or pecked at unopened buds, as though trying to find food for themselves before they knew what was edible.

At least one sanate, easy to recognize because she had somehow lost her tail and gotten a piece of bright red tape entangled around her right leg, built a second nest and hatched a second brood after successfully rearing her first. How many others were double-brooded I could not learn.

By the first of July the nesting season was drawing to a close. On the night of July 8 the grackles roosting in the palms were restless, shifting their positions and often crying out in the dark. After this, most of them abandoned the hilltop that had so long been their home, leaving only a few sanates with nestlings, one whose eggs were just hatching, and two faithful clarineros. The early mornings were strangely silent without the grackles (Skutch 1954).

11

The Small Cowbirds

Among the most deeply rooted and persistent of the activities of birds are the incubation of eggs and the nurture of young, both indispensable for their perpetuation. Of about three thousand species of birds in the Western Hemisphere, only ten are known to neglect these activities totally, foisting their eggs upon unwilling or unsuspecting foster parents. One of these brood parasites is the Black-headed Duck of Argentina. The other nine are altricial birds, including three species of cuckoos and six of cowbirds. One of these, the Giant Cowbird, I shall turn to in the following chapter. The five smaller parasitic cowbirds belong to the genus *Molothrus,* which now claims our attention.

The decay of parental instincts is so surprising, the shifts adopted by brood parasites to compensate for this loss so amazing, the effects upon their victims so various and often so disastrous, and the origin of the parasitic habit so obscure, that brood parasites have invited a vast amount of investigation and speculation. Few New World birds have been the subjects of so many studies as have the Brown-headed Cowbirds, which have become one of the most widespread and abundant of avian species in temperate North America, where so many professional ornithologists and amateur bird watchers reside.

The Bay-winged Cowbird

Only one cowbird, probably the ancestor of all the others, rears its own young: the Bay-winged of South America. Among the idiosyncrasies of this eight-inch (20 cm) bird is its coloration; the sexes are alike in their grayish brown attire, paler below than above, blackish tail, rufous-chestnut wings, black bill and feet. Adult males resemble females and retain their juvenile colors all their lives. The species ranges from eastern Brazil to Uruguay, Bolivia, and central Argentina, and from sea level to at least 9,000 feet (2,750 m) in the Bolivian Andes. It does not migrate but in winter wanders locally in flocks of about ten to thirty individuals, who often frequent gardens and dooryards. Like other cowbirds, Bay-wings forage walking over the ground in open places or light woods, gathering insects, seeds of wild plants, and fallen grains, mostly maize and rice. Hudson (1920) found them singularly inactive birds, slow and deliberate in their movements, spending much of their time resting in the trees.

Throughout the year, flocks of Bay-wings sing together in a leisurely manner that seems to express a calm temperament. Although their soft, sweet, clear notes are delivered with no attempt to keep time, they compose a pleasant chorus. In cold weather, a flock seeks a sheltered sunny spot on the northern side of a wood pile or hedge, where the birds loiter several hours each day, sitting inactive and chanting in their usual quiet, subdued style. With trilling notes they express alarm or curiosity. When about to leave a tree where they have been resting, they announce their intention to fly with long, clear notes, audible a quarter of a mile away.

In keeping with their tranquil, unimpassioned disposition, Bay-wings form monogamous pairs without evident courtship displays. They breed late, chiefly from November to early March, when other birds' nests, which they covet, have become more abundant and readily available because the young of the original occupants have flown. Unable to build the covered nests that they prefer, they claim the massive structures of interlaced sticks made by birds of the ovenbird family, especially Firewood-gatherers, Rufous-fronted Thornbirds, Short-billed Canasteros, and castle-builders or spinetails (*Synallaxis* spp.). Lacking these, they will occupy the clay ovenlike structures of Rufous Horneros, the domed nests with a side entrance of the Great Kiskadees, or a birdhouse. If they cannot find an empty nest that they like, they may fight to evict the builder, occasionally injuring severely if not killing a defender. Mostly they appear to prefer unoccupied nests that they can obtain peaceably. Hudson reported that Bay-wings open a gap in the side of a Firewood-gatherer's nest for the better illumination of the interior, but this appears to be unusual. After taking possession of a covered nest, they add to whatever lining they find already present and make minor alterations. If no roofed nest is available, a pair of Bay-wings, both sexes working together, will build an open cup of straws and weed stems in a tree, in a hole in a wall, or on the crossbars of a telegraph pole, at heights of six to twenty-five feet (2 to 7.5 m) above the ground.

After a nest is lined with horsehair or feathers and ready for occupancy, the female sleeps in it while she lays four or five eggs, which are grayish white and heavily marked with grayish or reddish brown. Larger numbers up to twenty in a nest have been reported, but most appear to have been laid by parasitic Screaming Cowbirds. It is not surprising that visiting females of a species so social as the Bay-wing may add an egg or two to the clutch of the resident female. During the incubation period of twelve days, the female alone covers the eggs, and only she broods the nestlings, but the male helps to feed them during the fourteen or fifteen days that they remain in the nest and after they fledge. He is the nest's chief guardian, mobbing predators and chasing intrusive Screaming Cowbirds.

In the province of Buenos Aires, Argentina, Rosendo Fraga (1972) found six to eight pairs of Bay-wings each year in an area of about one and three-quarters acres (0.7 hectares) where Firewood-gatherers abounded. The resident population of Bay-wings consisted of eight females and about fourteen males, of whom most individuals of both sexes were captured in mist nets and color-banded. Most nesting pairs were assisted by one or two helpers, who not only fed the nestlings but cooperated in defending the nests from predators and other trespassers. Although some nests lacked such auxiliaries, as soon as the young fledged extra individuals began to feed and protect them. After raising his own young, a Bay-wing might help at the nest of another pair. Young Bay-wings were often surrounded by a noisy, excited crowd of parents and their assistants. Once Fraga counted eighteen individuals in such a gathering, but with so many milling around, he could not ascertain how many were attending the fledglings. These aggregations included nonbreeding birds that wandered about his study area and often roosted there. Young Screaming Cowbirds, raised by the Bay-wings and difficult to distinguish from their own young, also benefited from the helpers' activities.

After her first brood fledged, a female deserted her mate to join one of her helpers, with whom she raised another brood. Both the change of partners and the second brood were exceptional in this population of Bay-wings, which are usually monogamous and have neither time nor enough available nests to raise more than one brood in a season. In the northwest Argentine province of Catamarca, Orians and his family (1977) found more than two adults in attendance at each of the three nests with young that they watched.

The Screaming Cowbird

Closely associated with the Bay-wing is the Screaming Cowbird, which parasitizes it. This bird resembles its host not only in size but in the absence of the sexual differences in plumage that we find in other cowbirds; however, in this case both sexes have male rather than female coloration. Their silky black plumage, glossed with purplish blue, is variegated by only a small chestnut spot beneath each wing, which is never prominent and sometimes lacking. Their bills and feet are black like their plumage. In southern Brazil, Bolivia, Uruguay, and northern Argentina, this species follows the range of its host, but it is absent from eastern Brazil, where the northern race of the Bay-wing resides. It does not migrate.

Less abundant and less social than their hosts, Screaming Cowbirds live at all seasons in pairs or, in winter, in parties of seldom more than half a dozen individuals. They eat small seeds and tender buds, and where horses and cattle are fed augment their diet with fallen corn. They accompany grazing quadrupeds to catch insects stirred up from the herbage by their movements. The Screaming Cowbird's temperament contrasts strongly with that of the placid Bay-wing, for, as Hudson (1920, vol. 1, p. 97) wrote,

> One of its most noteworthy traits is an exaggerated hurry and bustle thrown into all its movements. When passing from one branch to another, it goes by a series of violent jerks, smiting its wings loudly together; and when a party of them return from the fields they rush wildly and loudly screaming to the trees, as if pursued by a bird of prey. They are not singing-birds; but the male sometimes, though rarely, attempts a song, and utters, with considerable effort, a series of chattering, unmelodious notes. The chirp with which he invites his mate to fly has the sound of a loud and smartly given kiss. His warning or alarm note when approached in the breeding-season has a soft and pleasing sound; it is, curiously enough, his only mellow expression. But his most common and remarkable vocal performance is a cry beginning with a hollow-sounding internal note, and swelling into a sharp metallic ring; this is uttered with tail and wings spread and depressed, the whole plumage raised like that of a strutting turkey-cock, whilst the bird hops briskly up and down on its perch as if dancing.

Herbert Friedmann (1929) regarded the last of these vocal performances as the rarer of the Screaming Cowbird's courtship displays, at least in the more northerly parts of Argentina where he studied this bird. On one occasion, the male approached closer to the female each time he jumped, until she flew away. Usually he stood so near the female that he could not hop closer without bumping against her. In the display that Friedmann most often saw, the bird delivered, as with an effort, the explosive, squeaky *dzee* without hopping up and down on his perch. The monogamous, constantly paired Screaming Cowbird displays less frequently than do the other, more widespread, parasitic species. Friedmann saw its courtship displays only in the trees, never in the air or on the ground, where other cowbirds perform. On the single occasion when he saw a Screaming Cowbird give the widespread agonistic display with bill pointing straight upward, it was directed toward a male Shiny Cowbird while the two ate close together.

Until recently, the Screaming Cowbird was known to intrude its eggs only into nests of the Bay-wing, to which it is closely related. Like this host, it breeds much later in the year than do most other passerine birds in subtropical and temperate South America, waiting until the preferred nests of both species, especially the massive structures of Firewood-gatherers, are no longer occupied by the builders' own broods and are available without fighting for them. The Screaming Cowbird is generally credited with laying five eggs, but the number is difficult to determine in this, as in other, brood parasites. Because of its territorial habits, only one female is likely to lay in the same Bay-wing's nest, but she might distribute her eggs between two or more of the hosts' nests that happened to be in the territory claimed by her and her mate. However, in northwestern Argentina Gunnar Hoy (Hoy and Ottow 1964) found nests with six to twenty eggs that he attributed to the Screaming Cowbird. Most of these eggs were laid before the resident Bay-wing started to lay and were thrown out by her. Although the Screaming Cowbird's eggs closely resemble those of the Bay-wing, in any one nest it is usually possible to distinguish them; just as we can frequently distinguish the eggs of two females of the same species who happen to lay in the same nest. However, so great is the overlap in the appearance of the eggs of these two cowbirds that if a large number of both kinds are mixed together, they cannot with certainty be separated. The Screaming Cowbird's eggs have harder shells than those of the Bay-wing—a difference between the eggs of cowbirds and those of their hosts that we shall notice again in other species. The Screaming Cowbird's eggs are marked with purplish brown on a white ground, whereas the spots and blotches on the Bay-wing's eggs are mostly reddish or dark grayish brown, but the difference is not constant enough for certain identification. Incubated by a female Bay-wing, the Screaming Cowbird's eggs hatch in twelve to thirteen days, like her own eggs.

In Paraná (Rolândia) in Brazil, where the Screaming Cowbird is a recent immigrant and the Bay-wing appears to be absent, Helmut Sick (1993) found the former parasitizing the Chopi Blackbird, which nests in a diversity of holes, crannies, and other covered spaces. The Chopi rears the fosterlings with its own young. One Chopi's brood consisted of one of its own nestlings, one Screaming Cowbird, and three Shiny Cowbirds, five in all. In another Chopi's nest were three Screaming Cowbirds and one young Chopi. One Screaming Cowbird and three Shiny Cowbirds occupied a nest without a host's nestling.

Nestling Screaming Cowbirds closely resemble the nestling Bay-wings that share a nest with them. Friedmann noticed one difference: their fear reaction is weaker than that of young Bay-wings; a week-old Screaming Cowbird begged for food in his hand as a Bay-wing of the same age would not do. However, the fear reaction is even less developed in other parasitic cowbirds raised by a diversity of foster parents. After leaving the nest at about the same time, the two kinds of fledglings are too similar in appearance, voice, and behavior to be distinguished by ornithologists or, we may confidently assert, by the foster parents who attend them. Only when, in the postjuvenal molt, black feathers begin to appear in the grayish brown plumage of the young parasite's juvenal attire does it reveal its identity. Finding some of the changelings in flocks of Bay-wings, and finally seeing one of them fed by a Bay-wing, led Hudson to the disclosure of the Screaming Cowbird's parasitic habits, a discovery that pleased him as much as if he had "found a new planet in the sky"—all of which he relates in detail in the longest chapter in *Birds of La Plata.* He rightly concluded that the close resemblance of the Screaming Cowbird's eggs and young to those of the Bay-wing was not a case of protective mimicry, developed by natural selection to make the intruders more acceptable to their foster parents, but a consequence of their common ancestry. As the postjuvenal molt progresses, the young Screaming Cowbirds become curiously attired in a motley plumage of brown and black, still with chestnut on their wings, well illustrated in a painting by H. Gronvold in *Birds of La Plata.* After the completion of this molt, Screaming Cowbirds of both sexes resemble adults in their black plumage glossed with purplish blue but are less shiny.

The Shiny Cowbird

Far more widespread than either the Bay-winged or the Screaming Cowbird, and far less restricted than the latter in its choice of hosts, the Shiny Cowbird differs from both in the contrasting plumage of the sexes. The glossy black eight-and-a-half-inch (22 cm) male gleams with purple or purplish blue on his head, neck, breast, and upper back; blue on his lower back and abdomen; and greenish blue on his wings and tail. The slightly smaller female is grayish brown, darker on the upperparts than below, with whitish throat and eyebrows. In both sexes the short, stout, conical bill and the legs are blackish, the eyes brown. Shiny Cowbirds spread over almost the whole of South America except its southern tip, and from lowlands upward to at least 11,800 feet (3,600 m) in the Andes of Bolivia. Like a number of other icterids, they have been expanding their range as closed forests, which they avoid, are destroyed to make farms and pastures. Since the beginning of this century, they have been spreading northward and westward through the Antilles, reaching Puerto Rico in midcentury and Hispaniola in the 1970s (Post and Wiley 1977a). In 1982 they were first reported in Cuba, and a few years later stragglers appeared in the southeastern United States (Hutcheson and Post 1990; Post, Cruz, and McNair 1993). They are also established in eastern Panama. Although resident through most of their range, in Argentina Shiny Cowbirds migrate in large flocks that travel chiefly in the early morning and late afternoon.

In small parties, Shiny Cowbirds frequent cultivated fields, pastures, other open or lightly wooded country, and suburban gardens. They forage mainly on the ground, over which they walk, especially the males, with strongly uptilted tails. Like chicks around a scratching hen, they gather around the heads of grazing horses and cattle, and between bouts of foraging they rest upon the quadrupeds' backs, sometimes a dozen or more lined up together. In spring they follow the plow to pick up worms and grubs. Although animal matter predominates in their diet, they eat many weed seeds and grain spilled where domestic animals are fed. Occasionally they descend in damaging hordes upon fields of ripening maize or rice; but by devouring great quantities of insects injurious to crops, they compensate, at least in part, for these depredations.

Gregarious at all seasons, even while they breed, Shiny Cowbirds sing in concert. Hudson (1920) told how at their roosts in trees they continue to chant together until it is quite dark—much as I have heard Melodious Blackbirds singing in their evening gatherings. When disturbed in the night, male Shiny Cowbirds often burst into song while they flee. On rainy days, when they seek shelter in trees, they often continue to sing together for hours, the blending of innumerable voices making a rustling sound, as of a gale. In different parts of the Shiny Cowbird's range, its song has been described as "warbling, pleasing, musical," or "a melodious warble, suggestive of an oriole," or "like the twittering of some finches such as the Goldfinch in North America or the *Spinus icteria* in Argentina." The Shiny Cowbird sings well enough to be kept in cages for its voice, and the escape of some of these captives appears to have helped the parasites' spread through the Antilles.

Apparently, none of the foregoing descriptions refers to the courtship song, which, according to Friedmann (1929), begins rather indefinitely, well down in the throat, with a curious combination of purring and bubbling sounds. With quivering body, the bird belches forth three of these guttural notes, which sound as though they were forced up through water. Closely following these are three high, thin, rather glassy notes, the first two short and the last prolonged. The whole performance sounds like *purr-purr-purr-pe-tsss-tseeeee* and resembles that of the Brown-headed Cowbird of North America.

The male delivers this song while displaying to a female on the ground. He stands in front of her, fluffs out his feathers, lowers his head, and rapidly beats his arched wings three or four times, often striking the ground with them. If the female turns away, he jumps to a new stance before her, all without altering his posture. At highest intensity, the ardent male rises into the air and circles widely twice or thrice around the female, singing all the while. Both Hudson and Friedmann noticed that the spectator of these engaging performances appears to be unmoved by them. However, at their termination she often flies to join her wooer on the perch where he has alighted. Less often than on the ground, the cowbird displays in similar fashion in a tree, where he bows more deeply and fans the air with his wings instead of striking the ground with them.

Of the Shiny Cowbird's three frequent call notes, two are used by both sexes, one mainly by the male, and one by the female. Both sexes often utter a *chuck* not unlike the calls of other icterids. The male alone voices a prolonged, thin, plaintive, glassy *pseeeee*, which usually follows a *chuck* and is repeated once or twice. The female makes a rattle or chatter, a rapid sequence of clicking notes that usually announces impending flight.

Males of the smaller cowbirds are more numerous than females. Hudson surmised that males outnumbered females because the eggs from which they hatch have thicker shells, more resistant to the indiscriminate pecking by which cowbirds destroy so many eggs of their own and other species, but this conjecture might be hard to prove. Apparently, there are at least three male Shiny Cowbirds for every two females, but the numerical disparity may be even greater. Despite the imbalance, Shiny Cowbirds exhibit a tendency toward monogamy and territorial fidelity that is often weakened by the excess of males, or by a female's failure to find enough hosts for all her eggs within her mate's home range, leading her to seek nests beyond it. The abnormal concentration of cowbirds in cultivated areas where food is abundant is especially likely to thwart any tendency toward monogamy and territorial exclusiveness that might exist in the species.

Shiny Cowbirds have not wholly lost the tendency to build nests that their ancestors surely had in more than rudimentary degree. Hudson twice saw them trying to make nests, which they ultimately failed to finish. In an aviary, a female built in a large hollow log, while her companion brought materials and looked on, a nest of grass and sticks in which she never laid an egg. Shiny Cowbirds depend wholly on other birds to provide nests for their eggs, hatch them, and rear their young. Friedmann and Lloyd F. Kiff (1985) listed 201 species known to have been parasitized by Shiny Cowbirds throughout their range. Three of these hosts were pigeons, one a woodpecker, the remainder passerines from woodcreepers to finches. The pigeons' method of feeding their young is so different from that of most passerine birds that none reared the parasites' nestlings, but 53 passerine species are known to have raised one or more cowbirds at least until they fledged. The parasitized nests are of diverse forms: open cups of thrushes and tanagers, roofed nests of Great Kiskadees and wrens, nests in tree holes of tityras and certain flycatchers, the deep pouches of oropendolas and caciques, the clay ovens of horneros, the bulky constructions of interlaced twigs of other ovenbirds that are preferred nest sites of Bay-winged Cowbirds and the Screaming Cowbirds that parasitize the former.

Hudson was impressed by the Shiny Cowbirds' strong attraction to closed nests though they seldom lay in them. He wrote (1920, vol. 1, pp. 90–91),

> Before and during the breeding-season the females, sometimes accompanied by the males, are seen continually haunting and examining the domed nests of some of the [Furnariidae]. This does not seem like a mere freak of curiosity, but their persistence in their investigations is precisely like that of birds that habitually make choice of such breeding-places. It is surprising that they never do actually lay in such nests, except when the side or dome has been accidentally broken enough to admit light into the interior. Whenever I set boxes up in my trees, the female Cow-birds were the first to visit them. Sometimes one will spend half a day loitering about and inspecting a box, repeatedly climbing round and over it, and always ending at the entrance, into which she peers curi-

ously, and when about to enter starting back, as if scared by the obscurity within. But after retiring a little space she will return again and again, as if fascinated by the comfort and security of such an abode. It is amusing to see how pertinaciously they hang about the ovens of the Oven-birds [Rufous Horneros] apparently determined to take possession of them, flying back after a hundred repulses, and yet not entering them even when they have the opportunity. Sometimes one is seen following a Wren or a Swallow to its nest beneath the eaves, and then clinging to the wall beneath the hole into which it disappeared.

Hudson viewed the cowbirds' interest in closed nests that they feared to enter as a vestige of an ancestral habit of breeding in such nests, much as the Bay-winged Cowbird now does. Since he left his native Argentina to pass the remainder of his long life in England, numerous records of Shiny Cowbirds laying in closed nests of ovenbirds and others have accumulated, and cowbirds have been reared in such nests. In Trinidad, Southern House Wrens, who nest in the most diverse assortment of holes and crannies, are the cowbirds' most frequently recorded hosts (ffrench 1973). Nevertheless, it remains true that throughout the cowbirds' range open nests, nearly everywhere the most abundant kind, are by far the most common recipients of their eggs.

Female cowbirds seem to find nests of their hosts chiefly by watching the behavior of the latter from convenient perches. The first cowbird who locates a nest may be joined by other nest-searching individuals. Fraga (1985) occasionally observed two, and once as many as five, female cowbirds showing interest in a Chalk-browed Mockingbird's nest. On one occasion, two females who had discovered a recently built mockingbird's nest perched nearby, giving calls that within two minutes drew three more females and three males. Probably such cooperation in finding hosts' nests is unintentional, since it is to the female's advantage to lay only her own egg in a nest. However, female Shiny Cowbirds appear not to fight over territories or nests.

Unlike Brown-headed Cowbirds, who regularly lay at dawn,

Shiny Cowbirds deposit their eggs throughout the forenoon until about midday. The number of eggs laid in a season by a Shiny Cowbird appears to be unknown. The number intruded into a single host's nest, perhaps by several cowbirds, varies greatly. One of the cowbirds' most frequent hosts is the Rufous-collared Sparrow, widespread in South and Middle America, where it inhabits farms and towns as well as wild country (Fraga 1978). In widely scattered regions of the southern continent, more than half of the sparrow's open nests, on or near the ground, contained eggs of these cowbirds; often 75 percent or more of them were parasitized. By placing only one egg in a host's nest, the cowbird gives her prospective progeny the optimum conditions for survival and growth, and apparently this is what she tries to do by distributing her eggs among a number of nests; but her efforts are frequently defeated by other individuals who continue to add their eggs to the same nest.

In Fraga's sample of twenty-nine parasitized nests of the Rufous-collared Sparrow, half contained only one cowbird's egg in addition to those of the host, but the number ranged up to five. A Rufous Hornero's nest held the amazing number of thirty-seven cowbird eggs, of which only five were fresh. This heap of eggs was obviously the product of a number of females, up to thirteen of whom have been known to lay in the same nest. Nests with so many eggs are abandoned, as no bird could incubate them. Although such large accumulations of eggs are exceptional, Shiny Cowbirds do indeed waste many of their eggs, often dropping them on the ground, or in abandoned nests.

Shiny Cowbirds may lay in a nest still without an egg of its builder, during the interval when the host is depositing her eggs, or after incubation has begun. Fraga found that most cowbirds' eggs were laid in Rufous-collared Sparrows' nests during the days when the sparrow was laying, and this is certainly the most favorable time. If a cowbird's egg appears in a newly finished but still empty nest, the owner is likely to abandon it. If the cowbird's egg is laid after incubation has started, it may hatch after the owner's eggs, and the young cowbird will be handicapped when competing with the legitimate offspring for food. A cowbird may remove an egg of its host on the day before she lays her own egg, on the same day that she lays, or on the following day.

In addition to removing hosts' eggs, cowbirds peck into eggs indiscriminately, those of cowbirds as well as those of their hosts, and sometimes eat them. Males also indulge in this wasteful practice, which is most frequent where cowbirds teem, as where human activities provide much food. While an incubating Blue-and-Yellow Tanager was absent from her nest, a male Shiny Cowbird went directly to it and promptly pecked into and ruined all three of her eggs, for no apparent reason. Egg pecking may help to keep the cowbirds from becoming so numerous that they seriously reduce populations of their hosts, which in turn would diminish their own opportunities for reproduction.

Shiny Cowbirds' eggs show an extraordinary variety in shape, which ranges from almost spherical to oval or elliptical oval, and in coloration. Many eggs are immaculate white. Others are white, bluish white, bluish green, creamy, yellowish, pale pink, or light brown, most diversely marked with minute specks of pink or gray, or larger spots and blotches of chocolate or rufous, which on some eggs densely cover the whole surface. One might suppose that with such a great diversity of coloration cowbirds would match their eggs with those of their victims for the eggs' readier acceptance. This would require the development of different strains or "gentes" of the parasites, each specializing on a host with eggs like its own, as the European Cuckoo has done; but no such tendency has been detected among the cowbirds. That such specialization would be helpful to the cowbirds is evident from the behavior of the Chalk-browed Mockingbird, which lays blue-green eggs spotted or blotched with shades of brown. It accepts the cowbird's spotted eggs but ejects the white ones from its nest. Similarly, the Brown-and-Yellow Marshbird leaves spotted eggs in its nest and removes plain ones. But the Yellow-winged Blackbird, whose white eggs are spotted only on the thicker end with dull brown and black, appears to accept the cowbird's plain white eggs and reject spotted ones.

Among Argentine birds, Tropical Kingbirds and Rufous Horneros commonly but not invariably eject all varieties of cowbird eggs, and the Rufous-bellied Thrush apparently does the same (Fraga 1985). Although only a few South American species are now known to reject cowbird eggs, the number will doubtless increase as the habits of the birds of the southern continent are more thoroughly investigated, as has happened in North America. Nevertheless, in both continents so many species accept and incubate the eggs foisted upon them, whatever their coloration, that cowbirds of several species have no lack of victims.

Other birds adopt different procedures to avoid the burden of rearing alien young. Some abandon the nest in which a foreign egg appears, which they are most likely to do if the cowbird lays in a nest that is still empty. Less frequently, they cover the intruded egg(s) with a thick new lining, almost a nest upon a nest, thereby insulating the foreign eggs from the heat of the incubating parent. If cowbirds continue to lay in this nest, a second or even a third floor may be added, resulting in an unusually tall structure. Among South American birds troubled by Shiny Cowbirds, this expedient is used by the Yellow-browed Flycatcher. By breeding earlier than the cowbirds, or by consistently guarding their nests, potential hosts may reduce or avoid parasitic intrusions.

Shiny Cowbirds' eggs hatch in eleven and a half to twelve days, an incubation period several days shorter than that of some of their hosts, especially flycatchers. Unlike some parasitic cuckoos, nestling cowbirds neither attack the host's young nor heave them from their nests, but by hatching first they gain an initial advantage, which is often augmented by their larger size at birth. By stretching their gaping, red-lined mouths higher when a parent arrives with food, they may claim so much of it that their nest mates starve. Nevertheless, the legitimate offspring sometimes grow up with the voracious fosterlings. A pair of Rufous-collared Sparrows raised two of their own nestlings in the same nest with one of the cowbirds.

Between nestling and fledgling cowbirds and their foster parents there is no such rapport as between young birds and their own parents (Gochfeld 1979). Before they are fledged, nestlings of many kinds respond to the warning notes of their parents by crouching down in the nest or by jumping from it prematurely and hiding in the ground cover, but cowbirds do not heed these cries. Newly hatched nestlings rise gaping for food when their nest is shaken or a hand or some other object is waved above it, but after they can see well such disturbances make them shrink down and cling tightly to their cradles. Young cowbirds lack such defensive reactions and appear to be fearless; since they may be raised by a variety of parents, they expect food from the most diverse sources: a human hand or a bird larger then their foster parents. Hudson recounted how a young cowbird in a Double-collared Seedeater's nest cried for food on seeing his hand approach. The naturalist took the nestling and twirled it about, making it scream, so as to inform the absent foster parents of its plight, then replaced it in, or rather upon, the little cup that it more than filled. Then, plucking half a dozen measuring worms from a nearby twig, he offered them, one by one, to the cowbird, who eagerly devoured them all, despite the ill-treatment it had just received and in utter disregard of the wildly excited cries of its foster parents, who had just arrived and hovered close above the nest.

In nests of Rufous-collared Sparrows, Shiny Cowbirds remained for twelve or thirteen days, and in the roomy cups of Chalk-browed Mockingbirds they stayed for fourteen or fifteen days; but often, especially if disturbed, they depart much sooner, when they can scarcely fly. After quitting the nest, they do not like many fledglings hide amid dense vegetation; often they perch conspicuously, with fluttering wings pleading for nourishment from every passing bird. Hudson told how in a stubble field he noticed a fledgling cowbird perching at the summit of a slender dry stalk, clamoring for food at short intervals. Presently a diminutive flycatcher arrived and, alighting on the back of its large fosterling, dropped a caterpillar into its upturned open mouth. Six or seven times more the busy little flycatcher returned, each time standing on the young bird's back to feed it. This habit of awaiting its meals in an exposed situation often costs the incautious fledgling its life. On the pampas of Argentina, the Chimango Caracara carries off many of them to feed its own brood. Chalk-browed Mockingbirds were seen to continue attending their fosterlings for at least twenty or twenty-one days after they quitted the nest.

Shiny Cowbirds seriously depress the reproduction of their dupes. In the Andean foothills of northwestern Argentina, James R. King (1973) found that two-thirds of all nests of their principal host, the Rufous-collared Sparrow, received the parasites' eggs, and at the height of the breeding season every nest was parasitized. When spared the intrusions of cowbirds, the sparrows reared an average of 1.11 fledglings per nest, but parasitized nests yielded only 0.50 fledgling sparrows. A pair of sparrows could successfully rear a maximum of 4 sparrows or 2 cowbirds, or the equivalent thereof in various combinations. On the opposite side of Argentina, in the province of Buenos Aires, Fraga (1978) reached similar conclusions about the number of nestlings Rufous-collared Sparrows could raise. Here, where 72.5 percent of forty nests received cowbird eggs, their nesting success was very low. Unparasitized nests produced an average of 0.55 sparrow fledglings; parasitized nests only 0.24. In the same locality, 78.0 percent of sixty-five nests of the much larger Chalk-browed Mockingbird were parasitized. The mean number of mockingbirds reared in unparasitized nests was 0.93; in parasitized nests only 0.28. In all these studies, predation caused a large proportion of all losses of eggs and nestlings, but the harmful consequences of parasitism by cowbirds are clearly evident.

Since a small race of the Shiny Cowbird invaded Puerto Rico about forty years ago, it has heavily parasitized its preferred host, the Yellow-shouldered Blackbird, and appears to be largely responsible for the alarming decline in the population of this endemic bird (Post and Wiley 1977b). In Barbados, the survival of the local race of the Yellow Warbler is also threatened by cowbird parasitism (Cruz, Manolis, and Wiley 1985). In contrast to this lamentable situation, the colonial, polygynous Yellow-hooded Blackbirds, studied in a marsh in Trinidad by Alexander Cruz and his companions, were only lightly affected by Shiny Cowbirds (Cruz, Manolis, and Andrews 1990). Parasitized nests produced as many of the host's fledglings as unparasitized nests did, but nests of both categories suffered heavily from predation.

I have seen little of Shiny Cowbirds, chiefly in Venezuela, where I found one of the parasite's nestlings in a well-enclosed nest of

Buff-breasted Wrens. My impression gained from reading is of a pacific bird that seldom fights with others of its own kind or with its hosts, but also of a bird with a rather disorganized life, maladapted for its parasitic habits. Unselective of its victims, it foists its eggs on some that will rear its young poorly or not at all. It wastes its eggs by dropping many on the ground or in abandoned nests, or by loading an active nest with more than the host can incubate or more young than it could rear if all the eggs hatched. It indiscriminately punctures eggs of its own species or its hosts, throws them from the nests, or eats them. Unattuned to their foster parents, young cowbirds are unresponsive to the adults' concern for their safety.

Despite all these shortcomings, Shiny Cowbirds remain a successful, abundant—often overabundant—species, thanks to the rapid development of their loudly and aggressively begging young and, above all, their great fecundity, which (as in many invertebrates) compensates for high egg losses and neglect of offspring. Relieved of strenuous nest building, time-consuming incubation, and the exhausting labor of rearing young, cowbirds can spend their days foraging for food to form their eggs. Hudson estimated that a Shiny Cowbird produces from sixty to a hundred eggs in a season, a number that Friedmann reduced to one-tenth. In light of recent studies of Brown-headed Cowbirds, I surmise that the Shiny Cowbird's productivity lies between these extremes.

The Bronzed Cowbird

We pass over the little-known, red-eyed Bronze-brown Cowbird, confined to a small area on the Caribbean coast of Colombia, to continue northward in our survey of the cowbirds. The next species we encounter is the Bronzed Cowbird, also known as the Red-eyed Cowbird, from the color of the iris of adults of both sexes. The eight-inch (20 cm) male has a black head, body, and neck ruff strongly glossed with greenish bronze. His black tail and wings have a bluish sheen. The slightly smaller, duller female is browner below. Both sexes have black bills and legs.

The Bronzed Cowbird ranges along the southern fringe of the

Bronzed Cowbird, *Molothrus aeneus*, male

United States from extreme southwestern California to southern Texas, thence southward over Mexico (except Baja California) and Central America to central Panama. Like many other icterids, this cowbird is expanding its range; an isolated breeding population has become established around New Orleans. Northern populations are partly migratory, but some individuals remain in the southern United States through the winter. Within the tropics, Bronzed Cowbirds migrate altitudinally. In western Guatemala, I found them in the mountains between 8,000 and 9,000 feet (2,240 and 2,740 m) during the breeding season from March until July, after which I met them only on the plateau around 7,000 feet (2,130 m). Similarly, during the year I passed in a clearing in heavy forest at 5,500 feet (1,675 m) on the stormy northern slope of Costa Rica's Cordillera Central, these cowbirds first appeared as the birds were beginning to nest at the end of March. By July they had vanished.

Avoiding rain forests, Bronzed Cowbirds spread over open country in small or large compact flocks. They frequent cultivated fields, pastures, roadsides, towns, and urban parks. Walking over the ground, they gather many wild seeds and fallen cereal grains. Occasionally they attack ripening ears of maize, distressing the small farmer. In pastures they catch grasshoppers and other insects, often in company with Giant Cowbirds and Groove-billed Anis. All three of these black birds find hunting most productive close to the heads of grazing cattle, snatching up insects aroused by the quadrupeds. The Bronzed Cowbirds also alight on the backs of horned cattle and mules to vary their diet with ticks and insect pests plucked from the animals' skin. On the shingly floodplain of the Río Morjá, a tributary of the Río Motagua in Guatemala, Bronzed Cowbirds joined Giant Cowbirds, Melodious Blackbirds, and Great-tailed Grackles in shifting small, water-worn stones to disclose edible creatures lurking beneath them, in the manner already described for the Melodious Blackbird. In these stone-turning parties, male Giant Cowbirds chased male Bronzed Cowbirds, but never far. I watched these gatherings chiefly in the late afternoons. As night approached, the Bronzed Cowbirds and Melodious Blackbirds finished their supper gleaned from the stony floodplain and retired to roost in a dense stand of young giant canes adjoining it. Until they fell asleep, the blackbirds continued to sing delightfully, as told in the account of that species, but their companions the cowbirds were rarely heard (Skutch 1954).

In mid-March, I watched about fifty Bronzed Cowbirds foraging in a compact flock around a straw pile beside a granary, in the highlands of Guatemala, apparently picking up spilled grain. From time to time, for no evident cause, they all took wing together, wheeled around in close formation, then descended to continue their gleaning. The breeding season was approaching, and the males were already courting the females. Now and again one would rise a few feet into the air to hover prettily on beating wings above a busily eating female. Other males rested in neighboring pine trees, where each spread and raised his cape until it surrounded his head like a glossy black halo, in the midst of which his red eyes gleamed, and sang with squeaky whistles.

As the sun set on an evening in July, I watched a male and

female who walked, foraging, on a lawn beside the Río Ulua in Honduras. With a stiff, ungraceful gait, he followed her, his head thrown back, chest puffed out in front, and wings quivering. Suddenly he jumped into the air and for about a minute hovered like a miniature helicopter about a yard above her. Next, he fluttered to the ground in front of her, and with fluffed plumage bobbed up and down by flexing his legs. She considered him for a moment but was apparently unimpressed by this gallant display, for she rudely flew off and left him to deflate himself alone before he flew in pursuit of her. In a variant of the courtship display watched by Friedmann (1929) a male bent down his head until the underside of his bill touched his breast feathers, then bounced up four times about an inch above the ground. While bouncing, he emitted three deep, guttural, bubbling sounds, followed by two short notes and one long, high, thin, squeaky note, not unlike the Brown-headed Cowbird's song, but shorter and wheezier.

As in other cowbird species, male Bronzed are more numerous than females, resulting in irregular marital relations. This is the only small cowbird that consistently lays unmarked eggs, pale bluish green, bluish, greenish, or nearly white. They are deposited at daily intervals. Contrary to earlier reports, Michael Carter (1986) found that Bronzed Cowbirds frequently pierced earlier-laid cowbirds' eggs as well as hosts' eggs, particularly the latter. Friedmann and Kiff (1985) listed seventy-seven species in whose nests the cowbirds' eggs have been found, of which twenty-eight species are known to have reared the intruded young to the point of fledging. The hosts include flycatchers, wrens, thrushes, mockingbirds, thrashers, vireos, wood warblers, and tanagers, but chiefly sparrows and other finches, and, above all, orioles and a few related birds. Nests of nine species of orioles, from cups to long, pensile pouches, have received the cowbirds' eggs, and the swinging bags in colonies of Yellow-winged Caciques often contain them. The partiality of Bronzed Cowbirds for nests of orioles and finches contrasts with the less restricted selection of hosts by Shiny and Brown-headed Cowbirds, and reminds ornithologists of the narrow specialization of the Giant Cowbird, the Bronzed's closest extant relation.

How many eggs a Bronzed Cowbird lays in a season appears not to be known. Most often she deposits only one in a host's nest. When two or more eggs are present, they may have been contributed by different females. A small, platformlike nest of unknown origin, found near Brownsville, Texas, contained fourteen cowbird eggs and one of a host. Whether deposited in an old, abandoned nest or piled too deeply in one newly built, such large accumulations of eggs are always wasted.

After an incubation period of ten to twelve days, nestling Bronzed Cowbirds hatch with gray down sparsely shading orange-pink skin. The inside of the mouth is reddish and the flanges at its corners white or cream. Two cowbirds may grow up in the same nest; a pair of Audubon's Orioles was observed feeding three of the young parasites, almost fully grown. Likewise, nestlings of both the cowbird and its host may lie together in a nest. On the Isthmus of Tehuantepec in southern Mexico, I found a Green Jay's nest with a single Bronzed Cowbird, well feathered and nearly ready to fly, sitting awake and alert between two young jays, bigger than itself but naked and slumbering. In Guatemala, I was shown a Yellow-green Vireo's nest with three nestlings of the vireo and one of the cowbird.

When eleven or twelve days old, cowbirds leave their nests. Although here in the Valley of El General in southern Costa Rica cowbirds are rare, a few years ago I watched a pair of Orange-billed Nightingale-Thrushes feeding two young Bronzed Cowbirds. As often happens, the thrushes had divided their brood, each parent attending one of the fosterlings. The female fed one in a guava tree while the male took care of the other on the ground. While he hopped over the close-cropped pasture grass gathering food, the cowbird walked after him with fluttering wings. When he caught an insect, he placed it in the fledgling's red-lined mouth, although the inside of a nightingale-thrush's mouth is yellow. Reports of foster parents feeding fledged Bronzed Cowbirds are not rare, but, strangely, I have found none of adults simultaneously attending the fosterlings and their own fledged young, who while still in a parasitized nest often give promise of growing up to leave it. Can it be that naturalists have not paid attention to this point or have deemed it not worth reporting; that fledgling cowbirds expose themselves more than other birds of similar age; or that the more active and importunate young cowbirds claim so much of the foster parents' attention that the legitimate young perish of neglect? Possibly competition between the young of cowbirds and those of their hosts hurts the latter after they all leave the nest more than while they lie together in it. This question deserves further attention.

In the Guatemalan highlands, I watched a pair of the spotted race of the Rufous-sided Towhee attend a young cowbird. While the youngster clamored loudly for food, an adult cowbird alighted beside it and, after an interval, touched its open mouth with her(?) bill. Unfortunately, I was too distant to see whether she fed the fledgling. The incident appears to reveal that at least a vestige of parental attachment persists in the brood parasite, which is significant in the light of rare reports of Brown-headed Cowbirds feeding young of their kind, and of glossy cuckoos of the parasitic Old World genus *Chrysococcyx* occasionally giving food to fledglings.

The Brown-headed Cowbird

Including the dwarf race in Mexico and the southwestern and south-central United States, Brown-headed Cowbirds range in length from about six to eight inches (15 to 20 cm). The male is glossy iridescent black, with a contrasting brown head and neck. The female is grayish brown. The short, conical bills and the legs of both sexes are black or blackish; the eyes are dark. The cowbirds' original home was the short-grass plains of the midcontinent, where great herds of bison roamed. As men of European origin felled the forests that covered much of the region east of the Mississippi to make room for pastures and grain fields, the birds spread eastward in increasing numbers, finding new and more receptive hosts, as they continued to do well into the present century (Mayfield 1965). The species now breeds from southeastern Alaska and southern Mackenzie across southern Canada to Newfoundland, throughout the United States (except southern Florida), and southward to central Mexico. In the Rockies it is

Brown-headed Cowbird, *Molothrus ater,* male

occasionally found as high as 10,000 feet (3,050 m). Partly migratory, these cowbirds withdraw from the northern parts of their vast territory as winter approaches, some traveling southward as far as the Isthmus of the Tehuantepec, but many remain as far north as northern California, Missouri, and southeastern Canada (Bent 1958).

Brown-headed Cowbirds commonly migrate by day, chiefly in the early morning and late afternoon, often in flocks with Red-winged and Rusty blackbirds, Common Grackles, less frequently with Common Meadowlarks and American Robins in the east, and Brewer's and Yellow-headed blackbirds in the west. In western New York, Friedmann (1929) recognized five categories of migrant cowbirds. First to appear in early March were vagrant birds of unknown origin and destination. Next, when skunk cabbage was flowering in wet woodlands but most vegetation was still dormant, came transient males on their way to breeding grounds farther north. These were followed by males who would remain to breed. About the same time, but continuing longer, transient females passed through the region. Resident females did not appear until early April, about ten days after the first resident males arrived. Last of all, toward mid-May, males and females that seemed to be immature settled in the area. The early transients foraged in the uplands and in the evenings joined noisy Red-wings to roost in the marshes, which still offered them little food. Resident males went promptly to the trees, where they would sing and display during the first few hours of the morning, before they descended to seek breakfast in the fields.

Brown-headed Cowbirds forage on the ground, over which they walk or run but seldom hop, with tails held almost upright and wings drooping low. Sociable birds, they usually forage in flocks of their own kind or mingle with blackbirds, Starlings, and other ambulatory birds. Stubble fields yield them fallen grains and weed seeds, while freshly plowed fields supply larvae and other invertebrates. Their diet changes with the seasons. In summer, and especially while they breed, cowbirds eat much animal food—beetles, caterpillars, bugs, grasshoppers, and spiders—often amounting to half or more of their total intake, the remainder being seeds, chiefly of yellow foxtail grass *(Chaetochloa glauca)* in western New York, but also knotweed *(Polygonum),* ambrosia *(Ambrosia),* and the grass *Echinochloa.* From late fall until the following spring, the cowbirds' diet is almost wholly vegetable, including seeds of many plants and above all yellow foxtail grass, which Friedmann called the cowbirds' "staff of life." Although fruit is at best a minor component of their meals, they have been known to gather blackberries, huckleberries, wild cherries, grapes, and cedarberries. While plucking dandelion seeds, a cowbird held down a stalk with a foot.

Like other cowbirds, the Brown-headed is strongly attracted to grazing quadrupeds, who stir up insects hiding in the herbage. Although often the birds rest in numbers on the backs of horned cattle and probably relieve them of ticks or other parasites, or catch biting flies, these pests appear to be minor elements in their diets. Grasshoppers set in motion by the grazing animals yield them more nourishment. Although domestic cattle and horses are now the cowbirds' chief mammalian associates, on the Great Plains of the North American Midwest they were in former centuries so strongly attached to the great herds of bison that they were commonly known as buffalo birds. They moved with the herds, dining along the way on the insects aroused by their advance, and between meals resting together on their shaggy backs. Ernest Thompson Seton (in Friedmann 1929) told of a cowbird who flocked with others around a small group of bison confined in a park near Winnipeg, Canada. After the cowbirds flew southward in autumn, this particular individual remained behind, possibly because it was crippled and unable to undertake the long journey. All winter it stayed close to the buffalos, especially the biggest, fiercest bull, sharing his food, warming its toes in the wool of his back, and at night snuggling into a hollow it had made in the hair just behind the great animal's horns. The food and protection the lone cowbird found among the bison helped it survive the severe winter of this northern land until its companions returned from milder climates in the spring, finding it fat and fit.

It has been surmised that Brown-headed Cowbirds developed the habit of foisting their progeny upon other birds to leave themselves free to follow the moving herds of bison. Although this can hardly be the origin of their parasitism, which they evidently inherited from ancestors in South America, where bison were absent, attachment to nests and young would have limited the range of their wanderings. Associated so closely with huge quadrupeds, buffalo birds were almost fearless of relatively puny bipeds, gathering close around the horses they mounted, even alighting upon a human shoulder. Not only are Brown-headed Cowbirds strongly attracted to the horses and cattle that have replaced the bison over much of North America, they accompany sheep and pigs and alight upon their backs as well. In zoological gardens they are equally intimate with such exotic animals as African antelopes, Asian yaks, and Australian emus.

Brown-headed Cowbirds have limitations other than their breeding habits; lacking the ability to build nests, they are likewise deficient in melody. Two of their utterances serve as songs. The first, sometimes called the male's "true song," consists of two or three low, guttural or gurgly notes that do not carry far, followed abruptly by a few high, squeaky ones. It has been variously transcribed as *glub-glub-kee-he-heck* or *bub-ko-lum-tsee,* with the last shrill note often prolonged. Courting males utter this song while posturing before females or in intimidation displays before rival males, most frequently in spring and early summer, when the cowbirds are preparing to breed or seeking nests for their eggs. It is also used by males displaying alone on a high perch. The second, more frequently heard song is composed of a prolonged, high-pitched, squeaky sound followed by two or three sibilant notes, each shorter and lower in pitch: *wheeeee tsitsitse* or *whsss pseeee.* Usually a male starting to fly or alighting after a flight will sound these notes. Perhaps the utterance should be considered not a song but a call by which a male announces to his companions his intention to take wing, or that he has arrived at his destination, and when alone uses to make contact with them. A bird foraging by himself repeats it until he hears an answering call, then flies in its direction. The cowbirds' repertoire also includes a throaty *chuck* given by both sexes, a *tic* or *kek* by females foraging alone, and a rattle uttered in flight.

Of all the displays of cowbirds, the strangest and most puzzling is that called by its discoverers, Robert K. Selander and Charles J. La Rue, Jr. (1961), the "interspecific preening invitation." A cowbird about to solicit preening bows its head until its bill points vertically downward or inward toward its breast. The feathers of its head and nape are conspicuously ruffled, but the plumage of its body is usually slightly compressed or sleeked, and its wings and tail are held in the usual resting positions—features that distinguish this display from others, differently motivated, in which the head is bent low. In this posture the cowbird silently approaches some other individual—the recipient—by sidling along its perch or in a more direct, head-forward manner, until its head is about one inch from that of the recipient or until its crown touches the recipient's breast. It may also perch side by side with the recipient, turning its head toward the latter. Sometimes it solicits preening while both birds cling to the wire side of an aviary or cage or while they stand on the ground. Whatever the site of the display, the cowbird directs the back of its head toward the recipient's head. Silent and rigid, it avoids looking directly at the recipient and making sudden movements that might alarm it; but it may move its head slightly to help the recipient preen different regions of its fluffed plumage.

Captive Brown-headed Cowbirds invite small birds of a number of species, even stuffed specimens, to preen them. When first approached by a displaying cowbird, the recipient usually responds by either pecking or fleeing; but finally it often obliges. Surprisingly, even birds not known to engage in allopreening—the technical term for preening other individuals—are induced by this display to preen the cowbirds' heads. The longer captive cowbirds are deprived of contacts with birds of other kinds, the more eager they become to be groomed by them; and they can be extremely importunate.

Red-winged Blackbirds were the species that Selander and La Rue most frequently used in their study of the interspecific preening invitation. When a displaying cowbird approached a female Red-wing, she retreated a few steps along the perch, followed by the cowbird repeating his solicitations. When she flew, the persistent cowbird pursued her, and when she alighted he advanced toward her until his head pressed against her breast, whereupon she either retreated again, pecked at him, or yielded to his solicitations. In one hour, a male cowbird demanded preening from a female Red-wing ninety-three times, and a female cowbird did so sixty times.

A cowbird's insistence upon being preened was well demonstrated by a female placed in a small cage with a Budgerigar, a parrot that habitually engages in mutual preening with its mate. Neither the cowbird nor the little parrot had been together before. During eight minutes, while the episode was filmed, the cowbird displayed fifty-eight times, while the addressee pecked thirty-nine times at her head, on three of these occasions making contact and pulling out feathers before she could escape. After a while the Budgerigar's response to the displays became ambivalent; it began to preen the cowbird, then desisted and pecked her. The cowbird's insistence gradually wore down the parrot's resistance until the number of positive responses increased while the rebukes became less frequent. The confrontation ended with a bout of preening that continued for thirty-seven seconds. In a later test, the parrot preened a cowbird's plumage for nearly five minutes—the most prolonged preening episode recorded in this study.

In the experiments with captive birds, female Red-winged Blackbirds received more displays than male Red-wings did, and they preened much more frequently. Meadowlarks were often invited and became dependable preeners of cowbirds. Male Great-tailed and Common grackles did not receive displays; female Great-tails seldom did; no grackle preened. House Sparrows of both sexes were often approached and responded favorably. Inca Doves were invited to preen but larger Mourning Doves and Feral Pigeons were not; none of these pigeons acceded to invitations to preen. Dummies of a number of small passerines, from wood warblers to American Robins, excited displays, but some other stuffed birds did not. The dummies' passivity sometimes aroused attacks by thwarted cowbirds, who pulled cotton from their "eyes." Similarly, living birds who ignore repeated invitations to preen may be pecked or otherwise abused. Little Inca Doves who refused to preen cowbirds were injured by them.

Among unconfined birds that cowbirds have been seen to invite preening, the most frequent are House Sparrows, who occasionally comply but mostly do not. For more than ten minutes a free female Red-wing preened a female cowbird while they perched among canes. Apparently in the wild, as in captivity, Red-wings are not infrequently solicited by cowbirds of both sexes, whom they sometimes preen. Northern Cardinals and Dickcissels are also invited to preen but may not comply (Dow 1968). After a male cowbird had displayed for ten seconds to an unresponsive female Scissor-tailed Flycatcher on a telephone pole, he pecked at her, whereupon she flew back to her nest.

In the experiments of Selander and La Rue, the cowbirds rarely requested preening of other cowbirds confined with them, and then chiefly as a sort of redirected activity, when birds of other kinds had rejected their invitations. The cowbirds remained unresponsive. In a differently designed experiment by Stephen L. Rothstein (1977a), captive cowbirds frequently directed this same display to other cowbirds, who often received it passively but sometimes reciprocated the attention, making the display mutual. Both male and female cowbirds directed significantly more displays to female than to male cowbirds. Cowbirds display more frequently to new associates of their species than to those with whom they have been confined for several months, which may explain the different results of the experiments of Rothstein and those of Selander and La Rue.

In situations exceptionally favorable for observation, as when birds perched on overhead wires near feeders, Rothstein saw free male cowbirds invite other male cowbirds to preen them, but less often than they solicited the House Sparrows that rested near them. Because this display is often directed to other cowbirds of the same species, Rothstein found the designation "interspecific preening invitation" inappropriate and called it simply the "head-down display," but this description does not distinguish it from different displays with bowed heads. Briefly called the "preening invitation," it is sometimes given to juveniles a month or less old (Lowther and Rothstein 1980).

This preening invitation is widespread among cowbirds; both the Bronzed and the Shiny perform it just as the Brown-headed does. Similarly, Shiny Cowbirds do not respond by preening a displaying bird and may peck it for violating the "individual distance" that cowbirds of several species, like swallows perching on a wire, insist upon preserving between themselves and their nearest neighbors. Giant Cowbirds invite preening by birds of other species and even by people. Moreover, even the Bay-winged Cowbird—so different in many ways from other species of *Molothrus* that it was until recently classified in a different genus, *Agelaioides*—solicits preening from birds of other kinds. Confined in a cage with three male Chestnut-capped Blackbirds, a male Baywing would approach one of them by sidling along a perch with body almost horizontal and bill directed downward. But as the cowbird neared the blackbird, the former raised its head until its bill pointed almost straight upward, widely spread the feathers of head and neck, and leaned away from the blackbird, who preened it for an interval of twenty seconds or less (Selander 1964).

The significance of these unusual performances is far from clear. They appear to have both interspecific and intraspecific relevance, but what can it be? The displaying cowbird's approach with averted beak suggests that this might be an appeasement ceremony, a gesture to soften the resistance of birds whose nests the cowbirds are about to violate. Long ago, Frank M. Chapman (1928) twice saw female Giant Cowbirds display in this fashion to a female Chestnut-headed Oropendola on or near her nest, and Neal G. Smith (1968) has witnessed both inter- and intraspecific preening invitations in nest trees of Chesnut-headed, Montezuma, and Crested oropendolas and Yellow-rumped Caciques. If the invitation is ever given in the less exposed situations where most of the cowbirds' victims nest, it has apparently not been reported. However, Raleigh Robertson and Richard Norman (1976) found that models of cowbirds, set close to nests of host species, were attacked by these birds less often, or less violently, when posed in the preening invitation display than when normally upright. Moreover, the bowed models tended to elicit displacement behavior, such as bill wiping, self-preening, and pecking the ground, perch, or grass stems, more frequently than the upright cowbirds stimulated these activities that substitute for attacking.

From a prolonged study of Shiny Cowbirds in Puerto Rico, Post and Wiley (1992) concluded that the head-down display is often an aggressive pose by which the displayer supplants the recipient at a preferred spot in a diurnal or nocturnal mixed species roost, thereby increasing the cowbird's chances of survival. After watching free and captive Brown-headed Cowbirds display and examining four hypotheses to explain this behavior, T. W. Scott and J. M. Grumstrup-Scott (1983) called it "an appeasing, aggressive behavior" that mitigates the antagonism of the recipient to the posturing cowbird, who is generally dominant over the bird so addressed. The head-down display helps the cowbird to obtain food, conserve energy while roosting, and (or) establish its station in a flock. Despite many fine studies, we need to know more about the motives of cowbirds presenting their heads to be preened by other birds before we can reach firm conclusions.

Some instances of the invitation display, such as that of the captive Giant Cowbird who solicited preening from Payne (1969b) in the Fort Worth, Texas, zoo, suggest that the cowbird enjoys having its head scratched and, since others of its kind will not gratify it, seeks the attention of birds of different species, even of humans. I dwell on this behavior because I find it so extraordinary, another reminder of how opaque to humans the minds of nonhuman animals are.

More elaborate than the preening invitation, more often witnessed, and of less doubtful interpretation is the courtship display, well described, and portrayed by a series of photographs, by Friedmann (1929) in *The Cowbirds.* Arching his neck, puffing out his plumage, spreading his wings and tail, a male cowbird begins to bow forward. He may start this performance while perching on one foot, but as he tipples downward he needs to cling with both to save himself from falling. He also folds and elevates his tail. When he displays on a branch he may end with his bill touching it, sometimes pecking it. When he performs on the ground, he seems to fall on his head. While preparing for this striking act by fluffing out his plumage, the cowbird utters the low *bub-ko-lum* song already described. As long as he falls forward, he continues to emit a high, shrill, squeaky *tseee.* In silence he regains his upright stance. If another male cowbird is nearby when a male is about to start his display, the latter may admonish the former to keep distant by standing upright and pointing his bill toward the zenith.

While watching cowbirds eating millet seed that she spread over the ground beside her home on the outskirts of Nashville, Tennessee, Amelia Laskey (1950) saw hundreds of such displays, which she designated as "toppling over" because they terminated with the male bird's bill or head on the ground, and which she interpreted as intimidation or threat. In the somewhat similar "peck-gesture," the cowbird thrust his head forward rather than bending it down while he ran or flew toward another. Both of these aggressive displays, mainly toward other male cowbirds, were occasionally directed to a Mourning Dove, Common Grackle, Brown Thrasher, or House Sparrow sharing Laskey's bounty. The gesture might end with pecking or actual fighting. Contrary to Friedmann's experience of cowbirds' pacific temperament, she saw five fights between male Brown-headed Cowbirds, as did Margaret Nice (1937) while engaged in her classic study of Song Sparrows at Columbus, Ohio.

As in other cowbirds, male Brown-heads are more numerous

than females, sometimes in the ratio of three to two, apparently because adult males survive better than adult females; in James Darley's (1971) study in Ontario, Canada, the annual mortality of adult females reached 55 percent, that of males only 38 percent. The relation of the sexes has been variously interpreted by different observers. Many hours of watching her color-banded cowbirds convinced Laskey that they were monogamous, the dominant male among the visitors to her feeding area mating with the dominant female, whom he guarded from surrounding males by interposing himself between them and her. She also prevented the close approach of other females to him in much the same fashion but less frequently.

Years earlier, Friedmann (1929, p. 171) had written: "My experience has been that if the birds are not really strictly monogamous, at least a tendency towards monogamy is very strong." In support of this view, he cited observations of Alexander Wetmore, who, in an open area in Utah with excellent visibility, watched six pairs of cowbirds, of which each remained apart from the other pairs. More recently, Darley (1982), after a study of 154 banded male cowbirds in Ontario, concluded that "dominant males mated with resident females in mostly monogamous relationships." They guarded their mates, and the females sometimes protected their consorts from other females, much as Laskey had described. However, excess, unmated males took advantage of a dominant male's temporary absence to court his spouse and sometimes copulate with her. The presence of these unattached males, and the females' occasional need to seek nests for their eggs beyond the range patrolled by their mates, made strict monogamy difficult to preserve and supports the long-prevalent view that Brown-headed Cowbirds are promiscuous or polygamous. Apparently, too, their sexual relations vary from region to region, and they are least strictly monogamous where most abundant. In New York State, Alfred M. Dufty, Jr. (1982), learned that monogamous cowbirds often returned to the same mate and territory in successive years.

On a male cowbird's arrival in the spring, he selects a conspicuous post or "singing tree" that he will, if he can, hold throughout the breeding season. From this point of vantage he patrols a definite area of one to sixty-two acres (0.4 to 25.0 hectares), usually about twenty or thirty (8 or 12 hectares). This range is sometimes called his "territory," which is inaccurate, for territory is defined by ornithologists as a defended area, whereas the cowbird does not defend his range from other male cowbirds. As a result, the ranges of mated resident males often overlap, or one male's range may fall almost wholly within that of another. A male's mate tends to remain within his range. Cowbirds are strongly attached to the localities where they breed; when removed in homing experiments, they have returned from distances up to 1,200 miles (1,900 km). Migratory cowbirds tend to settle in the same area year after year.

In 1963 Friedman reviewed the host selections of parasitic cowbirds. In 1985 Friedman and Kiff listed 220 species parasitized by Brown-headed Cowbirds, of which 144 are known to have reared the cowbirds' young at least to fledging. This figure exceeds slightly the number of hosts recorded for the Shiny Cowbird, which spreads over a continent much richer in birds than that occupied by the Brown-head but poorer in people who take an interest in birds and report what they have learned.

Female cowbirds commonly locate the nests in which they will foist their eggs by finding birds building them. Nests in open fields devoid of trees or shrubs are more likely to escape their notice than those near hedgerows, fences, or woody plants that serve as observation posts. After finding a nest under construction, they sit, alone, inconspicuous, and silent, amid the foliage of a shrub or tree, often for a quarter of an hour or more, intently watching the builder at work—an experience that apparently stimulates the development of their eggs. Less frequently they walk slowly over the woodland floor, stopping at intervals to watch birds in the trees above them. In the dense shrubbery of the forest edge, or of hedgerows in more open places, they fly noisily into the foliage. If a frightened bird bursts out, they investigate the spot from which it arose, perhaps finding a nest.

If any of these procedures discloses an appropriate nest, the finder follows its progress by visits of inspection. To lay her eggs at daily intervals in different nests, she must simultaneously keep several under observation, and at appropriate times she returns to lay in them. Aside from the production of her eggs, a female cowbird's chief contribution to reproduction is the task of finding suitable nests, then laying in each at the proper time—neither too soon, which may cause the builder to desert, nor too late, which will handicap her progeny if they hatch. Successful parasitism depends upon the proper timing of laying (Norman and Robertson 1975).

Brown-headed Cowbirds deposit their eggs at dawn, before their hosts lay (Hann 1937; Scott 1991). Stealing silently into a nest, a cowbird drops her egg rapidly, often in less than a minute, then promptly departs; she does not, as small birds often do when they lay eggs in their own nests, linger in a nest for many minutes, perhaps as much as half an hour. There is good reason for her haste; if she delays long in the nest she has invaded, she will probably be attacked by the resident pair and often by their neighbors as well. Group defense prevents Red-winged Blackbirds' nests in marshes from becoming parasitized as frequently as are Red-wings' nests more widely scattered in upland fields. George M. Sutton (in Bent 1958) watched a flock of marsh-nesting Red-wings pursue a female cowbird until she was exhausted and plunged into the water to escape. By fastening dummy cowbirds beside nests, Robertson and Norman (1976) learned that the more heavily parasitized a species is, the more aggressively it confronts the dummies—a trend most evident among sparrows.

While sitting in a blind watching a Scarlet Tanager's nest, at dawn of a June day in Michigan, Kenneth Prescott (1947) witnessed a dramatic episode at a Red-eyed Vireo's nest, five feet (1.5 m) up in a white oak sapling and also visible from his enclosure. While a vireo scolded and struck with its wings a female cowbird on the ground near its nest, the male tanager flew down and dived at the intruder's head. When she retreated, the assailant withdrew; but in less than a minute the persistent cowbird returned and alighted close to the vireo's nest, which already contained four of her own eggs and two cowbird eggs. Undeterred by the attack of a vireo, who apparently struck her with a loud slapping noise of its wings, the invader advanced to the nest, and with a peck drove out the other vireo sitting there. While the cowbird rested in or on the nest, both vireos, continuously scolding, struck her with their wings as they darted past her. Joining the attack with a high *chip* alarm note, the male tanager soundly pecked the sitting cowbird's head. After enduring these assaults for about a half a minute, long enough to lay her egg, the cowbird flew away. The nest still contained six eggs, but Prescott was not sure whether they were the same as the eggs present when she entered it.

Unlike the European Cuckoo, cowbirds do not remove a host's egg when they enter a nest to lay, but they may carry out an egg one or more days before they lay, later on the day they lay, or on the following days. In the second and third cases, a cowbird risks removing a cowbird's egg, her own, or that of another cowbird who had laid in the same nest. Often she deposits the extracted egg on the ground and eats it, but, according to D. M. Scott and his collaborators (Scott, Weatherhead, and Ankey 1992), up to 40 percent of the eggs removed may remain uneaten. Although punctured eggs are sometimes found in nests parasitized by Brown-headed Cowbirds, they are less frequent than in nests visited by Shiny Cowbirds, and the holes may be made accidentally by the birds' toenails.

Brown-headed Cowbirds' eggs are white, grayish white, or less often bluish white, speckled or blotched all over with chocolate, light brown, tawny, or cinnamon-rufous. These markings are usually heaviest around the thick end of the egg, where they may form a wreath. Sometimes they are so thickly aggregated that they almost obscure the ground color. No tendency of cowbirds to lay eggs that match those of their hosts has been demonstrated. Apparently, cowbirds with enough hosts' nests try to place only one egg in each of them. When two or three cowbird eggs are present in a nest, they are often so different that they are obviously the products of more than one female. Occasionally a nest contains seven or eight cowbird eggs, but the much larger accumulations of wasted eggs laid by Shiny Cowbirds are not known in this species. Likewise, the Brown-headed Cowbird does not drop many eggs on the ground, as the Shiny Cowbird does; the Brown-head appears to be more advanced and better adapted to its parasitic life.

As to the number of eggs laid by a single female in a season, careful estimates range from about twenty-five (Walkinshaw 1949) to forty. In contrast to other passerine birds, which if they lose a nest wait at least five or six days before they begin to lay again, cowbirds can lay sequences of eggs or clutches with an interval of only a day or two, thereby producing about three dozen in a two-month breeding season (Scott and Ankey 1980). In this almost continuous laying, cowbirds resemble domestic hens that have been selected for egg production and rarely, if ever, become broody. They have been called "passerine chickens." A captive yearling kept by Nadine Jackson and Daniel Roby (1992) laid thirty-two eggs on consecutive days.

Since the presence of parasitic young in nests seriously depresses the reproduction of the hosts, it is not surprising that they have evolved various means to combat this loss. As we have seen, the efforts of the victims to hold the adult female parasites aloof are often ineffective; the very species that behave most aggressively toward the cowbirds tend to be the most heavily parasitized. A cowbird can easily slip into a nest and quickly deposit an egg while the owners are briefly absent or inattentive, or even while they attack her! In this situation at least, an ounce of prevention is not worth a pound of cure. Potential hosts have developed such cures as abandoning the nest with an alien egg, which they are most likely to do if it appears before they have laid one of their own. Of all the remedies, this is the most costly in energy and time, for the birds must seek a new site and build again. This delay may result in a brood that hatches after the most favorable time for rearing it has passed.

Hosts have resorted to the somewhat less costly recourse of covering over the unwanted egg with a new floor, upon which the nest's builder will lay her own eggs. This step is often taken by the Yellow Warbler, one of the Brown-headed Cowbird's frequent hosts. If the remodeled nest receives another cowbird's egg, the owner may cover this one over in the same manner, if necessary repeating the procedure until she has a nest four to even six stories high, with cowbirds' eggs sandwiched between its floors, and perhaps, after all her work, a few foreign eggs lying beside her own in the open cup at the top. About thirty species, including some like the Eastern Kingbird that usually throw out cowbirds' eggs, at least occasionally cover them over.

By far the simplest, most economical way to thwart cowbirds is simply to remove their eggs from the nest. Thanks to Rothstein's research (1975, 1977b), we now know that at least seven North American birds regularly follow this course: Eastern Kingbird, Blue Jay, Gray Catbird, Brown Thrasher, American Robin, Cedar Waxwing, and Baltimore Oriole. All but the last of these grasp the eggs in their bills and carry them away. Probably because its bill is small but sharp, the oriole strikes the cowbird's egg with parted mandibles, piercing it to lift it over the high rim of its deep pouch. This procedure is not without danger to the oriole's own eggs. The exceptionally strong, thick shell of the cowbird's egg, combined with its roundness, sometimes deflects the oriole's stroke to one of its own adjacent eggs, damaging the shell and making it unhatchable (Rohwer, Spaw, and Røskaft 1989). In Puerto Rico, Red-legged Thrushes, Pearly-eyed Thrashers, Greater Antillean Grackles, and other species eject eggs of Shiny Cowbirds experimentally placed in their nests. Since these birds have been exposed to brood parasitism for only a few decades, it appears that the habit of casting out foreign eggs is readily acquired, probably because it is derived from some widespread avian behavior, such as the removal of the white fecal sacs of nestlings, which nearly all passerine birds do. The ejection of nearly all intruded eggs by these rejecter species has not caused cowbirds to desist from laying in the rejecters' nests. Moreover, all the North American rejecter species except the jay are known to be occasionally careless and to rear the parasitic young.

When cowbirds outwit the spirited defenses of Red-winged Blackbirds and slip eggs into their nests, these birds accept them, thereby exposing their eggs to a special danger. Contact with the thicker shells of the parasite's eggs occasionally damages the blackbird's shells. Even a slight crack or dent may cause an egg to spoil (Blankespoor, Oolman, and Uthe 1982; Picman 1984).

When I reflect that a newly hatched European Cuckoo can eject hosts' eggs from nest cups, I find it incredible that any bird, except possibly a tiny, weak-footed hummingbird, is physically incapable of shoving out or otherwise removing a cowbird's egg from its open nest. Measurements of bills and eggs led Rothstein to a similar conclusion. However, fumbling attempts by small-billed Cedar Waxwings to throw out cowbirds' eggs break many of their own eggs as well as those of the parasites (Rothstein 1976a, 1976b). Other birds with small bills are more skillful.

Why, then, do so many birds accept and incubate the intruded eggs, to the great detriment of their own broods? Long ago Friedmann observed that "they act as though they are proud of the large egg and are as much concerned over its welfare as they are over their own." He wrote this statement before Niko Tinbergen (1951) made naturalists familiar with the effects of supernormal stimuli, which incite Oystercatchers to try to incubate eggs so much bigger than their own that they cannot sit on them, or to prefer a clutch of five eggs to their usual three. Similarly, Ringed Plovers neglect their own eggs to incubate artificial eggs that are more boldly spotted; and Piratic Flycatchers occupy the great swinging pouches of oropendolas, where they usually fail to rear a brood, apparently because they lose contact with their eggs in these nests much too wide for them.

Probably the success of cowbirds' unconventional way of life owes much to the unrelated fact that many small birds are captivated by supernormal stimuli, such as those provided by eggs larger or more heavily marked than their own. If we ask why natural selection has not suppressed a tendency so injurious to a species, the answer appears to be that changes in structure or innate behavior depend upon random mutations that are not always forthcoming when they would be most helpful. Birds' persistent reception of eggs unlike their own points to the relative recency of the cowbirds' impact on some of their hosts, especially those of the eastern forests, such as the Wood Thrush, Red-eyed Vireo, and many wood warblers, which have been exposed to cowbirds in considerable numbers only since European settlers decimated these forests a few centuries ago, a brief interval in evolutionary time. With longer exposure to cowbirds, their hosts would probably develop better discrimination against dissimilar eggs, which in turn might lead to a situation such as we find in certain Old World cuckoos, whose eggs closely resemble those of the species that they regularly parasitize (Rothstein 1974).

The cowbird's incubation period has not, as some have supposed, been substantially curtailed to give their nestlings a head start over those of their hosts. It is usually eleven to twelve days, about the same as that of nonparasitic small icterids. It averages about half a day less than the incubation period of the Ovenbird of North America, a frequent host.

In appearance and behavior, the nestling Brown-headed Cowbird differs little from the nestling Shiny Cowbird. Like nestlings of all kinds, it tries to elicit from its attendants all the food it needs, and because of its greater size and more rapid growth, it requires more than most of those that grow up with it. On the whole its relations with its nest mates, whether the host's own young or other cowbirds, are amicable (Eastzer, Chu, and King 1980). Parent birds are stimulated to bring food by the begging of their nestlings, and up to a certain point can adjust the amount to the needs of their brood. Frequently the two kinds of young grow up together, as they are most likely to do if the host's eggs hatch first. Two cowbirds were raised with two Song Sparrows, one with four Song Sparrows, and two in a nest with four Ovenbirds—an extreme example of compatibility. Ovenbirds reared with cowbirds leave their nests at about normal weight; but often the host's young in parasitized nests are underweight when they fledge, which reduces their chances of survival.

Cowbirds spontaneously leave their nests when nine or ten days old, earlier than the young of some of their hosts but later than those of others. As fledglings they are as heedless of their foster parents' warning cries and as careless about exposing themselves as young Shiny Cowbirds are. Fledgling cowbirds reared by Ovenbirds persist in rising to a perch, although their foster parents keep their own young well hidden on the ground. Their habit of begging noisily from sundry passing birds sometimes wins them a meal but may lead to their destruction. With gaping mouth and fluttering wings, a fledgling cowbird on a lawn thrice approached a Common Crow, who each time retreated a few feet. Finally, the crow seized the clamorous fledgling, killed it, and carried it off, as reported by Millicent Ficken (1967). Foster parents continue to feed fledgling cowbirds until they are twenty-five to thirty-nine days old and can forage for themselves.

Adult cowbirds retain vestiges of parental behavior, three cases of which Bent recorded in his *Life Histories.* In Nebraska, a female came nearly every evening to feed a cowbird nestling in a nest with Rose-breasted Grosbeaks. If the young grosbeaks opened their mouths for food, she pecked their heads and refused to give them any. Another female cowbird fed a nestling of her kind in a Yellow Warbler's nest. Laurence B. Fletcher trapped a female cowbird with a fledgling whom she fed. He banded and released both birds and later saw them together. The same adult continued to feed the banded fledgling but not the others who were nearby. More recently, in Tennessee, M. D. Hernandez (1986) watched a male and female cowbird feed a single young cowbird who was also attended by a female Northern Cardinal, apparently its foster mother. I earlier mentioned that in Guatemala I saw an adult Bronzed Cowbird touch the open mouth of a fledgling of her species who was being fed by a pair of Rufous-sided Towhees. In none of these instances was there proof that the cowbirds were parents of the young they fed. An instinct so deeply rooted as the feeding of young remains latent in passerine birds long after it has ceased to contribute to the survival of a species.

For a bird that cares for its own young in a nest that it has built, rearing a brood is a straightforward, uncomplicated sequence of activities. After innumerable repetitions by its ancestors, it proceeds from nest building through laying, incubation, brooding, and feeding young as smoothly as an actor in a well-rehearsed play —assuming, of course, that no predation, inclement weather, or other extraneous circumstance interrupts the sequence. A brood parasite like a cowbird faces difficulties that an attendant parent does not know. It must find a nest at an early stage of a host species that will be a good fosterer for its progeny. It must lay its egg neither too early nor too late, often against the active resistance of the nest's owners. The chances of a poor choice of nests or poor timing of laying are many. A wrong choice of nests may result in the prompt rejection of its egg, burial beneath a new lining, or the nest's abandonment. Premature laying increases the probability of desertion; late laying, after the host's eggs have been incubated for several days, as occurs not infrequently when cowbirds can find no better nest for eggs ready to be laid, may prevent hatching or jeopardize the young cowbirds if they hatch. Occasionally, cowbirds lay in nests that already contain nestlings, or they may find no better place for their eggs than an abandoned nest—or the ground.

Birds that accept cowbirds' eggs differ greatly in their competence as foster parents. In a compilation by Howard Young (1963), 30 percent of 223 cowbird eggs in Song Sparrow nests yielded cowbird fledglings, as did 23 percent of 121 cowbird eggs in Red-eyed Vireo nests, 28 percent of 67 eggs in Common Yellow-throat nests, and 23 percent of 57 eggs in Ovenbird nests. But another frequent host of the cowbird, the Yellow Warbler, raised young from only 8 percent of 75 cowbird eggs. Red-winged Blackbirds, infrequent hosts, reared fledglings from 10 percent of 48 cowbird eggs.

Often hosts rear young from a larger proportion of their own eggs than from the eggs foisted upon them, despite the frequently larger size and more rapid development of the parasitic nestlings. In nests of the highly parasitized Eastern Phoebe, 46.0 percent of 1,203 phoebe eggs in unparasitized nests yielded fledglings, as against 41.7 percent of 139 cowbird eggs in parasitized nests. Song Sparrows raised fledglings from 35.8 percent of 906 of their own eggs and from 31.9 percent of 113 cowbird eggs. Of 161 Ovenbird eggs, 43.5 percent produced fledglings, while in the same study only 25.0 percent of 40 cowbird eggs yielded fledged young. In Russell Norris's (1947) study of 237 nests of fourteen species of passerine birds in Pennsylvania, 37.7 percent of 668 host eggs and 26.8 percent of 108 cowbird eggs produced fledglings. However, of the host eggs that hatched, 64 percent yielded fledged young; and of the cowbird eggs that hatched, 63 percent yielded fledged young.

Although the cowbird's breeding system is inefficient and wasteful, it is not ineffective. If we make the minimum assumption that, on average, a breeding female lays thirty eggs in a season and lives to reproduce through two seasons (a free cowbird lived to be nearly sixteen years old), her lifetime production of eggs will be sixty. To maintain at a steady state a population with a sex ratio of one and a half males to one female, less than three (5 percent) of the young from these eggs need survive to breed. Cowbirds' ability to expand their range is proof that, overall, their rate of reproduction is more than adequate.

Cowbirds remove the eggs of their hosts, causing many of them to abandon their nests and build again with loss of time and energy in a breeding season that may be short; and young cowbirds compete for food with nest mates that are often smaller and grow more slowly, occasionally trampling or smothering them—by these means, their intrusion into a finely adjusted sequence of interactions between parents and proper progeny causes losses difficult to assess. Nice (1937) concluded that with the Song Sparrow as host, each cowbird is reared to fledging at the cost of one young sparrow. Likewise, in Norris's study of fourteen host species, each parasite was raised at the price of about one of the hosts' young.

Other hosts suffer greater losses. Val Nolan, Jr. (1978), compared successful unparasitized and parasitized nests of the Prairie Warbler in Indiana, finding that the production of 0.91 cowbirds was at the cost of 2.45 warblers, or a 73 percent reduction from mean brood size in unparasitized nests. None of these estimates takes account of the loss of hosts' young after fledging, which might be attributed to the parents' simultaneous attention to cowbird fledglings, and which I suspect is considerable. When I reflect upon the Brown-headed Cowbird's vast range over almost the whole of the temperate zone of North America, and the great diversity of its victims, I am astounded by the magnitude of the losses of the productivity of a whole avifauna caused by this parasite's failure to rear its own progeny. Nevertheless, this loss is apparently far less than that attributable to nest-pillaging predators, above all snakes.

The reproductive rate of successful species of birds is adequate to confront all these sources of loss. Despite heavy parasitism, widespread birds like Song Sparrows and Eastern Phoebes have in the past managed to remain abundant. The situation with rare species is different. Kirtland's Warbler, which breeds in a small area of regenerating Jack Pine woods in the north of Michigan's southern peninsula, was so severely affected by cowbird parasitism that it appeared to be on the road to extinction until it was saved by reducing the numbers of the dusky pests. In the North American Midwest, cowbirds so heavily parasitized Dickcissels, invading 95 percent of their nests in the tall grass prairies of Kansas, that it has been listed as "endangered" in *American Birds*. Although, in the past, populations of widespread species were not greatly affected by brood parasitism, this may not long remain true. Today, when destruction and fragmentation of habitats, encouragement of raptors, widespread use of pesticides and other noxious chemicals, and fatal collisions of nocturnal migrants with the increasing number of television towers and other high constructions are causing alarming decreases in the population of many species of birds, cowbirds may help drive some of them to extinction.

The Evolution of Cowbirds

After reading—often for the second time after a lapse of years—some of the vast literature on cowbirds, finding certain speculations about the origin of their parasitic habit that do not appear sound to me, and after writing this long chapter, I cannot, as I finish it, refrain from presenting my own views for what they may be worth. (For other interpretations, see Friedmann 1929 and Hamilton and Orians 1965.) Briefly, I believe that brood parasitism arose, at least in this family of birds, not from the decay of any phase of parental behavior in either sex (that came later) but from the penchant of certain birds to breed in nests more elaborate than they could make. Many icterids nest in structures that more or less completely conceal the incubating or brooding parent and her nestlings: oropendolas, caciques, and many orioles in long, woven pouches; meadowlarks in domed nests; Chopi Blackbirds in nooks in buildings, holes in trees, or old nests of other birds; Bolivian Blackbirds in clefts in cliffs; Common Grackles occasionally in old woodpeckers' holes or other cavities in trees; Troupials and Bay-winged Cowbirds in enclosed nests built by other birds.

When occasionally the last-mentioned species builds in the open, it makes a roofless cup.

That the Bay-winged Cowbird and its parasite, the Screaming Cowbird, are closely related is indicated by the close resemblance of the juveniles of these two species, which, as Hudson recognized, can hardly be attributed to a mimetic function; for birds the most diverse feed the young they have reared, no matter how greatly these fosterlings differ from their own progeny in size, appearance, or voice. I surmise that the common ancestor of these two cowbirds already nested in closed structures built by other birds, in southern South America where so many members of the ovenbird family make elaborate nests. Geographically separated populations of this ancestor retained the habit of occupying nests of other birds, while they diverged so greatly in appearance, voice, and habits that when range expansion brought them into contact, they had become different species. These closely related birds began to compete for nests of the same kinds, which may not have been available in sufficient numbers for all of them. In these contests Bay-wings usually won, possibly because they are more sociable and have helpers to assist in nest defense. Rather than persist in confrontations that they usually lost, ancestral Screaming Cowbirds adopted the expedient of depositing their eggs in nests still held by Bay-wings, which they might do by stealing in at opportune moments, thereby avoiding waste of time and energy, minimizing risk of injury, and saving resources that they could apply to forming more eggs than their hosts, the Bay-wings, could produce. Thereby began the parasitic habit in cowbirds.

The increased fecundity that brood parasitism permits made an offshoot of the Screaming Cowbird stock so numerous that these birds could not find enough closed nests, occupied by pairs of Bay-wings or other species, for all their eggs, which perforce they deposited in open nests or dropped on the ground, as their descendants, the Shiny Cowbirds, still do. Shiny Cowbirds' continuing interest in closed nests, noticed by Hudson, and their occasional parasitism of such nests, point to their descent from the stock of which the Screaming Cowbird is the least modified surviving descendant.

Despite the geographical separation of the Brown-headed from the Shiny Cowbird, it appears to be more closely related to the latter than is the Bronzed Cowbird, which occupies intervening territory and appears to be somewhat closer to the ancestral Screaming Cowbird. The red-eyed Giant Cowbird might be a descendant of the red-eyed Bronzed stock, grown larger the better to parasitize nests of the big oropendolas and caciques. They retain the ancestral habit of parasitizing well-enclosed nests.

Neither adult nor young cowbirds exhibit such evident structural modifications as the hooks on the tips of bills that enable nestling honeyguides and Striped Cuckoos to murder their nest mates. The changes that make cowbirds successful brood parasites are not anatomical but behavioral, such as may spring from individuals' choices. When such choices are frequent and increase fitness to survive and reproduce, they may become innate and heritable by the process known as organic selection or genetic assimilation—the firm establishment of the acquired habit by appropriate mutations—which is apparently what has happened in cowbirds. Although most contemporary cowbirds lay their eggs in open nests, their parasitic habit appears to have come by a circuitous route from the widespread icterid preference for enclosed nests, manifesting itself in species that cannot build such nests. Might we say that the parasitic habit of cowbirds sprang from something akin to covetousness of nests more commodious than their ancestors could build?

12

The Giant Cowbird

The Giant Cowbird, the only species of the genus *Scaphidura,* is widespread in tropical America from southern Mexico to western Ecuador, Bolivia, and northeastern Argentina, and from lowlands upward to 5,500 feet in Costa Rica and 7,000 feet in Colombia (1,675 and 2,135 m). The plumage of the fourteen-inch (35.5 cm) male is wholly black, glossed with purple on the body and blue on wings and tail. On his head he wears a large erectile ruff. His bill and legs are black; his eyes bright red. The female is similar but smaller, with only a faint bluish gloss on her black feathers and a rudimentary ruff. In large flocks these birds frequent open country with scattered trees, plantations, light woodland, gaps in heavier forest, and the shores of streams. In the breeding season they are found mostly at or near colonies of oropendolas and caciques, which they parasitize, but at other times they scatter

Giant Cowbird, *Scaphidura oryzivora,* male

widely. The big male flies with resonant wing-beats, apparently caused by the passage of wind through his stiff primary feathers. After five or six rapid beats, both sexes fold their wings momentarily—a habit that at a distance distinguishes them from associated male Great-tailed Grackles, which are about the same size and look equally black against the sky but fly with continuous strokes.

Despite their greater size, the Giants are true cowbirds, often foraging with cattle. They settle on the backs of cows and mules to pluck vermin from their skins, as I have often seen in Guatemala and W. Goodfellow (1901) observed in Ecuador long ago. It is amusing to watch the big cowbirds examining the sides of a sleek-haired cow reclining in a pasture, for their long toes are not well adapted for clinging to such a surface, and if they venture too near the edge they slip clumsily down. Sometimes Giant and Bronzed cowbirds and Groove-billed Anis forage together in a pasture, verily a black assemblage. All these birds stay close to the heads of grazing cattle to snatch up insects that the quadrupeds stir up from the grass, and the two cowbirds fly up to the big animals' bodies for different kinds of arthropods.

Giant Cowbirds forage much on the ground, walking instead of hopping. In western Panama, I watched them settle in flocks in the late afternoon to hunt over golf links, where they were shy and difficult to approach. On a broad bare floodplain covered with small, water-worn stones beside the Río Morjá in Guatemala, I could depend on seeing the cowbirds forage almost every evening in the dry season, from an hour or so before sunset until long shadows spread over the shore. The Giant Cowbirds who formed the nucleus of these assemblages were joined by Bronzed Cowbirds, Great-tailed Grackles (principally males), and Melodious Blackbirds. Often a few Muscovy Ducks would be foraging nearby in the shallows of the river. From a distance all five appeared equally black. For some reason, the Giant Cowbirds were antagonistic to the male Bronzed Cowbirds, who also have red eyes, and sometimes chased them.

The cowbirds chiefly engaged in turning stones, for which their strong black bills were well fitted. They moved stone after stone, overturning the smaller ones, pushing aside those slightly larger, and only slightly lifting a side of the biggest that they tackled, all to disclose anything edible that might be hidden beneath. The grackles and blackbirds joined in this pursuit, but less energetically than the cowbirds. All turned stones in the same way, inserting the tip of the bill beneath the near edge of a stone and shoving forward. As they pushed, they dropped the lower mandible and held the bill slightly open. The Giant Cowbirds picked up a few small spiral snail shells but mostly found them empty (Skutch 1954).

In southeastern Peru, Scott Robinson (1988) saw other aspects of Giant Cowbirds' foraging. Instead of cattle, they regularly stood on the backs of capybaras, largest of rodents, in marshes beside an oxbow lake. The cowbirds caught the biting horseflies (Tabanidae) that plagued the capybaras by day. In twelve minutes, a male captured twenty-four of these large insects. A capybara inclined its head upward and closed its eyes while the bird walked on its nose; but the rodents would not permit cowbirds to probe into their fur. In this wild region, Giant Cowbirds also hunted on the backs of tapirs. They ate figs *(Ficus trigona)* and fruits of *Coussapoa* (another tree of the mulberry family), sipped nectar from flowers, searched for insects on branches and foliage up to thirty feet (9 m) above the water, and probed curled leaves for hidden prey. Males tore strips of bark from tree limbs. As farther north, Giant Cowbirds in Amazonian Peru walk over shores exposed in the dry season, up to eighty together on the beaches of the wide Alto Madre de Dios River, hunting for food. They glean from the sand, turn over fallen leaves with their bills, and rummage in piles of driftwood. Apparently, small stones are absent from these shores. Giant Cowbirds also eat rice, a habit responsible for their specific name *oryzivora* as well as the older English name rice grackle.

While female Giant Cowbirds diligently turned stones on the shingly beach of the Río Morjá, the larger males strutted among them, apparently too obsessed with their own importance to participate in so humble an occupation. Fortunately for the tranquility of the females, they were always much more numerous than the larger sex. A male walked with a sort of goose step, carrying himself stiffly, his chest inflated and his head drawn back like that of a horse with a cruelly tight checkrein—all of which imparted an air of ludicrous pomposity. His long hind toes, conspicuous as he walked, seemed to impede his steps. Beautiful purplish reflections played over his glossy mantle and chest; his ruby eyes shone like gems in his black head; he was handsome without being in the least graceful, if one can imagine such a combination of attributes; he seemed "swollen with insolence and pride."

As with stilted gait a male Giant Cowbird approached an intently foraging female, she would often retreat, whereupon the would-be suitor would often forget his dignity to hop stiffly in pursuit of her. When he reached a female, he would plant himself squarely in front of her and draw himself up until he towered above her, appearing three times her height. Arching his neck, he depressed his head until his bill rested among the puffed plumage of his breast, and he erected the feathers of his cape until they surrounded his head as an iridescent black ruff, in the midst of which his red eyes brightly gleamed. Sometimes, as though still further to accentuate his height and impress the indifferent female, the male with plumage all fluffed out bobbed up and down in front of her by flexing his legs. But the female's reaction to this display seemed mainly to be annoyance; she interrupted her foraging only long enough to rebuke her suitor with a peck and a sharp note. Sometimes he pressed his attentions upon her until she clashed with him, then fled while with sonorous wing-beats he pursued her across the river and beyond view. In these afternoon gatherings, I never saw a male favorably received by a female.

Giant Cowbirds are mostly silent. Courting males utter an unlovely spluttering, ascending screech. Females flee from irate oropendolas voicing harsh, nasal whistles. While cowbirds fly, they often sound a more pleasant *pernt*.

As far as known, throughout their great range Giant Cowbirds parasitize only other members of their own family, the Icteridae, and in particular the species that weave hanging pouches in colonies, the oropendolas and caciques (Friedmann 1929). Whether any species in these two groups escape the intruders' visits, I do not know. Of the behavior of female Giant Cowbirds at the nest trees of Montezuma Oropendolas, their persistence in trying to enter the nests, their contempt for the half-hearted efforts of the oropendolas to drive them away, and their mutual jealousy, I tell in chapter 16. In the community of Chestnut-headed Oropendolas watched by Chapman (1928), the cowbirds were just as conspicuous and just as unwelcome as I found them at colonies of Montezuma Oropendolas. Not only the female Chestnut-head whose nest was threatened, but other individuals of the same group, and even from other groups, joined in attacking the female cowbirds, who sometimes fought back, as I never saw a cowbird do to a bigger Montezuma Oropendola. Even the little Piratic Flycatchers, who attend their own eggs in nests stolen from oropendolas, assailed the cowbirds, sometimes with greater zeal than the oropendolas displayed.

Giant Cowbirds follow Chestnut-headed Oropendolas into the mountains. At the highest colony that I have seen, at 5,400 feet (1,645 m) in Costa Rica, a single cowbird surveyed the eighteen nests in the small community. Like the oropendolas themselves, the cowbird migrated altitudinally, following them upward when the nesting season began.

In Panama, where Smith (1968) studied colonies of Chestnut-headed, Crested, and Montezuma oropendolas and Yellow-rumped Caciques, he learned that Giant Cowbirds, like European Cuckoos, lay eggs that mimic those of their hosts. By their eggs, he distinguished five types of females; three whose eggs resembled those of each of the three oropendolas, one for the caciques, and one that laid a generalized, nonmimetic icterid egg—the "dumper type." Moreover, different colonies of the same host species laid eggs differing in color and size, matched by further variation in the mimetic eggs.

Cowbirds with mimetic eggs usually laid only one of them in a host's nest with a single egg; infrequently they laid in a host's nest with two eggs, but never in a nest with more than two. Dumper females with nonmimetic eggs usually deposited two or three in a host's nest, occasionally as many as five. These dumper cowbirds laid in empty nests as well as in nests with full sets of the host's eggs. They often dropped their eggs in nests previously parasitized by mimics but never in nests that already contained eggs of other dumpers. The mimetic and dumper females behaved differently in the nesting colonies. The latter were aggressive, infested a nest tree in groups, and often drove oropendola or cacique females from their nests. Cowbirds with mimetic eggs skulked shyly through a colony, waiting for a chance to drop an egg into a nest in the owner's absence. Giant Cowbirds' eggs vary in color from white to pale blue, diversely marked with spots, blotches, and scrawls of black, to match the hosts' eggs; or, in the dumper type, they are scarcely marked. Oropendolas whose nests are protected by wasps or bees generally throw out cowbirds' eggs; those lacking insect guards accept the intruded eggs. This rule does not apply to Montezuma Oropendolas in Costa Rica, who neither nest near social insects nor willingly accept intrusions by Giant Cowbirds (Webster 1994).

The cowbirds' incubation period of about twelve days is much shorter than that of its hosts—seventeen or eighteen days for oropendolas and sixteen for the cacique. Cowbirds hatch with natal down, but host hatchlings are naked. Within forty-eight hours of hatching, cowbirds' eyes open, whereas those of oropendolas remain closed for six to nine days. The aggressiveness toward moving objects of cowbird chicks at least five days old probably accounts for the scarcity of botfly larvae upon themselves. When two of these chicks rest side by side in a nest, they preen one another, removing all the *Philornis* larvae. Mutual preening at so early an age, and active, directed snapping rather than passive gaping when an attendant brings food, are behaviors unknown in other passerine nestlings. In colonies where females opposed cowbirds because bees or wasps kept botflies away, 433 of 666 cowbird eggs (65 percent) yielded fledglings. In colonies without bees or wasps but receptive of cowbirds, 682 of 1,708 cowbird eggs (40 percent) produced flying young. As parasites, Giant Cowbirds are highly successful, thanks in part to the precocious activity of their nestlings.

Amid the forests of southeastern Peru, where the Russet-backed Oropendolas and Yellow-rumped Caciques studied by Robinson (1988) reared only a single young, instead of two as often was the case in Panama, parasitism by Giant Cowbirds brought no advantage (see chapter 16). Here, whether potential hosts nested near wasps or bees or in trees devoid of them, they resisted the cowbirds that tried to steal into their pouches. In this effort, caciques were more successful than oropendolas because they seldom left their colonies unattended by birds of both sexes, whereas female oropendolas and their male escorts foraged as a group away from their colony, leaving it unguarded. Moreover, female cowbirds often had difficulty entering the smaller hooded nests of caciques while male caciques attacked them. Accordingly, the cowbirds concentrated their visits on nests of Russet-backed Oropendolas, who sometimes reared the alien young. By contrast, during a five-year study Robinson found no cowbird eggs in the nests of caciques, who apparently did not rear a single young of the parasite.

After leaving the nest, a Giant Cowbird travels with its foster mother. I (Skutch 1954) was puzzled by my failure to see any young cowbirds in postnesting flocks of Montezuma Oropendolas with many young oropendolas still being fed. Finally, I found a solitary female oropendola who fed a juvenile Giant Cowbird. I surmised that she deserted her flock rather than abandon her fosterling, who was rejected by other flock members; but more observations are needed. However it may be with Montezuma Oropendolas, young cowbirds still receiving food were prominent in flocks of Crested Oropendolas in Venezuela. The bills of these juveniles were pale like those of the oropendolas, not black like those of adult cowbirds. Was this a case of mimicry that made the fosterlings more acceptable to their fosterers?

Like smaller cowbirds, Giants solicit preening from birds of other species. Presenting its head to the prospective preener, a cowbird bows so profoundly that its bill rests on the breast, while the nape feathers are erected but not spread into a cape, as in courtship display. In the gardens of the Zoological Society of London, a solitary male assumed this posture before a pair of African Black-headed Plovers, who were preening themselves and ignored the cowbird (Harrison 1963). In an aviary in Fort Worth, Texas, another solitary male Giant persistently solicited preening from people who neglected him until Robert B. Payne (1969b) pushed a finger through the screen and rubbed the bird's head. So eager was the cowbird for this attention that when Payne moved his finger sideward or upward, the bird followed to be scratched again. The significance of this behavior is considered in the chapter on the smaller cowbirds.

13

The Orioles

Orioles are delightful birds, colorful, songful, weavers of exquisite nests. Their twenty-seven species compose the largest genus *(Icterus)* in their family, distributed over the American continents from southern Canada to central Argentina and in the Antilles. Their center of distribution is Mexico, where sixteen species nest. Of the eight species in the United States, five are primarily Mexican species that range into the southwest, from extreme southern Texas to California. Another Mexican species, the Spotted-breasted Oriole, has been introduced into southern Florida and become established there. Only three species are widespread in the temperate zone of North America. The Baltimore Oriole in the east and Bullock's Oriole in the west differ enough in appearance to have been described as two species. Because they occasionally hybridize where they meet in the midcontinent, they were united

Baltimore Oriole, *Icterus galbula,* male

as a single species, the Northern Oriole. Now it appears that the older classification was sound. The east of the United States and Canada has only two indigenous orioles, the Baltimore and the Orchard. In all South America, only eight species of orioles nest. Five indigenous orioles inhabit the West Indies.

Orioles are slender, long-tailed, sharp-billed birds that range in size from the Orchard, only six to seven inches (15 to 18 cm) long, to a number of tropical species that measure eight or nine inches (20 or 23 cm). They are mainly black and yellow or orange, with white bars or edgings on the wings of many species. The distribution of these colors varies from species to species, serving for identification. One of the yellowest is the Yellow Oriole of South America, which is black only on the lores, throat, tail, and white-marked wings. The blackest are the Moriche Oriole, also of South America, which is yellow only on the crown and hindhead, rump, thighs, and wing coverts; and the Epaulet Oriole, one race of which has only yellow, chestnut, or tawny shoulder patches to relieve

Streaked-backed Oriole, *Icterus pustulatus,* sexes similar; hollow-thorned acacia

its deep black attire. In another dark species, the little Orchard Oriole, the adult male is deep chestnut on the rump and underparts, elsewhere black. Streaked-backed and Spotted-breasted orioles are marked with black on the yellow back or breast. Females of the migratory Northern, Orchard, and Scott's orioles are much paler than adult males, and young males are intermediate between adults of the two sexes. Among the permanently resident tropical orioles the sexes are usually alike, or the females are only slightly paler than their consorts. Males, and sometimes females, may take more than a year to acquire their splendid adult attire and nest with signs of immaturity. Seasonal changes in coloration are known neither in the resident species nor the highly migratory Baltimore and Orchard orioles, which travel to Central America and northern South America (Beecher 1950).

A General Survey

Less gregarious than many other members of their family, orioles live in pairs, family groups, or small, loose flocks when not breeding (Skutch 1954). At night some gather in communal roosts. They frequent light woodland, gallery forests along streams in arid country, savannas with scattered trees, gardens and plantations, groves of palms, arid scrub, deserts, and forest edges. They avoid dark lower levels of heavy tropical forests but may forage in the sunny canopy. They eat largely insects, fruits, and nectar. For spiders, caterpillars, beetles, and other insects they glean amid foliage and probe curled leaves, often hanging head downward. Baltimore Orioles tear open the webs of tent caterpillars to extract the larvae. They eat a great variety of wild and cultivated fruits and, in their winter homes, are readily attracted to feeders where bananas or halved oranges are offered (Bent 1958).

Almost any kind of flower with abundant nectar accessible to orioles attracts them. They extract the sweet liquid from a variety of leguminous trees, including *Erythrina, Gliricidia, Inga, Caesalpinia,* and *Calliandra;* from introduced eucalyptus and rose-apple trees with powder-puff clusters of stamens; from the vine *Combretum;* and from the nectar cups of epiphytes of the marcgravia family. I have often watched wintering Baltimore Orioles take nectar from the long, slender, red flowers of the poró trees beside our house. Plucking one of these blossoms with its bill, the bird holds it against its perch with either its right or left foot, but always with the base inward beneath its body. Then it pushes the sharp point of its beak through the tubular calyx that surrounds the base of the corolla, and by forcefully separating its mandibles splits the thick, fleshy tissue, to extract the nectar through the gap. Then it drops the flower to the ground, where multitudes drained by Orange-chinned Parakeets already lie. In the southwestern United States, Hooded Orioles drink nectar from the blossoms of agaves,

Hooded Oriole, *Icterus cucullatus,* male

aloes, hibiscus, and lilies. If the flower is large and tubular, they often perforate the base with their bills. They even split unopened buds to procure the nectar prematurely. When they indulge in such practices, they fail to recompense the plant for its bounty by pollinating its flowers (Beecher 1951).

It is difficult to decide which aspect of orioles wins greater admiration, the splendor of their plumage or the beauty and profusion of their singing. Their best songs consist of flutelike and whistled notes, incomparably full, smooth, and mellow. The rarity or absence of harsh notes, which such gifted singers as mockingbirds frequently interject into their recitals, maintains the high quality of orioles' songs. Among the most vocally gifted of orioles is the Yellow-tailed, whose singing delighted me while I lived in the Caribbean lowlands of Central America. In the mellowest of voices, he would repeat over and over, with hardly a pause, a verse of five, six, or rarely more notes, then choose a wholly different phrase, of which he had a great variety, and reiterate it in the same delightful manner. The female's voice was no less pleasing than her partner's, but her verses were shorter and less varied than his,

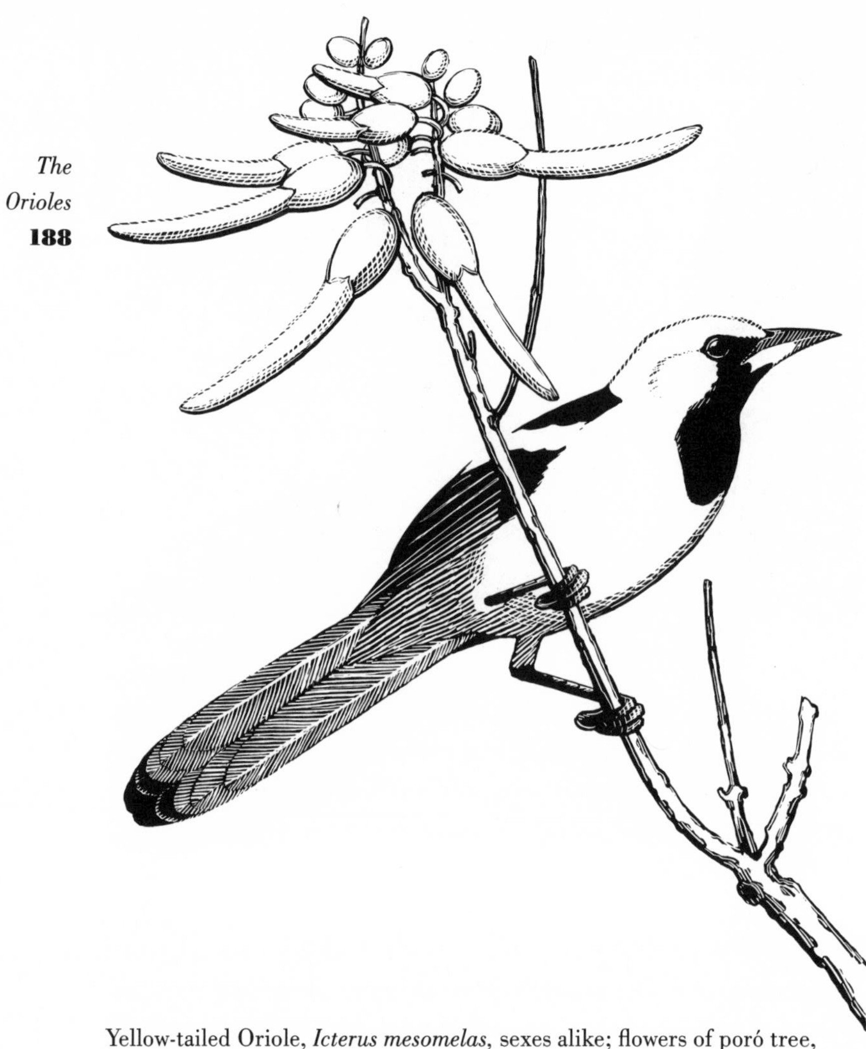

Yellow-tailed Oriole, *Icterus mesomelas,* sexes alike; flowers of poró tree, *Erythrina berteroana*

usually consisting of only two to four notes. While warming her eggs, she frequently answered the songs that he poured forth from a neighboring tree.

After listening to Spotted-breasted Orioles singing among high shade trees of a Guatemalan coffee plantation, I wrote in my journal that this was "the finest oriole's song that I know, a little better even than the song of the Yellow-backed Oriole, to which it is very similar." The male Spotted-breasted's song blended a series of the clearest, most mellifluous whistles into a continuous liquid stream of melody. His mate's simple song was composed of clear whistles distinctly spaced, a pleasant outpouring not unlike the song of the Altamira Oriole. Among orioles whose songs have won high praise are the Hooded, Audubon's, and Scott's, all of the southwestern United States and Mexico. Less dulcet in tone, the Baltimore Oriole's songs are admired for their vigor and variety. No two appear to sing the same verses, which often consist of about six notes. The song of Bullock's Oriole, which I have not heard, appears from written descriptions to be less pleasing than that of the closely related Baltimore. Quite different from all the foregoing is the Black-cowled Oriole's song, a varied whistle, sweet but hurried, often so low that one must stop and listen intently and at no great distance to appreciate it, for it lacks the mellow volume of the songs of slightly larger species.

Among the few orioles known to mimic notes of other birds is the Epaulet. In northeastern Argentina, Fraga (1987) heard it imitate the calls of Roadside Hawks, Guira Cuckoos, Green-barred Flickers, Rufous Horneros, Great Kiskadees, Chalk-browed Mockingbirds, and Solitary Black Caciques. Persistently repeated by both sexes of the oriole at all seasons, while they foraged or attended nests, these imitations were not included in their songs. Their significance is obscure.

Widespread among the vocalizations of orioles are churrs or rattles and nasal notes of complaint when their nests are or appear to be threatened. Even while they protest an intrusion at their nests, they seem unable to suppress their ebullient songfulness. Whenever I visited a low nest of a pair of Yellow-tailed Orioles, both parents approached as near as they dared and complained with notes almost as rich and mellow as their songs, sounding like *chup chup chup*. When my examination of a Black-cowled Oriole's nest suspended beneath a giant leaf of a traveler's tree frightened out the feathered young, the parents' nasal complaints were interspersed with snatches of low, whistled song, probably by the male. It was pleasant to hear melody burst forth in the midst of a scolding, like a beam of sunshine from a rift in dark storm clouds. Similarly, while I stood beneath the swinging pouch of a pair of Altamira Orioles, their churring protests mingled with the melodious notes that they constantly repeated while attending their nest.

In their winter homes in tropical America, migrant birds from the north are mostly songless, although a few sing while they settle their territorial claims upon their arrival in the autumn, and a few practice their songs as the date of their spring departure approaches. Most songful on their wintering grounds of all the birds I have heard were Orchard Orioles in the Caribbean lowlands of northern Central America. After their arrival in August, males often sang briefly. In September they were less vocal; but even in late October, two months after the appearance of the earliest migrants, I occasionally heard brief, subdued refrains. From October to early March, they were almost songless, but before the vernal equinox they began to sing sweetly again. In the cleared plantation lands of the Motagua Valley of Guatemala in April, where many Orchard Orioles wintered, they contributed more song than any other species. Until their departure late in the month, their songs increased in quality and profusion. As though eager to exercise their newly acquired singing voices, black-throated, yellowish young males sang as much, if not more, than mature males in chestnut and black. At dawn, young and old sang together in a many-voiced chorus of whistled notes delightful to hear.

These Orchard Orioles roosted gregariously in a dense stand of young Giant Canes that colonized a sandy flat left by the shifting course of the Río Morjá. In Honduras, many slept in a dense growth of elephant grass, six or eight feet (2 m) tall. They entered early, sometimes an hour before nightfall, and were soon completely hidden, but they revealed their presence by snatches of breezy song that rose above the chatter of the crowd of small seedeaters that joined them.

In this patch of tall grass the Orchard Orioles slept with seven other species of birds, including a few resident Black-cowled Orioles and numerous wintering Baltimore Orioles. In the evening the latter gathered in companies larger than one sees by day, sometimes a score or more, to roost amid the broad, close-set leaves of *Dracaena fragrans*—a small tree of the lily family—or in the abundant foliage of orange trees. Viewed in a flashlight's beam, with their heads turned back and buried in their fluffed plumage, they

Orchard Oriole, *Icterus spurius,* male

looked like brighter oranges scattered amid the dark, glossy leaves. Of three Baltimore Orioles who slept in a small orange tree beside my house in Costa Rica in the early months of the year 1943, one was a male in exceptionally deep orange plumage. At dawn on April 15, the three were present as usual, but that evening the brilliant male came alone, the other two having apparently begun their northward journey. For a week after the departure of his companions, he roosted in the same place; but after daybreak on April 22 he vanished, the last of his species I saw that spring. At dawn on October 10, an outstandingly handsome male appeared beside the house, and that evening he went to roost in the small orange tree, in company with an immature male. Although he was not banded, I had no doubt that he was the same individual who had slept in the same place the preceding spring. By feats of navigation that continue to baffle naturalists, despite all our sophisticated research, many birds appear to migrate yearly between two definite spots separated by thousands of miles, perhaps an orange tree in Costa Rica where they roost and a maple tree in Canada where they nest.

Unlike certain other members of their family, orioles breed in monogamous pairs. Winsor M. Tyler (in Bent's *Life Histories*, 1958) well described the courtship of a male Baltimore Oriole who had arrived from the south and claimed a territory. When a female alighted in a maple tree where the brilliant male sang loudly, he flew to her. Facing her a few inches away, he stood with body nearly upright, straight and tall, so that his glowing orange underparts filled her vision. Then, as he bowed swiftly forward until his head was level with his feet, his black foreparts eclipsed the orange, which flashed out again as his rump came into her view. Over and over, with a jerky movement, he repeated his changes of attitude, pausing briefly at the end of each downward or upward swing, presenting to the female a dazzling alternation of orange and black. Another description of the male Baltimore Oriole's courtship tells that his tail is spread and wings partly raised, while he utters his most dulcet and appealing notes. Bowing deeply, a courtship gesture widespread among icterids, has been reported also of a male Hooded Oriole slowly approaching a perching female. Courting Orchard Orioles sometimes rise high above the treetops, to descend with fervent song. Orchard Orioles feed their incubating mates, as few other orioles appear to do.

In the Western Hemisphere, pendent nests are built chiefly by certain tropical flycatchers and by orioles and their relatives, the oropendolas and caciques. The pouches of flycatchers are made by matting or felting fibers together; but with sharper bills orioles weave their nests, no less skillfully than the Old World weavers. The length of orioles' pouches varies greatly, being longest in the permanently resident tropical species, who have more time for such laborious undertakings, and likewise a wider choice of appropriate materials. All these constructions share two features: the woven texture and attachment by the rim rather than support from below.

Cups, about as deep as they are wide, are built among leafy twigs by Orchard, Audubon's, and Yellow-tailed orioles. The rims of these open nests are often incurved to prevent eggs from rolling out. Black-cowled Orioles suspend a shallow cup by fine fibers laced through perforations that they make in the broad leaf of a banana plant or traveler's tree, which spread green roofs above them. In the Antilles, similar nests are built by Martinique, St. Lucia, and Montserrat orioles. Likewise, some Hooded Orioles attach their cups beneath palm fronds by strands passed through holes that they pierce in the leaf tissue, or amid clusters of leaves bound together by fibers threaded through perforations. In the arid southwestern United States, Scott's Oriole sews its shallow nest to the edges of hanging dead leaves of Joshua trees (species of *Yucca*), which conceal it well.

The nests of Baltimore and Bullock's orioles may be rather shallow cups or pouches up to eight inches (20 cm) long, suggestive of the pensile nests made by a number of tropical species. The elongated, pyriform pouches of Altamira, Moriche, Yellow, Streaked-backed, and Spotted-breasted orioles, from one to two feet (30 to 60 cm) in length, are commonly attached to the ends of high, exposed twigs, from which they hang free, visible from a distance but almost impossible for ornithologists to reach. In arid regions where most trees are low, Altamira Orioles hang their nests from the tips of the highest exposed branches.

Most orioles' nests are well separated, but Orchard Orioles sometimes breed in colonies; nearly twenty of their nests were counted in one large live oak tree, as reported in Bent's *Life Histories*. On the Yucatán peninsula, Stephen Howell and his associates (Howell, Webb, and de Montes 1992) found two dense colonies of Orange Orioles, one with 30 to 35 nests in an area of 260 by 100 feet (80 by 30 m), the other still more crowded, with 26 to 30 pairs in a space of 196 by 82 feet (60 by 25 m). The deep pouches hung in bushes and low trees over water in flooded scrub, with up to 5 in a single small tree. Associated with the first colony were two pairs of Hooded Orioles, a building pair of Black-cowled Orioles, and at least three pairs of Black Catbirds. Orioles of several kinds

Baltimore Oriole, *Icterus galbula,* male, with nest

often build their nests near those of Eastern Kingbirds, Great Kiskadees, or other large flycatchers, who vigorously repel raptorial birds and other predators (Pleasants 1981).

For their nests orioles choose long, flexible strands that lend themselves to weaving. Orchard, Hooded, and Audubon's orioles prefer green grasses. Where available, Spanish moss *(Tillandsia)* is largely employed for nests well hidden in the midst of this dangling epiphyte. Thin fibers torn from decaying leaves or stems are important components of many nests; a Yellow-tailed Oriole made a nest of bright, clean fibers that the bird herself had apparently stripped from some suitable plant and woven into a fabric so open that the eggs were readily visible through the nest's bottom. Close to human habitations and on farms, Baltimore Orioles use string, bits of cloth, and horsehair. To line the bottoms of their cups or pouches, orioles gather down from willow catkins, pappi of dandelions or other composites, sheep's wool, hair, fine grasses, and long white wool from the tips of branches of certain organ cacti. The bottoms of long pouches may have only a thicker mat of the same kinds of fibers that compose the sides (Schaefer 1980).

Although male Orchard and Baltimore orioles have been reported to help, the female usually accomplishes the nest building with little or no assistance from her partner. To start their long pouches, Altamira, Streaked-backed, and Yellow orioles wrap fibers around the end of a high, often forked twig to form a loop. From this point they weave downward, hanging inverted inside a lengthening sleeve, until finally they close the rounded bottom, leaving the top open as an entrance. A female Spotted-breasted Oriole actively building was not accompanied by her partner as she flew back and forth gathering fibers, lengths of slender green vines, and cordlike roots of epiphytes. He was busy feeding fledglings of an earlier brood and singing superbly. While working she repeated a loud nasal call and, after weaving in her fibers, flew away uttering clear, musical notes. A Spotted-breasted Oriole's nest that I found when just begun appeared to be complete, externally at least, five or six days later. A Yellow Oriole, whose nest was in the open loop stage when first noticed, had closed the bottom only three days later, but her nest was exceptionally short. An Orchard Oriole built a nest in six days; nests of Baltimore and Hooded orioles took five or six.

Because so many orioles' nests hang inaccessibly high, we know too little about the size of their broods. Three appears to be the most frequent number of eggs laid by orioles at low latitudes, but sets of only two are not rare. Farther north, in Mexico and the southern United States, sets of four or five are more common. The orioles that breed at highest latitudes, the Orchard, Balti-

Altamira Oriole, *Icterus gularis,* sexes similar

more, and Bullock's, are the most prolific, laying three to six eggs and occasionally more. Orioles' eggs are white, sometimes lightly tinted with blue, gray, or buff, and more or less heavily marked, especially on the thicker end, with spots, blotches, or scrawls of black or shades of brown, purple, chocolate, or gray.

Probably all orioles help their partners to feed the young but not to brood them (Skutch 1954). Repeated passage through the opening at the top of a long, swinging pouch may tear it down along one side almost to the bottom. When this occurs, older nestlings stick gaping mouths through the enlarged aperture to receive food from a parent clinging to the outside. How long young orioles remain in such deep pouches hanging high appears not to be known. From an open cup, two Yellow-tailed Orioles departed when twelve and thirteen days of age. For Baltimore and Bullock's orioles, nestling periods of twelve to fourteen days have been reported. Beginning shortly after they fly from the nest and continuing while their parents feed them in trees, fledglings of these orioles are noisy, all day long clamoring for food with "a pathetic little childish cry or complaint, beseeching, yet insistent, half way between entreaty and demand, *dee-dee-dee-dee-dee*" (Bent 1958, 254). How long parents continue to feed these fledglings appears not to be known.

In Guatemala I watched both Altamira and Spotted-breasted orioles feed juveniles, already able to find food for themselves, while their mothers built nests for second broods. In the north, Baltimore and Orchard orioles appear to rear only one brood in a season.

Although the sexes of Altamira and Spotted-breasted orioles are alike, juveniles differ from adults, being much paler and less heavily marked. When, as in orioles that breed at high latitudes, the adult female is much duller than the adult male, juveniles of both sexes tend to resemble the former. Many male orioles do not acquire full adult attire until well over a year old, and yearlings (distinguishable from females by their black throats) may breed in immature plumage, as the Orchard Oriole does (Bent 1958; Sealy 1980). In one of the nesting pairs of Spotted-breasted Orioles that I watched in Guatemala, both members quite lacked the black spots that adorned the breasts of neighboring pairs who were apparently older.

Unlike some of the grackles and certain other icterids, orioles appear rarely to be troublesome to humans or (except the Troupial) to birds of other kinds. Although some may take cultivated fruits, they pay for this fare by ridding trees of injurious cater-

Spotted-breasted Oriole, *Icterus pectoralis,* sexes alike

pillars and other insects; on the whole, they are regarded as economically beneficial. Bright colors and cheerful songs make them welcome around our homes. Unfortunately, many pay dearly for being beautiful and melodious. In tropical America, they are captured and confined in cages that are often miserably small, so frequently that some, like the Yellow-tailed Oriole, have become rare because of this nefarious practice.

The Troupial

The large, black-orange-white, mellow-voiced Troupial, Venezuela's national bird, differs in several ways from other orioles. Its unique features include bare blue skin around its yellow eyes and long aculeate feathers on its throat, which stand out like a bristly beard when the bird sings. In a genus of skillful weavers, it makes no nest but pirates those of other birds. A frequent victim is the

Rufous-fronted, or Plain-fronted, Thornbird, a small brownish member of the ovenbird family, which in northern South America builds of interlaced twigs massive nests that hang conspicuously from high, exposed branches or service poles. These structures range in height from about two to six or seven feet (0.6 to 2 m) and contain from two to eight or nine chambers, one above the other, or rarely two side by side. These rooms do not intercommunicate; each has its separate doorway to the outside. In pairs or families of up to six fledged individuals, the industrious builders of these remarkable structures sleep throughout the year, and in them they lay sets of three white eggs and rear their young.

In northern Venezuela I (Skutch 1969) watched from beginning to end the capture of a thornbird's nest by a pair of Troupials and the rearing of their brood in it. They began by tearing open, in the side of the lowest of the five compartments, a doorway wider than those thornbirds use. In this low chamber the female Troupial slept, while her partner, who was repulsed whenever he tried to join her there, lodged in the room immediately above her. The six resident thornbirds continued to sleep in one of the upper compartments. This peaceable coexistence of builder and pirates continued for several weeks. Troupials are the only icterids known to sleep in dormitories rather than amid vegetation.

In the days when the pair of Troupials lodged in the thornbirds' nest before the female laid in it, they from time to time pulled sticks from the middle of the elongated structure, until it had an hour-glass shape. In this interval the thornbirds built a new nest on a neighboring branch of the same small tree, slept in it as soon as it was far enough advanced, and laid eggs in it. The male Troupial also moved to the new construction, to sleep in a room to which he had opened a wider doorway. The thornbirds, whose eggs had been eaten or thrown out, returned to lodge in their old nest, in the room immediately above that in which the female Troupial was now incubating. They entered late, in the twilight after their persecutors had settled for the night.

As the incubation of their eggs proceeded, the Troupials became more hostile to the thornbirds, chasing them from the captured nest, sometimes pursuing them mercilessly over the surrounding

Troupial, *Icterus icterus*, sexes alike, at nest of Rufous-fronted Thornbird, *Phacellodomus rufifrons*

pasture and through a neighboring thicket. The thornbirds responded to this situation by building a third nest on the far side of a nearby mango tree. The Troupials reared two young, who after fledging passed from my ken, while the parents continued to lodge in the thornbirds' nests. Not content with the two nests they had already captured, one of the adults invaded a neighboring nest belonging to a single pair of thornbirds, destroyed their nestlings, and slept in the lower chamber with an enlarged doorway, while the rightful owners continued to repair, and to lodge in, the higher of their two rooms. This pair of Troupials had pirated three thornbird nests, in all of which they slept and in one of which they nested. Their two young were reared at the price of two broods of thornbirds, probably six young. Fortunately for thornbirds, these brilliant rascals, the most aggressive of orioles, were rare in the locality where I studied them. I found only one breeding pair of Troupials but many families of their little brown victims.

Troupials invade nests other than thornbirds'. They have been reported to breed in the bulky nests of Great Kiskadees (Cherrie 1916) and in Jamaica (where they are introduced) in the long pouches of the Jamaican Oriole (Bond 1960). A frequent victim is the Yellow-rumped Cacique, which hangs its pouches in crowded colonies of a few to more than a hundred breeding females (Pearson 1974). In southeastern Peru, Scott Robinson (1985a) found Troupials extremely injurious to the larger caciques. To evict a female from her nest, a Troupial clung to the outside and pecked through the thin fabric with its sharp bill. In pairs, the pirates attacked caciques, often grappling with them until the combatants fell together into the water below the nest tree. They destroyed eggs and nestlings. In two days a belligerent pair evicted seven female caciques from a cluster of nests, and another pair seized five nests. Each time one of the invaders entered a stolen nest, its partner stood nearby and attacked any cacique that approached. (For more details, see chapter 15.)

The Troupials laid three eggs in the nest they wrested from the thornbirds, as did another pair in a nest found by George K. Cherrie in the Orinico region. While a Troupial incubated in the thornbirds' nest, I passed two mornings watching from a blind. Although I could not distinguish the sexes of this pair, my failure to witness a changeover during thirteen hours convinced me that only the female incubated, as in all other members of this family for which we have information. The similarity of the records I made on two sunny mornings five days apart is remarkable and suggests that, as in many other species, the length of an incubating bird's sessions and recesses is innately determined but influenced by external factors, chiefly temperature and availability of food. On June 10 I began my vigil at 5:55 A.M., while the waning moon shone brightly and the earliest birds were starting to sing. I did not see the male Troupial fly from his bedroom, but presently I heard him sing superbly. Then five or six thornbirds emerged from the compartment above the incubating female and for many minutes climbed over their damaged nest, passing from chamber to chamber (except that occupied by the female Troupial), until at 6:25 the last of them flew away. The Troupial did not leave her eggs until 6:32. By 12:43 she had taken nine sessions, ranging from 12 to 45 minutes and averaging 28.9 minutes. Her nine absences varied from 4 to 23 minutes and averaged 12.3 minutes. On June 15, the Troupial first emerged at 6:33 and by 1:06 had sat for nine intervals, ranging from 15 to 61 minutes and averaging 31 minutes. Her nine recesses ranged from 5 to 22 minutes and averaged 12.7 minutes. On the first morning she incubated for 70 percent of the time; on the second, 71 percent.

Sometimes, after flying from the nest, the female sang and was answered by her mate in the distance. Occasionally she sang loudly as she was about to enter. On some of her returns she was escorted by her partner, who rarely went as far as the doorway. But when, on the second morning, she entered with a small item, probably food, in her bill, he followed her into the nest and stayed for less than a minute. Apparently, the sight of something in the female's bill suggested to him that a nestling had hatched, and he entered to investigate. This observation reveals one of the ways that a sharp-sighted male bird who does not incubate learns that his progeny have hatched and it is time to begin feeding them. Since on this occasion the Troupial found only eggs in the nest, he did not return to it during the next four hours. Five more days passed before the eggs hatched. Unfortunately, I had not learned just when they were laid and could not determine the incubation period. At a Yellow-tailed Oriole's nest in Guatemala, this period was fourteen days. For northern orioles, including the Baltimore, Bullock's, Orchard, and Scott's, the incubation period is variously given as twelve to about fourteen days, but precise determinations are difficult to find.

Only two of the three Troupial's eggs hatched. The hatchlings were much like those of the Yellow-tailed Oriole. Their pink skins bore the sparse gray down usual in passerines; their eyes were closed; the interior of their mouths was red; and the flanges at the corners were white. Both parents fed them, chiefly on caterpillars and adult insects, including grasshoppers and crickets, with a liberal mixture of fruit pulp. This appeared never to come from the neighboring mango tree, although the parents stuck their bills far into the juicy pulp of its abundant fruits. During four hours of the morning when the nestlings were five days old, the parents brought food fifty-one times. When they were twelve days old, they were fed only thirty-eight times in four morning hours, but the items brought to them averaged larger. I could not distinguish the parents, but they appeared to take equal shares in nourishing their brood, often coming together and sometimes standing side by side in the doorway while delivering food. Often they sang while bringing food in their bills. Apparently, only the female brooded the nestlings.

At first excessively shy, the parent Troupials become bolder after their eggs hatched. When the nestlings were five days old, their mother entered the nest to brood them while I set up my blind in the open pasture about forty feet (12 m) in front of the nest; and later that same morning the parents continued to feed the nestlings while I folded up the blind. When I climbed a ladder to look into the nest, they alighted only a few yards above me, singing instead of scolding with harsh tones. Although their vocabulary included nasal notes similar to those of other orioles, I did not hear such notes on these occasions.

When only a few days old, the young Troupials were already heavily infested with *tórsalos,* white larvae of a dipterous fly that form relatively huge swellings under the skin of many tropical birds. One nestling bore ten of these parasites, including three on its head and three on one leg. At the age of a week, the nestlings had open eyes. At ten days, their plumage began to expand. The *tórsalos* that infected them had gone, leaving superficial scars that were already disappearing; but the nestling who had supported fewer of these larvae was far ahead of the more heavily parasitized one in size and the development of its plumage. When two weeks old, the more advanced of the young Troupials was fairly well feathered. Daytime brooding had ceased when the nestlings were twelve days old, but their mother continued to pass the night with them until they were at least seventeen days of age. During their last nights in the nest they were alone. The first departed when about twenty-one days old, the second at twenty-three days. In pattern their plumage resembled that of the adults, but their colors were less intense, pale yellow instead of deep orange, the prominent wing band impure white. Less careful of their young than are many other birds that sleep in nests, the parents did not lead them back to the thornbirds' nest or any other that I could discover, although an adult resumed lodging in the chamber where the young Troupials had been reared.

14

The Pouch-weaving Caciques

In appearance and nidification, caciques resemble some tropical orioles, but they are larger, noisier, and more social. Eight species of *Cacicus* inhabit tropical and subtropical South America. One of them, the Yellow-rumped Cacique, extends into Panama; another, the Scarlet-rumped, ranges as far as northeastern Honduras. Then, after a gap in their distribution, the ninth species, the Yellow-winged Cacique, appears on the Pacific coast of Mexico from southern Sonora to Chiapas. Most caciques prefer warm lowlands, but the Mountain Cacique lives high in the Andes from western Venezuela to Bolivia.

In size male caciques range from the eleven-and-a-half-inch (29 cm) Yellow-winged to the seven-and-a-half-inch (19 cm) Ecuadorian Black Cacique. Females, especially of the larger, polygamous species, are noticeably smaller than the males. The predomi-

Yellow-winged Cacique, *Cacicus melanicterus*, male

nant color of both sexes is black, with yellow on the wings, rump, and tail. Solitary Black Caciques and Ecuadorian Black Caciques are clad wholly in blue-black. The Yellow-winged Cacique wears on the back of its head long, slender plumes that wave conspicuously in a breeze. The eyes of several species are light blue or bluish, a color absent from the plumage of icterids. No less than their eyes, caciques' long, sharp, pale bills—whitish, yellowish, or greenish—contrast with their black heads and necks. Their legs and feet are dark.

In family groups or larger straggling flocks, often with oropendolas or in parties of mixed species of arboreal birds, caciques forage in the upper levels of heavy forests or through plantations and open country with scattered trees. Occasionally they descend low, even to the ground, to pick up food. Fruits and arillate seeds in great variety, as well as insects, compose their diet. Scarlet-rumped Caciques find many small invertebrates by rummaging among epiphytes that burden trees in rainy regions. They hang with head or back downward to reach tempting items. To investigate the interior of a curled leaf, they may draw it inward and place it under a foot; and in the same manner they hold a large insect while they detach and swallow fragments. I watched a Scarlet-rumped Cacique extract larvae and pupae from the cells of a small wasps' nest from which the covering had been removed. While this bird was so engaged, its companion rested nearby on the ground, appearing to eat earth or grit. On another occasion, I watched one of these caciques trying to remove a spittle insect from the large, frothy white mass, high in a tree, that the insect had secreted to protect itself. Drawing out long, viscid strands of spittle, the bird wiped them against a branch or held them beneath a foot. After continuing this for several minutes, the cacique departed, leaving the cercopid insect safe within its frothy citadel (Skutch 1972).

Like orioles, caciques enjoy nectar. Throughout the day, Scarlet-rumped Caciques came singly, or three or four together, to drink from the nectar cups of an epiphytic *Marcgravia* that spread over the crown of a tall rain-forest tree. They shared this feast with wintering Baltimore Orioles, Golden-naped Woodpeckers, Green Honeycreepers, and several kinds of hummingbirds.

Caciques roost gregariously. As night approached, Scarlet-rumps, who had foraged in small parties through the day, gathered in larger companies in the tops of tall trees at the forest's edge. Then, in smaller groups, they dived down to skim over the tops of banana plants in a great plantation to a tall, vine-draped tree standing alone, where they slept.

Sociable caciques use their voices freely to communicate with their companions and express their emotions. To convey some notion of their versatility, I shall tell of the notes of the Scarlet-rumped, the cacique that I have most often heard. Its most frequent call is a loud, bright note, suggestive of urgency and excitement. The bird often delivers this note in pairs or in a rapid sequence. What I take to be the cacique's song begins with a note hardly distinguishable from the call, which rises rapidly in pitch as it is repeated, to be followed at times by several similar notes in falling cadence. Scarlet-rumps also voice a liquid crescendo followed by a falling phrase of the same type. Another song is a sequence of high, bell-like notes, hurriedly poured forth, sometimes accelerated and attenuated almost to a buzz. A courting male emits a series of shrill notes, higher in pitch than the usual song, sometimes with a whining quality. The variations of the cacique's vocalizations are many, but nearly all are clear and bright, inclining to be piercing rather than mellow, fit expressions of the restless vitality of this trim scarlet-and-black bird that seldom remains long in one spot. Females sing, sometimes while sitting in the nest, much as males do but in voices that tend to be higher in pitch. Scarlet-rumped Caciques are songful throughout the year. Even at the height of the rainy season when scarcely any birds nest, their cheerful whistles alleviate the gloom of the darkest, wettest days.

Voices of the several species of caciques may be harsh as well as liquid. While a female Mountain Cacique built high on the slopes of Volcán Tungurahua in Ecuador, a male gave an amazing vocal performance, with loud squeals that resembled the cries of a protesting pig and sharp, excited notes of a different character. Then he called *wip wip wip wip wip*, and later accelerated these notes into a rattle. Amid his medley of notes soft and harsh, high and deep, I heard none that was mellifluous. Yellow-rumped Caciques east of the Andes imitate other birds as well as monkeys and mechanical sounds; but the race in northwestern Colombia and Panama is not known to mimic. While singing, caciques bow slightly forward, relax and vibrate their wings, shake their tails, and raise the bright yellow feathers of back and rump, all of which makes them more conspicuous. In this posture, widespread among icterids, they court females while pouring out streams of shrill notes.

Scarlet-rumped Caciques and Solitary Black Caciques are monogamous and scatter their nests. Yellow-rumped Caciques, Red-rumped Caciques, and Yellow-winged Caciques are promiscuous and nest in colonies, which in Yellow-rumps may become large. Possibly the Mountain Cacique is at least loosely colonial, for on Volcán Tungurahua I found three nests, the two most distant about a hundred feet apart. All appeared new, but only one was

Red-rumped Cacique, *Cacicus haemorrhous,* sexes similar

visited by a building bird during the short time in the afternoon that I could stay to watch so far above my lodging in the town of Baños. This and the remaining three caciques await study.

Built by the females alone, caciques' nests are pendent, skillfully woven pouches, resembling those of a number of tropical orioles, but often longer. They may be open at the top or hooded with a sideward-facing doorway; hoods may be added after the rains begin. Usually the nests hang free from high, exposed twigs, but sometimes they are low. If built inside the crowns of trees, they might become so entangled in the branches that the birds could not straighten them. Often they are beside the nests of wasps, which protect them by assailing animals that climb too near the vespiaries. When the wasps disappear, caciques' colonies tend to dwindle in size until they vanish.

Height and stinging wasps make it difficult to learn what caciques' nests contain without destructively shooting them down.

The better-known species lay two eggs, rarely one or three. White or blue-tinged, they are speckled, blotched, or scrawled with black, brown, chestnut, or a purplish hue on the thicker end or over the whole surface. The northernmost cacique, the Yellow-winged, lays two to four pale blue eggs streaked and spotted with black (Rowley 1984). As far as known, only females incubate. The incubation period of the Yellow-rumped Cacique is sixteen days; that of the Red-rumped, seventeen days. In the colonial species only females regularly feed the nestlings, but male Red-rumps occasionally help. Vigilantly guarded by male sentries, young Yellow-rumped and Red-rumped caciques remain in their nests for twenty-four to twenty-eight days (ffrench 1973).

The Scarlet-rumped Cacique

As an example of the breeding of a monogamous cacique with isolated nests, let us look at the Scarlet-rumped, which ranges from Honduras to northeastern Peru and in Costa Rica, where I have studied it, inhabits rain-forested regions at low altitudes, rarely ascending above 2,000 feet (600 m). The male is about nine inches (23 cm) long, the similar female an inch (2.5 cm) shorter. While perching with folded wings, their glossy blackness is relieved only by their light blue eyes and greenish white bills; only when they spread their wings in flight is the scarlet rump fully exposed. They wander through the rain forests and out into clearings with scattered trees in parties of about four to ten, often in company with such other treetop birds as Purple-throated Fruitcrows, Black-faced Grosbeaks, White-fronted Nunbirds, and Montezuma or Chestnut-headed oropendolas. Of their roosting, voice, and courtship I have already written.

In the Caribbean lowlands, Scarlet-rumped Caciques breed early; by mid-March of one year I saw them feeding juveniles almost as big as their parents. Nests that I found at this time were already abandoned. They hung conspicuously from exposed twigs, up to a hundred feet (30 m) high. One was fastened to a drooping branch of a slender vine dangling free in an open space, about sixty feet (18 m) up in a cacao plantation. Occasionally two unoccupied nests hung close together, both probably the work of the same individual. A female building a late nest worked at a leisurely pace, bringing material only six times in the hour when I found her most active. She gathered long fibers from epiphytes high in trees and spent much time weaving each of them into the lower end of the open sleeve that her nest had become. While she was so engaged, her mate always perched nearby, singing and plucking insects from neighboring branches. Usually he accompanied her trips for more materials, but sometimes he waited near the nest until she returned. Found at an early stage of construction on April 28, the nest appeared finished by May 15. Another late nest was built in about a week.

A Scarlet-rumped Cacique's fallen pouch measured twenty-nine inches (73.6 cm) from its attachment at the top to its rounded bottom. From the lower end of the elongated orifice in the upper part it was fifteen inches (38 cm) deep. Near the bottom it was five inches (12.5 cm) in diameter, and from this point it tapered upward to three inches (7.5 cm) in diameter at the lower lip of the entrance. This vertically elongated opening was ten inches (25.5 cm) high by one and three-quarters inches (4.5 cm) wide. The stiff, fibrous materials of which the pouch was woven included branched inflorescences and wiry roots of an epiphyte. These had been beautifully woven into a thin but strong open fabric, surprisingly uniform in texture, through which much light passed. Long strands dangled loosely beneath the nest. A thick pad of pale seed down covered the bottom where the eggs had rested.

The only nest that I could reach hung eleven feet (3.4 m) up, beneath a spiny frond of a pejibaye palm in a clearing beside the forest. At the end of April I spent six morning hours watching the female incubate her two eggs. Her six sessions ranged from 18 to 54 minutes in length and averaged 38 minutes. An equal number of recesses varied from 15 to 26 minutes and averaged 20.3 minutes. She attended her eggs with a constancy of 65 percent. Returning to her swinging pouch, she fluttered down from above and dived in headfirst, hardly pausing at the narrow entrance. Sometimes I could dimly discern her through the meshes of the nest fabric. Departing, she flew from her doorway toward the neighboring river, or less often into the still nearer forest. While covering her eggs she often sang, sometimes in a voice hardly to be distinguished from her mate's, sometimes at a higher pitch and more rapidly. On one occasion, 6 minutes of loud song preceded her departure. At times the male and his incubating partner sang responsively.

On five of the female's seven returns to her nest she was either accompanied by her mate or I heard nearby song that I took to be his. On two returns she seemed to come alone. Five times the female left her nest while he was singing nearby and they flew off together, or else she heard song, apparently his, in the distance and flew toward it. Although the male was attentive to his partner, he did not once go to the nest, and the longest time he spent in the surrounding trees was about five minutes. Mostly he was neither seen nor heard while she incubated but returned in time to accompany her when she went to seek food. In every respect, this pair belonging to a genus that includes a number of polygynous species behaved like a typical monogamous passerine. I fully expected to see the father feed his progeny.

On May 1 this nest held two hatchlings with pink skin devoid of natal down. The interior of their mouths was pale pink, much the shade of the skin on their bodies, and the flanges on the mouths' corners were pale yellow. On the following morning I watched this nest for three hours. On her five returns, the female dived into it so swiftly that I could not see whether she brought anything to her nestlings, although undoubtedly she did. She brooded them for intervals of twelve to sixteen minutes. The male was much less attentive than on the morning when I watched the female incubate. He did not approach the nest, and I saw and heard him only once.

When the nestlings were three or four days old their pinfeathers were sprouting and their eyes were partly open. While I examined them, their mother arrived with a white larva or pupa in her bill. Alighting nearby, she made no protest, and after I withdrew a short distance she carried in the food, then flew away with her mate. Later she took another meal into the nest while I stood only three or four yards away. While my son and I inspected the nest on May 5, the parents arrived together, the mother with a large spider, the father with a tiny object scarcely visible in his bill. She entered while three people stood close by, then flew off. He followed without having delivered what he held. Some male birds are so eager to feed their progeny that they offer food to the eggs, anticipating their hatching; others take hours or days to learn that the young have hatched and begin feeding them. This cacique was evidently one of the latter.

Three days later I arrived at dawn, confident that I would see the male cacique feed the nestlings. Alas, the nest was lying on the ground, sodden with rain, and the young caciques were dead. A limb falling from a neighboring tree had knocked down the palm frond from which the nest hung. The caciques' breeding season was waning, and late nests, laboriously built apparently for second broods, remained unoccupied. Well-grown juveniles had long been following their parents, from whom they differed chiefly in the duller red of their rumps and their dark instead of blue eyes. They were fed by their parents and other members of their flock (Skutch 1972).

The Solitary Black Cacique

From the brief available accounts, the Solitary Black Cacique appears to be less sociable than the Scarlet-rumped. These all-black, slightly crested, brown-eyed birds have pale greenish white or pea green bills, which contrast with their dark attire. The male is eleven inches (28 cm) long, the similar female slightly more than an inch (3 cm) shorter. Largely confined to lowlands below 1,300 feet (400 m), these caciques spread over much of tropical and subtropical South America east of the Andes from northeastern Colombia and Venezuela to eastern Brazil and northern Argentina. Alone or in pairs, they forage amid dense vegetation along rivers and at forest edges, in second-growth woodland, rank grasses, and even fields of sugar cane, usually remaining low.

The Solitary Black Cacique's large repertoire of curious sounds includes a squeaky, accordionlike series of ascending notes; a loud, resounding *tsonk* several times reiterated; a nasal, mocking *naaaaah;* and a strong, nasal, exhaling *wheeeah*. What appears to be its principal song is an odd sequence of "electronic" sounds (Hilty and Brown 1986).

The cacique's nest (illustrated in De la Peña 1979), fastened to the end of an exposed twig, is a pensile pouch, much like that of the Scarlet-rumped Cacique. Large examples measure twenty inches long by six inches thick near the lower end (50 by 15 cm). The two eggs are whitish, marked with chestnut spots or with blackish dots, blotches, and scrawls.

Another species that lives in pairs instead of flocks is the Ecuadorian Black Cacique, confined to the eastern forests and second-growth woods of that country. It resembles the Solitary but is smaller and has blue rather than brown eyes. Its nest appears to be unknown (Hilty and Brown 1986).

The Yellow-rumped Cacique

In April 1935 a small company of Yellow-rumped Caciques established a colony on a low, decaying trunk that stood, a hundred feet (30 m) from shore, in a narrow arm of Gatún Lake reaching far inland between wooded ridges of Barro Colorado Island. This stub was a solitary remnant of the forest that, twenty years earlier, had been drowned when the water of the Río Chagres was impounded for the Panama Canal. Although its own leaves had long since withered and fallen, the trunk was verdant with a diversity of ferns, orchids, aroids, bromeliads, and shrubs, whose roots draped and embraced it all around. Bees had built their hives and wasps their nests amid this aerial garden; and the caciques were weaving their pouches in the bush that flourished at the top of the stub, about twenty-five feet (7.6 m) above the water that might have protected them from flightless predators.

In the windless dawn, when only the little waves spreading from my dugout canoe, or cayuco, ruffled the smooth surface of the cove, I would paddle up to its head and sit in the boat to watch the caciques at work. The colony consisted of fifteen building females and at least eight males. Three of the females had brown eyes and were evidently younger than their blue-eyed neighbors. They were the last to start their nests, and at least one of them abandoned her work unfinished. Apparently, Yellow-rumped Caciques do not lay eggs until about two years old and as yearlings practice weaving.

From the wooded shore or deeper in the forest the females brought narrow strips of palm fronds, fibrous strands apparently pulled from vines or large leaves, and lengths of slender vines, dead or with small green leaves still attached. The first step in nest construction was the formation of a loop in a mass of fibers entangled around the terminal twiglets of a leafy branch. Sitting in this loop, the builder continued to weave more strands into the fabric, which lengthened downward, forming a sleeve. When this had become long enough to contain her, the cacique worked in an inverted posture, her widely spread feet clutching opposite sides of the tube in what appeared to be a strained, uncomfortable position, although probably she did not find it so. Each new strand was alternately pushed and pulled through the meshwork until it was firmly interwoven in the lengthening fabric. Although the caciques always worked inside, I could see their sharp yellow bills projecting through the nests' walls whenever they pushed a fiber through it from within or reached out to grasp a loose end and pull it inward. The builders always entered and left through the original loop at the top, never through the bottom of the sleeve, although at a certain stage of construction this would have been the most convenient exit.

After the sleeve attained its full length of twelve to eighteen inches (30 to 45 cm), the builder closed it by weaving a rounded bottom. Five or six days of hard work sufficed to weave a pouch, from wrapping the anchorage to the supporting twigs to closing the bottom. Then the cacique softly lined the bottom where her eggs would rest, chiefly with down that surrounded the seeds of the fat-boled barrigón trees that grew on the shores of the cove and were laden with heavy dehiscing pods. From a greater distance, the builders brought other soft and fibrous materials to mix with the seed down. They devoted two or three days to lining the nests, making seven or eight days the approximate time for finishing them. These structures had open tops without hoods, such as other races of Yellow-rumped Caciques add to their nests for better protection from rain and certain predators. While the female caciques built in this small group, I noticed none of the bickerings between near neighbors so frequent in crowded colonies of oropendolas and grackles, and no attempts to pilfer materials from each others' nests or bills.

While the female caciques went quietly about their tasks, the idle males claimed attention by their movements and, above all, their voices. They made the narrow cove resound with notes so bright and shining that they remain in memory for years. They had a marvelous repertoire of phrases, some arrestingly beautiful, others amusingly quaint. Most other vocalists of the caciques' excellence sing alone; no other colonial nester that I know equals them as musicians; to hear so many superb voices in the same tree made watching this colony of Yellow-rumped Caciques a unique experience. Among the many verses that I tried inadequately to paraphrase were: *chee chu chu chu*—series of full liquid notes descending in pitch, perhaps the birds' masterpiece; *key a woo woo a woo; key what; a woo woo; kee-ee;* and *key yo yu.* Like oropendolas, the caciques often poured out their liquid notes while clinging to the sides of their swinging nests.

A favorite position of a singing cacique was the broken-off top of the trunk that supported the colony. From this coveted station he was often driven by a rival who flew at him in full career, as though to knock him down into the water; but the target of this attack always flew just in time to avoid impact. Often the aggressor was content to occupy the perch that the other had just relinquished, but often he would dash in pursuit over the water. Pursuer and pursued flew rapidly, with resonant wing-beats. They might alight, not far apart and without evident animosity, in a tree on the shore, or the chase might continue back and forth among the trees. But however far the pursuit was carried, I never saw it end in combat, and rarely in contact. It appeared to be a game played by the idle male caciques, the chief rule of which was that the individual chased must flee and give the pursuer the pleasure of following. While the females built, the males almost continuously indulged in this game, and in singing.

No matter how gently one male flew toward another, the second invariably relinquished his place to the first. Often the individual thus supplanted would go promptly to claim the perch of a third male; I never saw such a claim contested. To give the claimant what he wanted appeared to be one of the caciques' most strictly obeyed rules of behavior. On only one occasion did I see two males come into contact. One had chased another from the nest tree and was pursuing him toward the shore. The pursued, feeling himself hard pressed, turned in the air to face his pursuer. The two collided but did not clinch. After this momentary encounter, they flew together to the shore and rested amicably not far apart. All the chasing back and forth reminded me of a good-natured game of tag, in which nobody lost his temper. Other watchers have reported a similar rarity or absence of fights among male Yellowrumps and other caciques, as well as among oropendolas.

Rarely a few wasps from vespiaries on the trunk attacked the caciques, who protected themselves by nipping them in the tips of their sharp bills and dropping them into the water.

The male caciques often followed the females from the nest tree to the shore, sometimes two or three with a female. Often one returned with a female bringing fibers, then rested above her nest and sang while she worked inside. Frequently males peered down into nests, especially those that had been finished, but I never saw a male enter a pouch. They were content to cling to the outside, in almost any attitude, and sing. I saw no indication that particular males were paired with particular females or that they were attracted to a certain group of them. While in the colony, the males rarely addressed any one female; if one did, the demonstration lasted only a few seconds and did not lead to mating. Coition probably occurred off in the forest, where the birds could mate in greater privacy, as Francisca Feekes (1981) found in Suriname.

By the first of May, when thirteen females had finished their nests and started to incubate, the colony in the cove became much quieter, presenting a strong contrast with the noise and bustle of the preceding weeks. The males now spent less time singing and chasing each other amid the nests and more in the woods along the shores. Occasionally one would visit the colony to perch near a nest or cling to its side and sing, but he rarely remained long. Even the Giant Cowbird and the Piratic Flycatcher who had been so conspicuous were losing interest in the nests that they had tried to exploit. The incubating females came and went in a direct, efficient manner, rarely loitering outside their nests. They tended to leave their eggs and fly off to forage in groups. A female departing her nest often started a general exodus of up to half a dozen of her incubating neighbors and such males as happened to be present. I could not decide whether females still in their nests learned of the first bird's departure by feeling the vibration of the supporting branches as she climbed out of her pouch, hearing her wing-beats as she took off, or perhaps even seeing her dimly through the meshes of their nests. At other times only one, two, or three females would follow the first, or one would leave alone. The sessions of four females that I watched simultaneously for two hours in the morning ranged from 10 to 30 minutes, their absences from 4 to 11 minutes. They incubated for 69 to 75 percent of the two hours.

On May 12 I first saw a female cacique enter her nest with food in her bill, and heard the weak cries of newly hatched nestlings. Three days later, over half of the nests held young, and the cries of the oldest were becoming loud. The males took no direct interest in their progeny. In Suriname, Yellow-rumped Caciques fed approximately equal amounts of fruits and insects to their nestlings (Feekes 1981).

One evening I stayed in my cayuco at the head of the cove until it grew dark, to watch the caciques' preparations for the night. During the late afternoon the community carried on its usual activities, the females flying back and forth between the colony and the shore and darting in and out of their nests, the males dashing to and fro and singing among the pensile pouches. At about six o'clock, long after the sun had fallen behind the western ridge, many of the females left their nests and flew to low trees on the nearest point of land, where they diligently preened all their plumage and shook out their wing and tail feathers. One of the males who rested among them courted a female, bowing up and down and striking her tail with his bill at each forward movement, continuing this until she sidled beyond his reach. Soon the males flew inland, and after finishing their grooming the females returned to their nests. Before it was too dark to write in my notebook or to distinguish the leaves on the neighboring shore, every cacique who remained in the colony was hidden in a nest; the males, and the few females who had not completed their pouches, roosted somewhere off in the dark forest on the island.

Soon after the eggs began to hatch, I noticed that some of the female caciques no longer visited their nests. Day by day the number of unattended nests increased, until half were deserted. Since none of the nestlings was yet old enough to fly, it was evident that something was preying on the eggs and young, probably by night, as I saw nothing menace the nests in the daytime. As daylight waned, I sat in my cayuco amid the marsh grass at the head of the cove to watch the blighted colony. The young in one of the nests cried out loudly and shook the swinging pouch. All the other nests appeared to be empty now. Suspecting that the attack on this colony surrounded by water would come from the air, I waited long in the darkness, looking toward the sky, against which a flying creature would be silhouetted. When after hours no aerial attack came, I paddled closer to the colony and threw my flashlight's beam upon it. Almost the first glance revealed a yellow-and-gray mica, a snake that I knew to be an insatiable looter of birds' nests, stretched in a sinuous line along a branch that supported several nests, which now hung empty and deserted.

Holding the flashlight in one hand, I fired my revolver at the serpent. Neither the sudden blaze of light nor the loud detonations seemed to make any impression on the snake, which slithered along from nest to nest. After I had exhausted my twenty shots to no apparent avail, the serpent continued its search among nests that it had already despoiled. Finally, it stuck its head into a pouch at the end of a branch and withdrew it with a bulge behind its mouth. Then it opened wide its jaws and spat up two eggs, which fell with a splash into the water and sank. After it had finished its search among the abandoned pouches where they hung closest together, it slid across to the opposite side of the colony toward the only nest that I knew to be still occupied by nestlings.

I had fired my last shot and hurled the last detachable object that I could spare. Unable to do more, with a heavy heart I paddled away through the night.

Next morning I returned to the cove, expecting to find the caciques' colony wholly devastated. As the cayuco approached it, I was amazed to see the mica hanging head downward beside an empty nest. On the shore I cut a long pole to pull down the snake. In its posterior half were three bullet holes. It is almost incredible that neither sounds or light or mortal wounds had availed to overcome the lust for prey that drove the mica from nest to nest. After swimming across the too-narrow expanse of water that separated the colony from the shore, it had easily ascended through the tangle of roots that enveloped the trunk, to hide by day in a nest and emerge at night to plunder more of them. For success, nesting colonies as conspicuous as those of the caciques and oropendolas need better isolation from all flightless predators.

Before I removed the snake, I noticed one of the caciques carrying food to young in a nest less than a yard from its dangling body. In one other nest the young had escaped destruction and were still attended by their mother. Females bereft of their nestlings continued to bring food, which after an interval they swallowed or carried away in their bills. Still others came to peer into empty nests or cling to their sides. Despite their losses, their attachment to progeny and nests remained strong. Nor did the males wholly desert the stricken colony. Until the last young bird flew away, they came to sing among the swaying empty pouches (Skutch 1954).

Often Yellow-rumped Caciques nest in colonies with other species that build hanging nests somewhat like their own. In Suriname, Feekes (1981) found mixed colonies of Yellow-rumped and Red-rumped Caciques, both of which build hooded pouches. Usually the colony was established by the Yellow-rumps and later joined by the Red-rumps, not without opposition by the founders. After this initial resistance, the two species dwelt in peace, each going its own way independently of the other, with rarely a tiff. Males of both served as sentinels, whose alarm calls when intense might make all colony members flee. More timid than Yellow-rumps, Red-rumps were often put to flight by disturbances, such as the approach of a human, that the former ignored. After an alarm, Yellow-rumps returned first to their nests. Both sexes of both species joined in attacks upon predatory birds.

Caciques and oropendolas, especially the Crested, sometimes nest together. This larger bird is more alert to sound the alarm when the colony is, or appears to be, threatened. Caciques tolerate unrelated birds in their colonies. While the Yellow-rumps nested at the summit of the stub in the cove described earlier, a pair of Tropical Kingbirds and a pair of Rusty-margined Flycatchers raised young amid the epiphytes that draped the trunk below them, all dwelling in harmony. On the *llanos* of Venezuela, I found two of the Rufous-fronted Thornbirds' many-chambered nests of interlaced sticks hanging among thirteen of the very different nests of Yellow-rumped Caciques in a wide-spreading tree (Skutch 1969).

In large colonies caciques tend to clump their nests, often so closely that contiguous pouches are woven together. At Cocha Cashu Biological Station in Manu National Park in southeastern Peru, nesting Yellow-rumped Caciques found safety on a tiny islet in an oxbow lake inhabited by black caimans and giant otters that preyed upon snakes trying to cross the water. Here Scott K. Robinson (1985b) counted seventy-six nests densely packed together. One group consisted of fifty-five nests in a space of approximately one cubic meter (35 cubic feet). These dense clusters of nests suggest the avian apartment houses of Sociable Weavers, Monk Parakeets, and Palmchats, with the difference that each cacique works only on her own nest, without cooperating with others even to build a base or superstructure for the aggregation.

Occupied pouches of caciques were often intermixed with empty pouches from earlier nestings in the same site or from more recently abandoned nestings. These may increase the safety of occupied nests by discouraging predators, which after examining a number of empty nests may desist from searching before they find eggs or young. Similarly, Marsh Wrens may benefit by the many "dummy" nests the males build.

Female Yellow-rumps competed keenly for the safest nest sites, such as those in dense clusters or on trees surrounded by water.

Long-term residents developed mutual tolerance and built in interwoven clusters from which they tried to exclude newcomers, who tended to weigh less than older individuals. Unlike the caciques in the small group that I watched in Panama, in these populous colonies females often struggled strenuously to possess a desirable site, to exclude a new arrival from a cluster, or to prevent one female from capturing a nest that another was building in a coveted safe situation. Losers excluded from prime sites perforce built in inferior ones, such as trees at the forest's edge or on branches overhanging water (Robinson 1986).

Caciques nesting simultaneously in clusters on islands enjoyed the highest success, rearing fledglings in about 60 percent of their nests. Success fell below 50 percent in island clusters where nesting was not simultaneous. Isolated nests on islands were less successful than massed nests. About 42 percent of clustered nests near wasps produced fledglings, but even when close to vespiaries isolated nests yielded few fledglings. Of nests at the forest's edge, only 20 percent were successful. Overhanging nests fared worst of all, producing fledglings from only 9 percent when they were clustered and synchronized, and failing totally when they were solitary. Although these Peruvian Yellow-rumped Caciques laid two eggs, they rarely reared more than one fledgling. They usually renested at least once during an eight-month breeding season.

Among the many enemies of caciques nesting at Cocha Cashu were at least three primates: white-fronted capuchins, brown capuchins, and squirrel monkeys. When not held aloof by stinging wasps, these mammals drove the caciques from their colonies and emptied every nest within reach by removing the contents through the doorway. Avian predators included Black Caracaras, who hung from a nest while they tore a hole in the side to extract eggs or nestlings; and Cuvier's Toucans, who tore a long, narrow gap in the side near the top for the same purpose. These predators were sometimes routed by mass attacks of the caciques. Great Black Hawks shook nests, then tore open those in which they heard nestlings. They were not mobbed by the caciques, who quietly abandoned the colony when they arrived; but empty nests reduced their ravages of occupied nests. Females who lost their eggs or young to any of these predators, or to snakes, nearly always sought a safer locality for their next nesting, usually another colony.

Piratic Flycatchers captured nests, probably by throwing out eggs or small nestlings while a parent was not within (as I have watched them do at covered nests of other birds), then moved in to rear their own offspring. In mixed colonies, Russet-backed Oropendolas sometimes destroyed the eggs or young in caciques' nests around their own. Of the fiercest, most destructive of the feathered pirates, the Troupial, I told in chapter 13, in the section on Troupial behavior at Rufous-fronted Thornbirds' nests. After inspecting a number of different colonies, this very aberrant oriole settled in one and systematically fought every female cacique and chased away every male who came near. Clinging to the outside of a pouch, it stabbed incubating caciques through the wall. One pair of Troupials consecutively treated seven females in this manner. When a persecuted cacique burst out of her pouch, one member of the Troupial pair either attacked and grappled with her or chased her from the colony.

In two days, this bellicose pair evicted seven caciques from their nests, tossed out fragments of their eggs, and chose a nest in the center of the group for its own use. Although in one instance a group of caciques, including one male, attacked the aggressors, they could not save the five nests of this group. However, one cacique, by sitting in her doorway and pecking the robbers whenever they approached, held them aloof and retained her nest. Troupials and Russet-backed Oropendolas apparently devastate nests clustered around the one in which they breed for the same reason that caciques often lay their eggs amid empty nests: because predators' failure to find food in nests that they investigate may cause them to abandon the search before they find an occupied nest.

Most land birds seek safety by hiding their nests, chiefly amid vegetation. This tactic is unavailable to caciques because their long pouches must hang free—and conspicuously—to avoid entanglement in boughs and foliage. Accordingly, they seek inaccessible sites, difficult for flightless creatures to reach, such as a tree with a long, clean trunk and an isolated crown or, better, a tree surrounded by water, or one inhabited by stinging wasps. The scarcity of such superior sites forces many birds to nest in colonies that are highly conspicuous and attract aerial predators. Compensating for this disadvantage, some colonial nesters, including caciques, join in mass attacks upon despoilers of nests, or their sentinels warn them in time to flee from enemies too powerful to be confronted. This appears to be the rationale of colonial nesting by pouch-weaving icterids. That caciques benefit by nesting in colonies is evident from the fact that of 1,129 nests watched by Robinson at Cocha Cashu and vicinity only 406 (36 percent) were looted by predators, a low loss for birds' nests in tropical rain forests (Robinson 1985a).

15

The Yellow-billed Cacique

Although by some authors classified in the same genus as the caciques that weave hanging pouches, this bird differs from them in so many ways that it is best kept in a genus of its own *(Amblycercus)*, of which it is the only species. The male is nine inches (23 cm) long, the female an inch (2.5 cm) shorter. Both have wholly black plumage with contrasting whitish or orange-yellow eyes and a yellow bill with a silvery sheen, earning for the bird the name *pico de plata* (silver bill) in Costa Rica. The rather long bill is conical and flattened at the tip, as recognized by the alternative name Chisel-billed Cacique. In its wide range from central Mexico to Peru and northern Bolivia, the yellow-billed bird shows interesting differences in its altitudinal distribution. In Costa Rica it extends from lowlands to 10,000 feet (3,050 m), but in Venezuela it lives only above 6,000 feet (1,800 m).

Yellow-billed Cacique, *Amblycercus holosericeus,* sexes alike

The denser and more entangled the vegetation, the more at home the Yellow-billed Cacique appears to be. It lurks in the riotous thickets that spring up in abandoned fields and other clearings in the humid tropics—thickets through which a human can scarcely take a forward step without cutting his way with a machete. Canebrakes, the bushy edges of forest, and in arid regions tangles of vegetation along waterways also offer enough cover for this secretive bird; grass if sufficiently tall and rank is also acceptable habitat. Dense vegetation near the ground is the cacique's prime requisite. In undisturbed forest at low altitudes, and even in taller second-growth woodland, the undergrowth is often too light and open for it. High in mountain forests it frequents the dense understory of tall bamboos and ravines choked with plants.

How reluctant this cacique is to quit low tangles of bushes and vines was strikingly demonstrated to me when an isolated patch of such vegetation was burnt. As the flames spread with loud crackling and a dense cloud of smoke, the more mobile and wide-ranging birds fled well in advance of them. Flycatchers and tanagers retreated first, then pigeons and doves. The more shy and secretive birds, who rarely expose themselves, were the last to appear at the edge of the burning tangle of plants. Finally, when the flames had come near, a skulking Plain Wren could endure it no longer and rushed forth to fly slowly and laboriously over an adjoining open field. But a pair of Barred Antshrikes and three Yellow-billed Caciques refused to abandon their sheltering thicket. Braving heat and smoke, they stuck steadfastly to a corner only a few yards across that escaped the conflagration. Seeing these three caciques at the same time, when the flames had driven them quite to the edge of the thicket at the roadside, was a memorable event (Skutch 1954). Although I had glimpsed lone individuals and occasionally a pair at many points in Central America and southern Mexico, I could not recall having seen three together.

Yellow-billed Caciques find most of their food in the depths of their thickets. Their yellow eyes, which from the edge of a dense tangle sometimes return the gaze of a prying bird-watcher, give them an inquisitive air that is intensified by their manner of foraging. They are constantly peering, prying, and tearing into the folds and crannies and hollow spaces of living and dead plants. They noisily ransack the rustling dead leaves that thickly drape the massive stems of banana plants in abandoned plantations overgrown with brush. Where a field of sugarcane adjoins a thicket, they enter it to pry up the edges of the sheathing bases of the leaves, searching for small creatures hidden between the leaf sheath and the stem. They break open the ends of thin dead twigs to extract larvae or ants from the pith, and they dig into rotting branches for the same purpose. On thin dead stems and vines they hammer vigorously with chisel-beaks, like a woodpecker, or perhaps more in the manner of a jay pecking into an acorn.

After they have pierced the harder outer layer of a stem, they insert the tip of the bill and by opening their mandibles spread apart the edges of the gap, laying bare any larvae or boring insects that might be hiding in the center. It is difficult to learn just how they extract their small victims; but since the gaping bill is fully occupied in holding the gap open, the tongue is apparently employed for removing the insects. When, while rummaging among the dead foliage of the banana or some other great-leaved plant, they wish to see what is on the other side of a leaf, they push the sharp bill through the tissue and by opening it make a gap through which they can peer.

Fruit seems to be a minor part of the caciques' diet. Among the few things that lure them out of their low thickets to exposed branches thirty or forty feet (9 to 12 m) above ground are the long, hard green pods of *Inga* trees. After searching until it finds one it can open, a bird will insert its bill and gape, then pull out green seeds enveloped in soft, sweetish white coats that children love to suck. Whether these or the larvae that often infest the pods are the chief attraction to the birds, I have not learned. The caciques demonstrate well the advantages of the special muscles that enable icterids to open their bills against pressure, whether to enlarge a hole in the ground to extract worms and grubs, in the manner of grackles, to help an oriole remove the juices of a pierced fruit with its brushlike tongue, or to open stems and tough pods, as the Yellow-billed Cacique does.

Often one hears, floating out of a thicket, a loud, clear double whistle, followed by a prolonged, rattling *churr.* If the listener is near enough, he or she may notice that the *churr* is preceded by a long, thin, ascending whistle, sounding as though it were produced by an inhalation—an impression strengthened by the knowledge that much breath is needed to make the protracted *churr.* From the repeated association of the two dissimilar utterances, I concluded that the mellow double whistle and the rattling *churr* were the responsive calls of a pair of caciques, and I suspected that the more melodious notes were the male's—a conjecture confirmed while I watched a nest. I never knowingly heard the female give the double whistle or the male the *churr,* although he utters a diversity of harsh notes that contrast sharply with his velvety whistles. Hearing these responsive utterances at all seasons convinced me that Yellow-billed Caciques live in pairs throughout the year—

a conclusion that might be hard to substantiate by visual evidence alone.

The two nests that I have seen in Costa Rica were found by a man cleaning small, neglected patches of his cane field, overgrown with vines and other growths, in the midst of thickets. One nest was forty inches (1 m) above the ground, supported upon vines and cane leaves between two upright canes. The other had been inadvertently cut down by the man, who propped it up in approximately its original position, thirty-nine inches above the ground. Both were well-made, thick-walled, open cups. The outer layer was composed of slender dry vines and strips from the fibrous leaf sheaths of banana plants, some of which had been looped around the supporting stalks in typical icterid fashion. The middle layer was principally of long, narrow strips torn from monocotyledonous leaves. Thin, dry vines lined the cup. The bulkier of these nests measured 6 by 4½ inches in overall diameter. The cavity was 3½ by 3 inches in diameter and 2½ inches deep (15 by 11.5 cm; 9 by 7.5 and 6.3 cm). When found, one nest contained a nestling and an unhatched egg; the other, two eggs. These eggs were pale blue, marked with large and small black spots, chiefly in a wreath around the thicker end. A recorded observation from Panama is also of a nest with two eggs.

The nest that had been cut down and propped up was much more exposed than when the birds had built it; its two eggs were slightly cracked. Nevertheless, the cacique continued faithfully to attend it. While I stood beside the nest with the boy who showed it to me, measuring and making notes, the bird circled around us just within the edge of the surrounding thicket, now and then peering out at us with keen golden eyes. My note taking done, I dismissed the boy and stood poorly concealed at the edge of the cane field, watching the bird, who must have seen me, return to her damaged eggs. On subsequent visits she continued to cover them while I approached through the thin stand of canes to within two yards, and later within arm's length, before she fled to the thicket's edge with harsh complaints that sometimes drew her mate. While at the nest one afternoon, I saw one member of the pair bill the feathers of its partner's neck while they watched me from the bushes. On a rainy afternoon the female continued to cover her eggs while I set up a blind only twenty feet (6 m) away. I was surprised to find how fearless these seclusive birds were.

At dawn on May 1, I stole into my blind to pass the morning with the caciques. As the sky brightened, the male appeared at the edge of the cane patch and repeated his full, soft whistle many times over. The female on the nest, who closely resembled him, replied with a long, sibilant whistle followed by a prolonged rattling *churr.* Then her mate, still in the thicket, voiced many harsh notes. Soon he flew to the canes that shaded the nest and rummaged among the dead leaves above it before he returned to the thicket. After I had seen more of these partners, I learned to distinguish the female by the frayed end of her tail. She alone incubated. During each of her sessions she answered her mate's whistles once or several times with a *churr.* While the female was absent for 20 minutes, the male continued to perch at the edge of the thicket nearest the nest, preening his long, soft black feathers. In the six hours of my morning's watch, the female's five sessions on the eggs ranged from 25 to 84 minutes and averaged 53 minutes. Her six recesses lasted from 11 to 24 minutes and averaged 18.6 minutes. She spent 74 percent of the morning on her eggs. Although she continued to incubate her cracked eggs for fourteen days after I first saw them, they failed to hatch. Her attachment to them gradually waned as the days passed.

The hatchling in the other nest had pink skin quite devoid of down and tightly closed eyes. The interior of its mouth was red. While I was at this nest a parent arrived and complained frantically with loud, harsh notes, only a yard or two from me. Soon its consort arrived and added its voice to the clamor. I departed, leaving the nest undisturbed. Young Yellow-billed Caciques still with their parents were only slightly smaller than the adults and equally black but less glossy. Their bills were as yellow as their parents', but their eyes were brown or blackish instead of yellow (Skutch 1954).

16

The Oropendolas

Largest of the icterids, the highly gregarious oropendolas range over tropical America from southern Mexico to northern Argentina but are absent from the Caribbean islands, except Trinidad. The twelve species are sometimes classified in five genera and sometimes lumped into one, *Psarocolius.* Males are thirteen to twenty inches (33 to 51 cm) long, the conspicuously smaller females nine to sixteen inches (23 to 41 cm); otherwise the sexes differ little. Their plumage is mainly black and chestnut in diverse patterns and shades; a few species are largely olive yellow or olive green and chestnut. A distinguishing feature of all oropendolas is the bright yellow of their outer tail feathers. Many wear short or rather long, narrow crest feathers, which, extending rearward, are usually inconspicuous unless tossed by a breeze or raised in courtship. Their stout bills taper to sharp points; the upper mandible of most species extends over the forehead as a frontal shield. These bills are often vividly colored: whitish yellow; bluish gray; black with a yellow, red, or orange tip; or, rarely, wholly black. A few species have bare pink or pink-and-bluish cheeks. The eyes may be blue, yellowish, or brown.

In straggling flocks of few or many birds, oropendolas range widely through the upper levels of tall forests and out over neighboring country with scattered trees, flying slowly with steady wing-beats, the males' often resonant, the females' usually silent. Largely frugivorous, oropendolas seek the fruits of forest and cultivated trees, the green spikes of cecropia trees, and bananas in neglected plantations. Arillate seeds, flowers, maize, citrus fruits, and occasionally cacao diversify their diet. Clinging head downward to the knobby stem of a banana inflorescence, they suck nectar from the white staminate flowers clustered beneath a fleshy, upturned, dull red bract. The large white flowers of balsa trees and nectar cups of epiphytes of the marcgravia family also offer them sweet drinks. For insects and other invertebrates they search through foliage and vine tangles, pry into bark crevices, and examine the undersides of branches and fronds. In the evening many gather from afar to sleep in trees or clumps of timber bamboos.

Despite their vocal flexibility, oropendolas are not known to sing sweetly. The utterances of males of the several species have been described as hoarse or rattling trills, harsh grating sounds, croaks, screeches as of a rusty hinge, the sound of ripping strong linen, and crashes. More pleasing are hollow gurgling notes or the tinkle of water dripping into a pool. The alarm call is a harsh, stentorian *cack*. To deliver the more musical utterances that might be called their songs, males commonly bow far forward, bringing the inverted head below the level of the feet, elevating the wings above the back and vibrating them, and slanting the tail upward. This is also the courtship posture. The quieter females, and less often the males, repeat a throaty *cluck*. When annoyed, as by others building too close to them, females complain with an expressive, high-pitched whine. Agonized screams reveal their concern for endangered offspring.

Oropendolas breed in colonies of few or many nests hanging

prominently high in trees that usually stand apart from others. The presence of wasps or bees that would attack prowling animals makes a tree more attractive to the birds. The same nesting tree may be occupied for many years. In these colonies the males, who are nearly always less numerous than females, do not pair with them but are polygamous or promiscuous. The females alone weave the pensile pouches that vary in length from about two to five feet (60 to 150 cm), rarely more. The roofless entrance is at the top of the pouch, which expands downward to the bulbous bottom, where the eggs lie upon a bed of loose leaves. Two is the usual number, although single eggs, possibly incomplete sets, have been reported. If a parasitic Giant Cowbird lays an egg in a nest less than twenty-four hours after the oropendola has deposited her first egg, she may fail to lay a second—an example of indeterminate laying. If a cowbird anticipates her in laying in her nest, she may not lay at all. In either case she absorbs her developing egg(s). Oropendolas' eggs are plain white, or pale pinkish, pale green, pale blue, or pale gray, lightly marked with reddish brown or blackish spots, blotches, or scrawls. Only females incubate. It is difficult to learn exact incubation periods without seeing the eggs, which is rarely possible in these well-enclosed nests hanging high above reach; but by watching the behavior of the attendant parent, the duration of incubation can be estimated. The incubation periods of Chestnut-headed, Crested, and Russet-backed Oropendolas are about seventeen to nineteen or twenty days. Nestling periods are more readily determined by noticing when the mother, who alone feeds the nestlings, begins and ceases to bring food to a nest, an interval that ranges from thirty to thirty-six days in several species.

The Montezuma Oropendola

The oropendola that I know best is the Montezuma, one of the largest species, the males being twenty inches (51 cm) long and weighing about one pound and two ounces (520 g); the females about fifteen inches (38 cm) in length and weighing about half a pound (230 g). The head, neck, and upper chest of both sexes are

Montezuma Oropendola, *Psarocolius (Gymnostinops) montezuma,* male

black, the remainder of the body and the wings deep chestnut, the central tail feathers brownish black, and the outer ones bright yellow. The long, sharp bill is orange over its terminal half and black basally. It extends over the forehead as a horny shield, which on the male is orange along the posterior margin. On each cheek is a patch of bare skin that is white faintly tinged with blue. The male wears an additional facial adornment in the form of an elongate warty protuberance of orange-colored skin on either side of his chin. The eyes of both sexes are dark and their legs are black.

From southern Mexico to central Panama these oropendolas range through the shrinking rain forests of the Gulf and Caribbean lowlands, and upward in decreasing numbers to about five thousand feet (1,500 m) in Costa Rica. Where the continental divide is high and broad, as in Guatemala and southern Costa Rica, Montezuma Oropendolas are restricted to the Caribbean side; where the barrier is low, as in northern Costa Rica, they spill through its gaps to the Pacific slope, where in this drier region a few are found around the bases of the volcanoes. Fortunately for these birds whose rain forests have long been diminishing, they adapt well to open country where scattered trees remain to provide fruits and nest sites.

Although when foraging in the forest Montezuma Oropendolas are so shy that I found it almost impossible to watch them, in their nesting colonies they are sometimes confident in the presence of humans. These colonies, conspicuous from afar, are among the most fascinating spectacles of tropical America. Usually they are in a tall tree with a long, clean trunk and wide-spreading boughs that do not come close to neighboring trees and therefore are not readily accessible to troops of monkeys, other mammals, or snakes that might devour all the eggs and young. Often the nest tree stands close beside a railroad, highway, or building. In the absence of a suitable dicotyledonous tree, the oropendolas may choose palms to support their nests. Attached to the rachises of seven tall, spiny pejibaye palms growing close together in a pasture hung sixty-one nests, from one to twenty-two in a palm. Three or four were in some cases suspended in a cluster, often in contact, beneath a single palm frond. Whatever the botanical classification of the tree, the nests, hanging like overgrown, gourdlike fruits, are rarely less than forty feet above the ground and often as high as a hundred feet (12 to 30 m).

One morning in late February, while I stood on a hilltop overlooking the Motagua Valley in Guatemala, a great whining and clucking of oropendolas drew my attention to a tall tree with an umbrella-shaped crown standing alone above a plantation of bananas. Although the tree was nearly half a mile away and several hundred feet below me, through my binoculars I could clearly see the flashing of yellow tails of many oropendolas clinging to branches and chattering loudly. Before I could reach the tree, the oropendolas had departed, leaving no signs of nests. Since this promised to be the founding of a colony, I kept careful watch of the tree. For twelve days the birds showed no further interest in it, but at the end of this interval I saw from the hilltop that they had gathered in the tree and were flying to and fro. Hurrying down, I found the very earliest stages of about a dozen nests, just a thin wrapping of banana fibers around twigs. While I watched, a few females flew up trailing fibers, but work seemed to have been suspended for the day. One bird who arrived with a billful of materials flew after her companions still carrying them.

For the next two days, the females continued to bring materials to this tree, but they worked little. They behaved like children on their first day at school who have brought their copybooks and pencils without knowing what to do with them. The birds flew around idly trailing fibers in their bills, perched on the newly wrapped twigs with an absent air, and after a while dropped their materials or carried them away. A male spent much time in the tree often alone, calling with or without bowing. Going from one to another of the slender twigs that were suitable sites for the nests, he called and bowed profusely. He seemed to be trying earnestly to persuade the females to return and continue to build. He made no nuptial demonstrations, at least none to any particular female. Despite all his efforts to hold the colony here, the nests never advanced beyond the earliest stage, probably because at this time the tree began to shed its foliage, soon leaving the limbs bare.

Another year I returned in late April to the Lancetilla Valley in northern Honduras and found oropendolas nesting in the same lofty tree that they had occupied in the preceding year. It stood above a dense thicket that had grown up in an abandoned banana plantation, between a well-worn path that led up to the thatched cabins near the head of the valley and, on the other side, beyond a grass-choked rivulet, a hillside pasture. Clustered at the ends of twigs hung about three score nests in which the females were incubating or already feeding clamorous young. But twenty-one nests, clustered too closely at the end of a living branch, had by their weight snapped it off at a point where it was two inches (5 cm) thick. It had crashed down to the brink of the stream, probably about a week before my arrival. Since I found no indication that parents or nestlings had perished, I was not sorry about the disaster, which gave me an opportunity to examine and measure nests and eggs that are usually inaccessible, and to watch the birds as, undaunted, they set about to replace their losses.

With tireless industry, the oropendolas worked at their new nests through much of the day, although, like most birds, they built most actively in the early morning. They used chiefly long, pliant fibers ripped from banana leaves, slender green vines with foliage still attached, and long, narrow strips of palm fronds. They procured banana fibers from a small plantation across the path from the nest tree, where the females went to gather them in small parties, usually escorted by a vigilant male. He stood sentry on a coconut palm or some other commanding position and with a loud *cack* of alarm put them to flight before I could learn how they gathered their fibers. Nevertheless, I persisted until I saw how they did it. Standing on the massive midrib of one of the huge leaves, a female, taking advantage of one of the transverse tears made by the wind in the broad blade, lowered her head through the gap and with her sharp bill nicked the smooth underside of the midrib, then pulled off a strand, sometimes as much as two feet (60 cm) long, from the fibrous outer layer. She doubled her harvest in her bill and returned to her nest, often with one end of the strand trailing far behind as she flew. The green midribs of the banana leaves were marked with long brown streaks where the fibers had been torn out, and many loose ends of fibers dangled beneath them.

The oropendolas crowded their new nests in two compact clusters close together, although half of the tree's spreading crown was unoccupied. Their recent calamity had not made them cautious, but perhaps this crowding had certain advantages. The pouches were sometimes fastened to thin unbranched twigs but more frequently to a fork and occasionally to three branchlets close together near the end of a bough. The twigs used for attachment were about the thickness of a lead pencil. They were always at the outside of the tree, never inside its leafy crown. Thus, the nests were more easily reached by flying oropendolas and less accessible to climbing animals.

The first step in nest construction was preparation of the anchorage by wrapping many fibers around the arms of a crotch or around the single twig chosen for support. The length of twig wrapped varied from eight to sixteen inches (20 to 40 cm). The oropendolas worked much as we would do if permitted to apply only one hand to such a task, pushing fibers over the twig and pulling them under it, intertwining and knotting them carefully. The second step was the formation of a loop. As the mass of fibers swathing the arms of the fork thickened, the bird stretched strands across the space between them. When this weft had become strong enough, on returning to work at her nest the female rested on this bridge instead of grasping the twigs themselves with her feet. Thereby the strands were gradually pushed downward, and as more fibers were stretched from arm to arm, a pocket was formed. Then she pushed apart the materials of the pocket, converting it into a loop, which would become the entrance of her nest.

I followed closely the procedures of a female who started her nest (number 17 on my chart of the colony) on a straight horizontal twig. After she had wrapped a thick mass of fibers around this support, she perched upon it and, lowering her head, tried to push it between the fibers and the underside of the twig. Then she hung in the strands below the twig, trying to force her way through them. After struggling strenuously for a short while, she flew off to rest. Returning, she again perched on the twig, bent down her neck, and pushed her head into the mass of fibers, then slowly swung down and pivoted around into the loop she was trying to form and into which she forced her way with difficulty. After she had repeated this several times, the loop became large enough for her to enter it directly when she returned with materials, but always it was a tight squeeze. Her somewhat violent methods had little by little torn one end of the loop until it was attached to the support by only a few strands. For a week this bird struggled, sometimes able to enter her loop and sometimes not, for it was constantly on the point of breaking loose at that end. But she never gave up, and at last, after adding many banana fibers, she managed to make her loop strong enough, and thereafter had no difficulty completing her pouch.

After an oropendola had made her loop, she always rested in it while she worked, invariably facing the center of the tree. As she wove more fibers into the loop, it lengthened into a sleeve, in which she hung head downward, her yellow tail projecting from the upper end and her head from the nether, as she intertwined more strands at the lower margin. Soon the sleeve became so long that she was wholly enclosed in it. Then I could see only her orange-tipped bill as she pushed or pulled a strand through the close-meshed fabric, or her head projecting briefly from the bottom. One might imagine that at this stage, when the nest was open at both ends, the oropendola would have found it easier to emerge through the lower end, which she already faced. On the contrary, she always climbed out at the top, as she had entered. Thus, the loop remained the doorway of the nest. After her sleeve had reached full length, the bird wove in the rounded bottom and thereby completed the pouch.

The construction of the nests did not proceed without frequent squabbles among the builders, most of which arose over nest sites that were superabundant. Two birds often started nests so close together that they were in each other's way as they worked, and they paused to express their annoyance in loud, high-pitched, irritated voices, like children who interfere with each other at play. Sometimes, completely losing temper, each menaced the other with open bill. Next, meeting face to face in the air, they fluttered downward until their proximity to the foliage below warned them that it was time to cease their dispute, whereupon they separated and flew up to resume their weaving side by side.

What surprised me most in these generally orderly, industrious birds was their propensity to steal nest materials from their neighbors. A weaver frequently attempted to pilfer a fiber that hung loosely from another's nest and incorporate it in her own. If the upper end of such a strand was attached more firmly than she had reckoned, the would-be thief grasped it in her bill and hung with half-opened wings below the nest until the coveted piece broke loose or the owner returned to drive her away. Sometimes, finding the desired fiber too firmly attached to yield, the frustrated filcher went straightway to another nest with dishonest intentions. After witnessing many such incidents, I concluded that they were not detrimental to the colony but, on the contrary, were in a measure beneficial as a mode of discipline, to discourage careless building. It was not easy to detach a fiber competently woven into the fabric of a pouch. Birds who built most carefully, leaving few loose ends, were not often troubled by pilfering neighbors and finished the strongest pouches.

One morning, when the builder of nest 22 was absent gathering material, another female went to her unguarded nest, then in the loop stage, and tore at it vigorously, pulling apart the fibers and destroying its neat appearance. The loose ends left by the first robber attracted others, whose repeated visits further damaged the loop until only a few strands held it together at the bottom. To repair the injured loop was difficult. Straddling the gap with a foot grasping each end, in what appeared to be an uncomfortable posture, the builder continued to knot fibers along the sides of the loop without at first succeeding in binding them together and closing the gap. While she struggled with her problem, neighbors persisted in stealing from her. For two days the loop lengthened without becoming sound; but finally the oropendola bridged the gap and finished a serviceable nest, which differed from others chiefly in its longer entrance. This and similar incidents showed that the oropendolas were not as proficient at mending as they were at weaving. With a better grasp of a situation, they might have repaired their nests—with knots they knew well enough how to tie—in an hour instead of fumbling along for days.

Not only did the oropendolas steal materials from neighbors' nests, they frequently snatched it from their bills. At times, returning with long strands trailing behind her, a bird rested briefly on a perch near her nest before proceeding to weave her new material into it. Seeing the strands hanging from her neighbor's bill, a second bird, perhaps just preparing to go abroad to gather material, would change her plan and grasp one dangling end, then hang from it with nearly closed wings, doing her best to pull away the strand. To utter a single note of protest would have been disastrous to the rightful owner, for to have opened her mouth would have resulted in the loss of all that she held, as happened to the crow with cheese in the fable when he succumbed to the fox's flattery and tried to sing. At other times, two oropendolas perching side by side on a branch would engage in a tug-of-war over a billful of fibers, and in the end perhaps unwillingly divide the spoils. Hens whose fibers had been torn from their nest or their bill usually took their loss without revealing emotion; if fights arose from this cause, they were less frequent and protracted than those over nest sites. On the whole, the oropendolas were too industrious to waste much time pilfering or quarreling, and building proceeded apace.

At the nest tree were usually several females for each male. The males gave no indication of claiming territories or being mated to particular females or groups of them; they were ignored by the females and mostly ignored them. While building continued, the males accompanied the hens on expeditions for foraging or collecting fibers. When present, they strutted around on branches of the nest tree with pompous gait and heads held high. Although idle, they never quarreled, as females did. Sometimes, indeed, one male dashed at another, who always retreated promptly and thereby avoided a clash, since the aggressor quickly forgot whatever pique he might have had. Such pacific comportment of males is general in colonial-nesting oropendolas, caciques, and grackles.

At intervals the male oropendolas delivered far-carrying calls. Bowing profoundly, until the raised tail stood directly above the inverted head, lifting the spread wings above the back, and fluffing all the body feathers, they uttered, or seemed rather to eject with exhausting effort, an indescribable liquid gurgle. Heard from afar, there is no sound, save possibly the ventriloquial call of the Short-billed Pigeon or the organ notes of the Great Tinamou, that to me is more expressive of the wonder of tropical American rain forests; but close at hand the effect is marred by screeching overtones, as though the machinery that produces this inimitable song were badly in need of lubrication. The male oropendolas did not bother females who were building, as male Great-tailed Grackles do; their bows and gurgles were not addressed to individuals so much as to the world at large.

The males guarded the colony. At the approach of danger, actual or apparent, and not infrequently when no cause for alarm was evident, they uttered a loud *cack,* which sometimes sent all the oropendolas within hearing dashing headlong into the nearest sheltering thicket but at other times was ignored by most of the community. Could this have been because some males were more trustworthy than others as watchmen? They generally signaled my approach to the nest tree by a few such *cacks*, but after they became accustomed to me, few or none of the building females heeded the warning. I could stand or sit wholly exposed and watch all the activities of the colony without causing the least unrest. How different from the behavior of oropendolas foraging in the forest, when the sentinel's warning sent the whole party into instantaneous flight!

Such was the patient application of the hen oropendolas that their great hanging pouches, three to four feet (90 to 120 cm) in length and seven to nine inches (18 to 23 cm) in diameter near the bottom, were completed in the average time of ten days. One bird finished in only seven or eight days a nest considerably shorter than the others. The female who had much difficulty starting her loop on the unbranched twig finished last of all, after working seventeen days; while number 22, who was so harassed by thieving neighbors, took fifteen days to build. Most of the weavers took from nine to eleven days. Probably these birds finished so promptly because they were experienced weavers who had lost their first nests and were building again. In other birds, replacement nests and those for second broods are built more swiftly than nests started early in the season. In a pioneer study of the Chestnut-headed (formerly Wagler's) Oropendola, Chapman (1928) found that they took three or four weeks to weave their shorter pouches, the season's first.

Sometimes two Montezuma Oropendolas, building side by side, wove opposite ends of the same strand into their nests. Occasionally two pouches were sewn together by birds weaving fibers through contiguous walls. If, as sometimes happened, one of these united nests broke away from its support, it might be saved from falling by its attachment to the other. A nest with nestlings tore away from its anchorage and hung for a week precariously tied by a few strands to an adjacent nest, which sustained it until the young could fly. Pouches closely sewn to others were sometimes abnormally long. One such nest, picked up from the ground, measured six and a half feet (198 cm) in length. The part in contact with the adjoining nest was open rather than tubular and was bound to the outside of its companion throughout its length; the pouchlike part of this very long nest began at the bottom of the other and hung completely below it. I do not know how contact with another nest made the structure grow in this strange fashion.

To appreciate fully the quality of the oropendolas' weaving, one must cut a fallen nest open and spread it to the light. It is a regular, uniform fabric, with meshes wide enough to admit air to the brooding female and her nestlings, yet strong and durable.

After the bottom of the pouch was closed off, the female remained away for a day or two, during which courtship and coition probably occurred in the privacy of forest trees (as in the Casqued Oropendola [Leak and Robinson 1989]), where it has been witnessed (Howell 1964); I never saw intimate relations between the sexes in the colony. Then the hens returned and labored assiduously for three to six days more, plucking dying or dead leaves from trees at a distance, tearing them between feet and bill into pieces an inch or two (2.5 or 5 cm) long, and carrying them into the pouch, where they formed a thick but loose and yielding litter in which the eggs rested. In addition to helping to conserve heat, they served to prevent the eggs from rolling together and breaking when a strong wind rocked the swinging pouches. Sometimes a bird brought fibers and leaves alternately, as though she had started to line the nest before she had quite finished the weaving. Even during the incubation period, or while the nest contained young, a female occasionally took pieces of leaf into it, and more rarely a fiber.

After the completion of the new nests, the colony in the Lancetilla Valley contained eighty-eight, but not all were occupied. I could not see the eggs in those high nests, but examination of the twenty-one that had fallen revealed that the usual set was two.

At the height of the breeding season, a bustling activity prevailed in the tall nest tree, with females hurrying back and forth bringing materials for nests or food for nestlings, or clucking while they rested on its branches. Their occasional quarrels and high-pitched remonstrances, the males strutting along the boughs or flying with noisy wing-beats from one to another and pouring out their liquid calls, the whining cries of nestlings in their swinging cradles, and the voices and movements of birds of many kinds that visited the nest tree to rest or search for insects—all created a continual commotion.

The disturbances caused by lurking Giant Cowbirds augmented this throbbing activity. As long as any nests in which incubation had not yet begun remained, the big black female cowbirds continued to haunt the nest tree. Of all birds, save of course dangerous raptors, they alone were opposed by the oropendolas. Continually chased by first one and then another of the larger yellow-tails, the parasites circled around and returned with undaunted persistence, wholly undeterred by repeated rebuffs. By skulking most of the day in the colony and repeatedly peering down through the entrances of the pouches, they apparently kept well informed about the contents of each of them, and at the proper time they tried to slip in and lay their eggs. Occasionally they miscalculated the hour of their visit and attempted to enter while the oropendola was at home, only to beat a hurried retreat when the irate owner emerged, as though to demand an explanation of the intrusion. Again and again the cowbirds were driven away as they were about to enter a nest; but never discouraged, a cowbird would eventually manage to elude many watchful eyes and steal into a pouch, where apparently she left an egg to be incubated by the female oropendola.

The Giant Cowbird who lays in an oropendola's nest meets far more opposition than do parasitic birds of other species that foist their eggs on small birds nesting alone. To deposit an egg in an oropendola's nest, a Giant Cowbird must not only dodge the watchful oropendolas but must also outwit jealous rivals of her kind, each eager to drop her own egg into a nest where incubation is just beginning and ready to drive away another cowbird who tries to get ahead of her. Contemplating how difficult it was for the cowbirds to impose their eggs upon oropendolas, I wondered why they did not build nests and rear their own progeny.

Had the oropendolas made a concerted attack upon these intruders, they might have driven them permanently from the colony; but they are mild-mannered birds and appeared content merely to prevent the cowbirds' entry into their nests. They rarely continued to pursue a cowbird for more than five or ten yards from a threatened nest, and never farther than the next tree. Long-suffering as they were, the yellow-tails were not always deceived by the alien eggs. Early on a morning toward the end of May, several cowbirds tried with exceptional persistence to enter nests in the group most recently completed, but, chased by the owners, they fled with harsh, nasal whistles. Soon several cowbirds were chased by as many oropendolas, and confusion prevailed in that part of the tree. Apparently, one of the cowbirds, taking advantage of the disorder, stole into a nest and left an egg.

What I actually saw, a few minutes later, was an oropendola emerge from one of these nests with a white object in her bill. Flying to a higher branch, she dropped what she held. With a *plunk* it struck a banana leaf in the thicket beneath the nest tree, rolled off, and lodged on dead leaves carpeting the ground. When I picked up the cowbird's egg, unbroken by its eighty-foot (24 m) fall, it was still warm. Thus, Montezuma Oropendolas are, at least in some colonies, rejecters of cowbird eggs, of which a number have been identified in temperate North America, where Brown-headed Cowbirds abound. That they do not consistently reject cowbird eggs is evident from an observation I made in a later year.

After incubation began, I could only conjecture what was happening inside those high, closely woven nests. Often a male flew with sonorous wing-beats to the side of a pouch, pouncing heavily upon it. One seeing him do this for the first time might fear that he had torn the fabric, for he alighted with a sound as of ripping strong linen. This noise, however, was a vocal utterance that is often heard away from nests. Clinging head downward, the big bird would raise his wings above his back and deliver his liquid gurgle, while the hen incubating or brooding within protested this rough treatment of her pouch with a high-pitched whine, in which she was often joined by the group of birds whose nests were jarred by this boisterous behavior.

The indefatigable female oropendolas continued their daily tasks after the sun had dropped behind the western hills. As dusk fell over the quiet valley, they wove a last strand into an unfinished nest or carried a final meal to waiting nestlings. Others sat among the high boughs of the nest tree, enjoying the refreshing cool of evening after a warm day, the hens clucking in contented tones, the males often bowing while they uttered their stirring songs. Then, in small parties, males and hens with neither eggs to incubate nor nestlings that needed brooding, flew toward the north, whence floated the calls, mellowed by distance, of males who had gone before. Finally, as tree frogs began their shrilling and the loud *kweahreo* of the Common Pauraque announced the end of the day, the last lone male oropendola jumped heavily on the side of a pouch, sang loudly while making his deep obeisance, then flew with noisy wing-beats to overtake his companions. Meanwhile, the remaining females retired one by one to cover their eggs or nestlings in their swinging cradles. Older nestlings slept alone until their mothers returned at dawn. Every oropendola who remained in the tree was inside a nest.

When I first arrived at Lancetilla, the oropendolas roosted in a double row of tall bamboos that shaded a path, half a mile from their nest tree. Here they were joined by a number of Chestnut-headed Oropendolas, whose liquid *plunk* mingled with the more prolonged and melodious calls of the male Montezuma Oropendolas, and by Garden Thrushes, who contributed their staccato *tock tock tock* to the medley of voices. At their roosts the oropendolas are much shier than in their nest tree. Apparently because I watched them retire, they moved to a site on a hillside about two miles from their colony. Thither they flew in successive flocks late in the evening; thence they returned to their nests at break of day, flying up the valley with wing-beats as regular as crows', and appearing as black in the dim early dawn. The mothers of nestlings stopped on the way to find food for them and arrived at the nests with the day's first meals in their bills.

How long a female incubated her eggs I did not learn, but the period was probably about seventeen days, as in other species of oropendolas. Finally came a day when a mother returned to her nest with fruit from some distant tree, telling me plainly that her nestlings had hatched. Soon their nasal, quavering whines left no doubt of what the pouches held. This was a busy time for the mother; for as she had built her nest and incubated her eggs alone, so she had to feed her nestlings unaided by their father, with food often procured at a great distance. These big birds breeding in populous colonies cannot find enough food near their nests, in the manner of smaller birds nesting singly. Returning with food in her bill, the hen flew directly toward the nest's doorway and when almost there folded her wings to glide deftly in, never pausing until the pouch had swallowed her form, just as she had done while she built and incubated. To delay with her head inside and body outside would needlessly expose her to the attacks of raptors whose approach she could not see. Coming out was different, and often she delayed with the posterior part of her body inside and her foreparts exposed, while she surveyed the outer world before launching forth.

By noting how long a mother continued to bring food, I learned that the young remained in their lofty pouches for about thirty days, a nestling period exceptionally long for a passerine bird. Before they flew, their nests showed signs of wear. The pouches became more bulbous at the bottom. The female's innumerable passages through the entrance tore it downward until it extended far on one side of the pouch. Before the occupants were ready to fly, several nests hung precariously by only a few strands and swung dizzily as the parent went in and out. The females did nothing to reinforce the attachment of these pouches, as they might well have done by lacing it with added fibers. The males failed to understand the delicacy of the situation but continued to jump on the damaged nests and sing, just as they did on sound nests. Repeated impacts by the heavy birds jeopardized the nestlings and hastened their fall. The males often cocked their heads and with evident interest peered down into pouches with nestlings, but they never entered, and it never occurred to them that the young might be fed.

Early on a misty morning at the end of May, I found the female oropendolas flying about the nest tree, clucking as though unusually excited, apparently to coax from the nest a young bird ready to leave, as House Martins do. Presently I saw a pouch jerk downward at short intervals, as though something heavy were climbing upward inside by intermittent advances. Soon the pale head of a fledgling appeared in the doorway. Perhaps startled by its first clear view of the landscape, the youngster paused briefly, as though uncertain what to do. The mother, who had been circling around the tree, made a turn close in front of the nest, apparently to encourage the fledgling, who now flew boldly forth. Without so much as touching the nest tree, it followed her across the rivulet to a small tree on the hillside, completing a flight of about two hundred feet (60 m) on a slightly descending course—not a bad achievement for a young bird who had never before been able to spread its wings fully! Several other females followed the young oropendola on its first flight and joined the mother in driving away a Brown Jay who entered the same tree. After feeding the fledgling there, the mother coaxed it still farther from the nest tree—her every effort seemed to be directed toward leading it to a distance from the colony.

Another newly emerged fledgling, whose mother was also feeding a sibling still in the pouch, delayed among the nests for at least two hours, but by noon it had vanished. It was quite unusual to see an exposed fledgling in or near the colony; most were probably led away as soon as they emerged from the nest, like the one whose departure I witnessed. The nest tree was too conspicuous and widely known for the young to remain there safely without the protection of the pouch. Even the adults would not pass the night in so exposed a situation.

The last oropendola to leave its nest that year was, to judge by his size, a male. For a full month I had watched his mother bring food to his pouch. Apparently he left prematurely; instead of flying strongly away, like the youngster whose departure I had watched, he fluttered down into the thicket beneath the nest tree, where despite his efforts to flap away I caught him. As I was about to pick him up he squawked madly and beat his wings frantically, while his mother and half a dozen other oropendolas flew around above us in the greatest excitement. When I offered the young bird a finger to perch on, his long toes gripped it so strongly that I winced.

Never have I had such an escort of birds as that which followed us back to the house. The one who appeared to be my captive's mother led, flying from tree to tree and keeping close to us, clucking vehemently, and repeating a scream that I had not previously heard. After we reached the building, the other oropendolas dropped away; but the faithful mother remained, perching high in a neighboring tree with an orange-colored fruit in her bill, while we photographed her fledgling. Then we placed him on the screened porch while I developed the films, to make sure they were satisfactory before we released the bird. All this while his mother flew around the house in great circles, answering her fledgling's calls. Satisfied with the negatives of his portraits, we returned the impatient youngster to the spot whence I had taken him a few hours earlier. Here in the thicket his mother attended him for the next three days, after which he vanished.

Although smaller, this young oropendola closely resembled an adult male. His body was duller chestnut, but he had the same bright yellow outer tail feathers. The orange of his bill, confined to the extreme tip, darkened to the black that covered most of the bill. His iris was dark brown. The bare skin of his face and forehead were palest pink. The swollen, bean-shaped area of bare orange skin on either side of the adult male's chin was lacking. Female fledglings are similar but much smaller.

From the start of building a nest to the fledglings' emergence from it, two months or more elapsed—about twice as long as small passerines take for a nesting. After the departure of the last fledgling at the end of June, the colony in the Lancetilla Valley stood like a deserted village. Already some of the earlier nests had decayed and fallen. Eight or ten of the later nests had been abandoned before I saw food for nestlings carried into them. The young of some nearby nests had apparently been taken by an undetermined predator, which might have stolen the eggs from these. In nests fallen from other colonies I found a hole in the wall near the bottom, one of them three inches (7.5 cm) in diameter, made by cutting the fibers and forcing them apart. Possibly an owl or a large carnivorous bat had opened the pouches to take eggs or young. Successfully fledged young followed adults of both sexes up and down the valley in flocks of usually less than a dozen birds and continued to be fed by their mothers for many weeks after their first flights. Females gave food to young males bigger than themselves.

In Caribbean Central America, nesting begins in January and continues until September, but I am aware of no single colony that was active throughout this long interval; nor have I a record of a second brood. Male oropendolas' interest in nests continues beyond the end of the breeding season. Long after the last young has flown, they return from time to time to examine deserted nests, perching above the doorway and peering down into a pouch, as they do while young are within. They jump to the side of a nest, making the sound of ripping cloth as they alight, and hang head downward to deliver their songs. During the rainy final months of the year, when females associate in flocks, one often sees a lone male perch at the top of a lofty tree and, scarcely bowing, pour forth his stirring liquid call (Skutch 1954).

The Chestnut-headed Oropendola

The first detailed study of an oropendola was made by Chapman (1928, 1929), who carefully watched black Chestnut-headed Oropendolas on Barro Colorado Island during the early part of their breeding seasons in 1926, 1927, and 1928. On the whole, the behavior of these smaller oropendolas closely conforms to that of the Montezuma. Chapman noticed that females of some groups started their nests with little friction and worked competently from the beginning, whereas in other groups of the same colony they might squabble for days over nest sites. These females would bring fibers for their nests without appearing to know what to do with them, and finally drop them unused. The former were apparently experienced birds who had known each other in an earlier year, the latter young females trying to start their first nests. Although young, inexperienced birds, such as Song Sparrows, may build their first simple nests as competently as their elders, to construct a more elaborate nest, like the closed structure of the Village Weaver, nest making must be perfected by practice; as Nicholas Collias and Elsie Collias (1984) have shown, the first nests of these African birds are less neatly finished than their subsequent nests.

Chestnut-headed Oropendola, *Psarocolius (Zarhynchus) wagleri*, sexes alike

Apparently, the same is true of oropendolas. Those that I watched at Lancetilla were experienced birds, replacing nests that they had lost. Chapman's oropendolas who started to build at the beginning of their nesting season might take a full month to weave and line a nest; some that began later did the same work in twenty-three to twenty-five days. These nests were twenty-two to forty inches (56 to 102 cm) long.

The most prolonged study of oropendolas was made by Neal G. Smith (1968, 1983) mainly of the Chestnut-headed in Panama, where he found colonies of four to over a hundred nests, most often of from thirty to forty. With a ladder and long, telescoping aluminum poles, he lowered high nests, examined their contents, and replaced them with sticky tape or contact cement. He found that members of a colony tend to nest together, in the same site, year after year, but a small exchange of individuals between colonies prevents excessive inbreeding. Colonies usually contain five times as many females as males. The unbalanced sex ratio originates during the nestling period, when males, who need about twice as much food as females, starve in greater numbers when it is scarce, resulting in an average sex ratio at fledging of three females to one male. When they leave the nest, males weigh twice as much as females. In these colonies, building takes ten to fourteen days, incubation seventeen days, and rearing the young to fledging thirty to thirty-six days. Males do not establish a dominance hierarchy, never fight among themselves, and do not mate with females at the colony, in all this agreeing with Montezuma Oropendolas.

Although Chestnut-headed Oropendolas' nests are ravaged by the same predators that harass other oropendolas, caciques, and many different birds—toucans, snakes, opossums, bats, and others—the chief cause of nestling mortality in most years is infestation by larvae of botflies (*Philornis* spp.). Adult flies deposit eggs or larvae directly upon the nestlings' skins, into which they burrow, making conspicuous swellings. Often these larvae emerge to pupate before young birds fledge; but small oropendola chicks with more than ten, a not uncommon number, usually succumb, especially if they are undernourished. Oropendolas nesting close to social wasps or stingless but biting bees (*Trigona* spp.) are less troubled by botflies because the adult flies are attacked by the wasps or bees, who react to them much as they do to the many parasitoids that try to enter their nests. (This raises the question of whether other birds, especially flycatchers, that build covered or hanging nests close to vespiaries do not gain a certain immunity from botflies, which might be more beneficial to them than the protection from nest pillagers that wasps give them.)

The young of oropendolas nesting away from protecting bees and wasps tend to be more heavily infested by botfly larvae but, paradoxically, they are often benefited by the very intruders that compete with them for food—nestlings of the parasitic Giant Cowbird. Young cowbirds preen their nest mates, removing eggs and larvae of botflies, but young oropendolas do not. The result is that more oropendolas fledge from nests with one, or if food is exceptionally abundant with two, cowbird chicks than from nests without cowbirds—"the advantage of being parasitized," as Smith wrote. Oropendolas in colonies protected by wasps or bees chase female cowbirds and toss out a high proportion of the alien eggs. Oropendolas in colonies at a distance from wasps or bees are generally more receptive of cowbirds and reject fewer of their eggs. Cross-fostering experiments over nine years demonstrated that the oropendolas' reaction to cowbirds— tolerating or chasing them—is learned, not innate; but how it is learned is unclear.

Over a ten-year period, oropendolas nesting near wasps or bees enjoyed slightly better nesting success than those depending upon cowbirds for relief from botflies. Nevertheless, success is low; although Chestnut-headed Oropendolas usually lay two eggs, like other oropendolas and caciques, in ten years the average number of young fledged per nest was only 0.40. Compensating for this low rate of reproduction, oropendolas live long; five females banded as chicks in one colony were actively building nests when twenty-six years of age.

The Crested Oropendola

Crested Oropendolas range from western Panama to Bolivia and northern Argentina. The male is glossy black, with dark chestnut lower back, rump, upper and under tail coverts, and the usual yellow outer tail feathers. Long, thin crest feathers adorn his head. His eyes are blue, his bill pale yellow. The much smaller female is similar but less shiny black. In northern Venezuela, where Crested Oropendolas were abundant in lightly wooded country, I found groups of only two, three, and seven nests. On the island of Trinidad, colonies average about a dozen nests and rarely contain as many as forty, which is half the size of some Montezuma and Chestnut-headed colonies in Central America (ffrench 1973). Apparently, male territoriality, which is absent in the Central American species, is responsible for the greater dispersion of Crested Oropendolas' nests.

The male Crested Oropendola's display, which I often watched in Venezuela, is spectacular. Bowing forward until his body is inverted, with bill pointing earthward and tail lifted toward the sky, he raises his wings over his back and at the same time makes a sharp, rattling sound, followed by a soft, liquid, soprano, far-carrying *whoo-oo,* often with a rippling or quavering quality. As he resumes his normal perching stance, he vigorously shakes his wings, audibly striking the relaxed primary feathers together.

Like other oropendolas, Cresteds breed chiefly in the drier months, in the northern tropics from January to May or June. By late June, all but a few belated hens had finished nesting, and the birds had begun to flock over the open or lightly wooded hills south of Lake Valencia. Many with ragged plumage were molting. The flocks, obviously composed of oropendolas from a number of the small nesting groups, continued to grow until, in July, I sometimes counted a hundred oropendolas, flying with steady wing-beats in long, straggling columns, high above the fields. They flew with clucking sounds that were dry or pleasantly melodious, especially when softened by distance. The abundant mango trees, great domes of dark green foliage scattered over the pastures, were now ripening their luscious fruits in such abundance that humans, domestic animals, white-tailed deer from the thickets, and birds, including the oropendolas, could not consume them all. The yellow-tails sometimes descended to the ground to peck into fallen mangos. They also ate the small yellow fruits of the

Cordia trees that grew almost everywhere, and they laboriously pecked open hard green pods of *Inga* trees, to reach the sweetish white pulp that surrounds the smooth green embryos. Well-grown juveniles traveling with the flock continued to receive food from their mothers. I watched a female attend a young male bigger than herself. Flapping his wings, he voiced raucous, groaning-moaning cries while she fed him, then he clumsily helped himself to *Cordia* fruits. Other female oropendolas fed clamorous young Giant Cowbirds (Skutch 1977).

Crested Oropendolas were studied intensively in Trinidad by Richard E. Tashian (1957) and William H. Drury, Jr. (1962). According to the latter, females of this species do not, like the Montezuma and Chestnut-headed, collect a mass of fibers, then force them apart to make a loop. They weave a hanging apron, then sew the edges together to form the beginning of a sleeve—much as was done by Montezuma Oropendolas whose loops were torn apart. Until the sleeve is closed, Crested Oropendolas may enter it from below, which the Montezuma and Chestnut-headed were not seen to do. Only two of ten matings seen by Drury occurred in the nest trees (which never seems to happen with the Montezuma and Chestnut-headed); the other eight occurred in trees without nests. The territoriality that this author ascribed to the Crested appears to be weak, unable to prevent promiscuous matings. As in all other studies of colonial oropendolas and caciques, males supplanted one another but were never, or scarcely ever, seen to fight.

Naturalists widely agree that the greater size of males than females in animals the most diverse evolved because large size gives males an advantage in struggles with other males for territories and (or) females. Since, in the most carefully studied species of colonial icterids, males scarcely ever fight and do not establish evident social hierarchies, why should the difference in the weight of the sexes be exceptionally great among these birds? Can it be that impressive size alone gives them an advantage in winning females? Or is it that bigness makes them more effective guardians of the colony against certain predators? The question merits further study.

17

Retrospect

Orioles, oropendolas, caciques, meadowlarks, Bobolinks, grackles, blackbirds, cowbirds, and their kin compose a heterogeneous family of about ninety species. Confined to the Western Hemisphere, they breed from the Arctic to the Cape Horn archipelago. Most abundant at low and middle altitudes, a few ascend high into the mountains. Long recognized as a separate family, the Icteridae, in recent years they have been demoted, together with the wood warblers, tanagers, and emberizine finches, to the rank of a subfamily, the Icterinae, of the cosmopolitan Emberizidae, a huge aggregation of songbirds. If we insist that every family of birds be as different from every other family as many nonpasserine families are, we may eventually find all passerines (half the world's avian species) in one or two unwieldy families. Therefore, I retain the Icteridae as a family.

In size icterids range from Bobolinks, hardly bigger than sparrows, to male oropendolas, among the largest of passerine birds. Their prevailing colors are black, often glossy and iridescent, brown, yellow, orange, and red. Green and blue are absent except in the blue eyes of certain caciques and oropendolas; yellow and dark eyes are more frequent in the family. In nonmigratory tropical species that are constantly mated, the sexes tend to be alike, whereas in polygamous and highly migratory species they may differ greatly in appearance; males that are largely or wholly black often have brownish or grayish, plain or streaked consorts. Polygamous or promiscuous males are often much larger and heavier than the females. Male oropendolas and Yellow-winged Caciques wear rather skimpy crests of slender, elongated feathers, and male cowbirds have ruffs. Male grackles bear notably long, keeled or boat-shaped tails, folded upward. Otherwise, ornamental plumes are absent in the family. Species that perform the longest migrations show the most conspicuous seasonal changes in plumage, especially Bobolinks, which fly from southern Canada and the northern United States to Argentina, and Rusty Blackbirds, which travel from the Arctic and sub-Arctic to the southern United States.

The food of orioles, oropendolas, and pouch-weaving caciques is largely fruits and insects that they gather from trees and shrubs. They sip nectar from flowers, including those of the banana plant. Meadowlarks, Oriole-Blackbirds, grackles, and cowbirds find much of their food on the ground, over which they walk or run with alternately advancing feet instead of hopping with feet together, picking up insects, worms, seeds of many kinds, and fallen grains. Grackles are omnivorous; to a diversity of invertebrates they add small vertebrates, including fish that they catch in the water. They plunder birds' nests, and at least the larger of them seize small adult birds. Their depredations on maize and other crops often distress farmers. Grackles and Giant Cowbirds pluck ticks and other vermin from patient cattle, capybaras, and tapirs.

A special mode of foraging of icterids is gaping. Other birds forcefully close their bills, as to crush seeds, but need to exert little force to open them. Icterids' mandibles are so muscled that the birds can spread them against resistance. Thrusting their sharp bills into the soil, the ground foragers gape, opening a little hole, and probably looking through their open bills, while they extract worms, grubs, and the like. By pushing their bills beneath the edge of a pebble, fragment of wood, or other small object lying on the ground and shoving forward, they lift or overturn the piece to reveal whatever edible lurks beneath it. By gaping, Yellow-billed Caciques open decaying canes and vines to extract larvae. Gaping also helps icterids to reach the nectar of flowers. I have often watched a wintering Baltimore Oriole pluck a long, slender red flower of a poró tree, hold it against a branch with a foot, pierce the thick calyx with its bill, and gape to extract the nectar. This activity reveals another of icterids' abilities. Not all birds use their feet for holding.

Voice is richly developed in icterids, which include many delightful singers. The Spotted-breasted Oriole, Yellow-backed Oriole, Yellow-tailed Oriole, and Melodious Blackbird, among others, sing superbly with full, mellow voices. Females of many tropical species also sing well, in tones somewhat weaker than their mates'. They duet with their partners or answer their songs, sometimes while sitting in their nests. The notes of the sexes may be similar or contrasting; the male Yellow-billed Cacique's loud, clear double whistle is answered by a long-drawn rattling *churr* from his spouse. Male Yellow-tailed Caciques of the Panama race have an amazing repertoire of brief, bright phrases that command attention. Great-tailed Grackles proclaim their presence with many strong notes that reveal the range and power of their voices. The utterances of these two birds are exclamations rather than sustained songs. Heard from afar, the male Montezuma Oropendola's liquid gurgle evokes a sense of the wonder and mystery of the tropical forest. The brief verses of Red-winged Blackbirds and meadowlarks bring cheer to northern meadows and marshes in early spring. In Argentina, the social singing of flocks of Brown-and-Yellow Marshbirds were to Hudson a "delightful hubbub," not devoid of soft, silvery notes.

The Epaulet Oriole of South America mimics the calls of other birds. At least in captivity, a Troupial repeats verses that he has heard. Bobolinks, meadowlarks, Red-winged Blackbirds, Red-breasted Blackbirds, and Jamaican Blackbirds are among the icterids that sing while flying, the last-mentioned above the woodland canopy, as few forest-dwelling birds do. Added to these songs, a diversity of rattles, squeaks, screams, and other queer notes demonstrate the range and versatility of icterids' vocal organs.

Among the displays of icterids, performed in courtship and otherwise, are several versions of the song-spread. The male bows profoundly, spreads his wings, depresses or elevates his tail, while he pours forth his notes. Towering pompously above a female of his kind busily foraging on the ground, the male Giant Cowbird bows his head to his breast, spreads his ample glossy cape around it with his ruby eyes gleaming through it, and bobs up and down by flexing his legs. Usually she is unimpressed. A Bronzed Cowbird hovers in the air above a foraging female. Of all the displays of icterids, the most curious and puzzling is the preening invitation of cowbirds. With ruffled plumage of the neck and bowed head, cowbirds solicit preening from a variety of birds, including other individuals of their own species, and are often obliged.

Nuptial feeding has been recorded in the Melodious Blackbird, Rusty Blackbird, Brewer's Blackbird, Yellow-headed Blackbird, Bolivian Blackbird, and apparently the Baltimore Oriole. Feeding of the female by the male appears to be rather rare in the family.

Monogamy, which is the prevailing relation of the sexes in the Icteridae, is practiced by orioles (the largest genus), Melodious Blackbirds, Yellow-billed Caciques, Yellow-winged Blackbirds, Yellow-shouldered Blackbirds, Oriole-Blackbirds, Jamaican Blackbirds, Brown-and-Yellow Marshbirds, Scarlet-headed Blackbirds, Scarlet-rumped Caciques, Solitary Caciques, Common Grackles, Bay-winged Cowbirds, and even the parasitic Screaming and Brown-headed cowbirds. Meadowlarks, Bobolinks, and Brewer's Blackbirds may be monogamous or polygynous, depending upon the ratio of the sexes, quality of territories, and perhaps other factors. Red-winged Blackbirds, Tricolored Blackbirds, and Yellow-headed Blackbirds are polygamous. In these species females nest when about a year old, but most males do not breed until their second year; the unbalanced sex ratio in the breeding population promotes polygamy. The number of a male's consorts depends upon the size and quality of his territory and doubtless also upon his own persuasiveness. Male Great-tailed and Boat-tailed Grackles likewise delay breeding until their second year (while females nest earlier), and they are promiscuous. Also promiscuous are colonial-nesting caciques and oropendolas, although many species of the latter remain to be studied.

Nests differ greatly in form and position. With the possible exception of the horneros and American flycatchers, no other family of New World birds exhibits such a diversity of architecture. Simplest are the nests of Bobolinks, little depressions in the ground scratched by the birds or found by them, loosely made of grasses and weed stems and lined with finer materials. Meadowlarks also make terrestrial nests but more substantial structures usually oven-shaped, with a roof and side entrance in the two northern species, often roofed or only half-roofed in the South American species. Another ground nester, the Red-breasted Blackbird, constructs a deep open cup of dry grasses lined with finer materials, which is sometimes approached through a tunnel in the herbage.

Blackbirds of the genus *Agelaius* and Yellow-headed Blackbirds attach substantial open cups to cattails, reeds, and other upright marsh vegetation, or to the branches of shrubs and trees. Nests of the Jamaican Blackbird are similar in form but composed of quite different materials—small epiphytic orchids or dark roots of air plants. The open cups in trees, palms, cattails, and buildings of Oriole-Blackbirds, Melodious Blackbirds, Brown-and-Yellow Marshbirds, Brewer's Blackbirds, and grackles are plastered inside with mud or cow dung, over which the birds provide a lining of fine vegetation. In northern coniferous woods where mud or cow dung may be hard to find, Rusty Blackbirds plaster their bowls with duff (the decaying vegetation from the forest floor), which when dry becomes as hard and stiff as papier mâché.

The most skillful nest builders in the family—orioles, caciques, and oropendolas—make pouches finely woven of leaf strips, grasses, rootlets, and other pliable strands. In length these structures range from the three- or four-inch pockets of Orchard Orioles through the one- to two-foot bags of some tropical orioles, to the wonderful pouches of oropendolas, which may reach two to five feet in length. From the top, where they are attached and the birds go in and out, all these pouches broaden to a rounded bottom, lined with vegetable down or leaf fragments that provide insulation and may prevent the eggs from knocking together and possibly breaking when wind sways the pouches that hang freely exposed, like gigantic fruits, from branches of tropical and subtropical trees. Black-cowled Orioles pass fibers through holes that they apparently make in the broad leaves of banana or traveler's trees to suspend their shallow pockets beneath green roofs, or they hang them from nails and service wires under the eaves of houses. Yellow-rumped Caciques weave a hood over the opening at the top of their nests when the rains begin.

In a family that contains so many skillful builders, species that depend upon other birds for nests are an anomaly. Troupials expel Rufous-fronted Thornbirds from their multichambered, hanging nests of interlaced sticks and build a cup nest in one of the rooms; or they invade colonies of caciques and oropendolas and evict them from one or more pouches. Bay-winged Cowbirds claim abandoned closed nests of horneros or other birds in which to breed; or, if no such nest is available, they build a simple one for themselves. Five species of cowbirds build no nests but foist their eggs upon many kinds of birds that incubate the intruded eggs and feed the alien young. The Giant Cowbird drops its eggs only in the swinging pouches of oropendolas and caciques.

Nearly always the female builds the nest alone, not only in the polygamous icterids but likewise in the monogamous orioles. Exceptions occur: the male Melodious Blackbird helps his mate to build, and when Bay-winged Cowbirds do make a nest, the male does most of the work. Other male icterids, such as the Oriole-Blackbird and the Yellow-winged Blackbird, may occasionally bring material to the nest their mate is building, only to drop it or carry it away. Unique in the family is the polygamous male Yellow-hooded Blackbird, who in a marsh builds a succession of unlined cups, then by a special fluttering flight invites females to come and line them for their eggs. When a female Common

Grackle delayed to start her nest, her impatient mate built for three days—most unusual behavior.

In the tropics, most icterids lay two or three eggs, while northern species have sets of three to five or six, occasionally seven. This "latitude effect" is evident in a single species. Thus, Red-winged Blackbirds in Costa Rica lay two or three eggs, northern Redwings three to five or even six. Great-tailed Grackles in Guatemala lay two or three eggs, Common Grackles in the United States, four or five and not infrequently more. Parasitic Brown-headed Cowbirds, which do not incubate, produce more eggs than other icterids, up to about forty in a season.

Icterids' eggs are white or more often tinted—grayish blue, greenish blue, reddish brown, cinnamon rufous—and either immaculate or spotted, blotched, or scrawled with black or shades of red, brown, or lilac. Cowbirds' eggs are extremely variable in color; those of the Shiny Cowbird may be pure white, greenish white, bluish white, or light brown, variously spotted or blotched. One might expect that a parasite with such diversely colored eggs would deposit them in the nests of hosts whose eggs they most closely resembled, thereby increasing the probability of their acceptance, as the European Cuckoo does, but such matching of parasites' eggs to hosts' eggs may occur only in the Giant Cowbird. Cowbirds' eggs have exceptionally thick shells, the better to resist the beaks of their hosts or rival cowbirds.

As far as known, only female icterids incubate, while their mates rarely feed them. A Jamaican Blackbird covered her eggs for 68 percent of the daytime, with sessions that averaged 26.5 minutes. One Yellow-shouldered Blackbird sat with a constancy of 72 percent and another for 77 percent of her active day. Thirty-seven sessions of these two females averaged 23.5 minutes. A Yellow-billed Cacique covered her eggs for 74 percent of a morning, with sessions averaging 53 minutes; a Scarlet-rumped Cacique for 65 percent of a morning, with sessions averaging 38 minutes; and a Melodious Blackbird for 80 percent of seven hours, her sessions averaging 62.5 minutes. Colonial nesting Yellow-rumped Caciques were less patient, remaining in their swinging pouches only about

10 to 30 minutes at a stretch. Incubation periods of nonparasitic icterids range from eleven to fourteen days for smaller species to seventeen or eighteen days for the big oropendolas. Eggs of the smaller cowbirds hatch in eleven or twelve days, which is substantially shorter than the incubation periods of many of their hosts, to the great advantage of their early hatching chicks.

Nestlings are born with closed eyes and usually sparse down, but those of the Yellow-billed Cacique and the Scarlet-rumped Cacique have naked pink skin. The interior of the hatchling's mouth may be pink or orange-yellow but in older nestlings it is usually red. Although a male Common Grackle was reported to brood briefly, brooding exclusively by the female is the rule among icterids. In oropendolas, colonial caciques, and Great-tailed Grackles, she alone feeds the nestlings, but in most icterids, she receives more or less help from their father. In polygamous Red-winged Blackbirds, Tricolored Blackbirds, Yellow-hooded Blackbirds, Yellow-headed Blackbirds, Brewer's Blackbirds, and Bobolinks, males usually feed nestlings, giving preference to the offspring of their first or primary consort, without neglecting later broods. However, attendance at a nest appears to be determined largely by the nestlings' needs. By experimentally increasing the number of nestlings, researchers induced Yellow-headed Blackbirds to give more food to later broods. Monogamous males of all species feed their young. Food is brought in the bill, mouth, or sometimes deeper regions.

Because so many icterids nest in high trees or amid dense, low vegetation, neither of which offers an appropriate stage for the broken-wing act, feigning injury is rarely witnessed in the family, but it has been recorded in the two northern meadowlarks, the Bobolink, and the Red-winged Blackbird. Red-wings spiritedly defend their nests, threatening or even striking intruding humans. By contrast, a pair of Melodious Blackbirds whose nest I visited watched calmly from a safe distance.

Nestling periods show a wide range, as we should expect in a family so diverse in size and habits. Brown-headed Cowbirds leave the nests of their foster parents when only nine or ten days old, and Red-winged Blackbirds depart at ten or eleven days, Great-tailed Grackles at twenty-one to twenty-three days, Montezuma and Chestnut-headed oropendolas at thirty to thirty-five days, sometimes more. After the young icterids abandon their nests, their parent(s) continue to feed them, Red-wings for about two weeks, or until they are about a month old. Montezuma Oropendolas continue to feed young males who have grown much bigger than their mothers.

Helpers feed, or otherwise serve, other birds who are neither their mates nor dependent young. Intraspecific helpers attend others of their own kind; interspecific helpers aid different species. The former attend nests and are conspicuous around fledglings of Bay-winged Cowbirds. At nests with young of Brown-and-Yellow Marshbirds they are active. Austral Blackbirds and Bolivian Blackbirds also have helpers, but the extent of their aid remains to be studied. Helping appears to be rare in Common Grackles, but a male who had lost his brood to a predator fed a neighbor's young. To another nest two males, one evidently a helper rather than the father of some of the brood, brought food throughout the nestling period. A Red-winged Blackbird who had lost her fledglings helped to feed those of a more fortunate neighbor. At nests of Bobolinks auxiliaries arrive to feed nestlings. At nests of cooperatively breeding Forbes' Blackbirds, two to four helpers assist in all operations except possibly incubation. The only report of interspecific helpers among icterids that I have found is of a Common Grackle who fed and protected orphaned Chipping Sparrows.

Finally, icterids are highly social. Oropendolas, caciques, Yellow-headed Blackbirds, grackles, and some orioles nest in crowded colonies. As night approaches, these birds often fly far to sleep in large aggregations. While nesting, males and females neither incubating nor brooding young leave the breeding colony to join an often distant communal roost. Even birds whose nests are widely spaced, such as Melodious Blackbirds, when not covering a nest sleep in roosts that may be far away. Several kinds of icterids may roost together, as in Puerto Rico, where Yellow-shouldered Blackbirds sleep with Shiny Cowbirds, which threaten their survival, and Carib Grackles in big communal roosts. Where they are not numerous, icterids sleep with birds of different families. In a patch of tall grass in Honduras, a few Black-cowled Orioles and wintering Orchard and Baltimore orioles passed their nights amid a crowd of small seedeaters and other birds, giving us a last glimpse of a fascinating avian family.

Bibliography

Allen, A. A. 1914. The Red-winged Blackbird: A study in the ecology of a cattail marsh. *Abstract of the Proc. Linn. Soc. N.Y.*, nos. 24–25:43–128.

Bancroft, G. T. 1984. Patterns of variation in Boat-tailed Grackle *Quiscalus major* eggs. *Ibis* 126:496–509.

———. 1986. Nesting success and mortality of the Boat-tailed Grackle. *Auk* 103:86–99.

Beason, R. C., and L. L. Trout. 1984. Cooperative breeding in the Bobolink. *Wilson Bull.* 96:709–710.

Beecher, W. J. 1950. Convergent evolution in the American Orioles. *Wilson Bull.* 62:51–86.

———. 1951. Adaptations for food getting in the American blackbirds. *Auk* 68:411–440.

Bent, A. C. 1958. Life histories of North American blackbirds, orioles, tanagers, and allies. *U. S. Natl. Mus. Bull.* 211.

Blankespoor, G. W., J. Oolman, and C. Uthe. 1982. Eggshell strength and cowbird parasitism of Red-winged Blackbirds. *Auk* 99:363–365.

Bollinger, E. K., and T. A. Gavin. 1989. The effects of site quality on breeding-site fidelity in Bobolinks. *Auk* 106:584–594.

Bollinger, E. K., T. A. Gavin, C. J. Hibbard, and J. T. Wootton. 1986. Two male Bobolinks feed young at the same nest. *Wilson Bull.* 98:154–156.

Bond, J. 1960. *Birds of the West Indies.* London: Collins.

Carter, M. D. 1986. The parasitic behavior of the Bronzed Cowbird in south Texas. *Condor* 88:11–25.

Chapman, F. M. 1928. The nesting habits of Wagler's Oropendola *(Zarhynchus wagleri)* on Barro Colorado Island. *Amer. Mus. Nat. Hist. Bull.* 43:123–166.

———. 1929. *My tropical air castle.* New York: D. Appleton and Co.

Cherrie, G. K. 1916. A contribution to the ornithology of the Orinoco region. *Mus. Brooklyn Inst. Arts and Sci., Sci. Bull.* 2:133a–374.

Collias, N. E., and E. C. Collias. 1984. *Nest building and bird behavior.* Princeton, N.J.: Princeton University Press.

Cruz, A., T. D. Manolis, and R. W. Andrews. 1990. Reproductive interactions of the Shiny Cowbird *Molothrus bonariensis* and the Yellow-hooded Blackbird *Agelaius icterocephalus* in Trinidad. *Ibis* 132:436–444.

Cruz, A., T. Manolis, and J. W. Wiley. 1985. The Shiny Cowbird: A brood parasite expanding its range in the Caribbean region. In *Neotropical ornithology,* ed. P. A. Buckley, M. S. Foster, E. S. Morton, R. S. Ridgely, and F. G. Buckley. Amer. Ornith. Union, Ornith. Monogr. 36: 607–620.

Darley, J. A. 1971. Sex ratio and mortality in the Brown-headed Cowbird. *Auk* 88:560–566.

———. 1982. Territoriality and mating behavior of the male Brown-headed Cowbird. *Condor* 84:15–21.

De la Peña, M. R. 1979. *Aves de la provincia de Santa Fe.* Santa Fe: Ministerio de Agricultura y Ganadería.

Dow, D. D. 1968. Allopreening invitation display of a Brown-headed Cowbird to Cardinals under natural conditions. *Wilson Bull.* 80:494–495.

Drury, W. H. 1962. Breeding activities, especially nest-building, of the Yellowtail *(Ostinops decumanus)* in Trinidad, West Indies. *Zoologica* (N.Y. Zool. Soc.) 47:39–58.

Dufty, A. M., Jr. 1982. Movements and activities of radio-tracked Brown-headed Cowbirds. *Auk* 99:316–327.

Eastzer, D., P. R. Chu, and A. P. King. 1980. The young cowbird: Average or optimal nestling? *Condor* 82:417–425.

Emlen, J. T., Jr. 1941. An experimental analysis of the breeding cycle of the Tricolored Red-wing. *Condor* 43:209–219.

Fautin, R. W. 1941a. Development of nestling Yellow-headed Blackbirds. *Auk* 58:215–232.

———. 1941b. Incubation studies of the Yellow-headed Blackbird. *Wilson Bull.* 53:107–122.

Feekes, F. 1981. Biology and colonial organization of two sympatric caciques, *Cacicus c. cela* and *Cacicus h. haemorrhous. Ardea* 69:83–107.

ffrench, R. 1973. *A guide to the birds of Trinidad and Tobago.* Wynnewood, Penn.: Livingston Publishing Co.

Ficken, M. 1967. Interactions of a crow and a fledgling cowbird. *Auk* 84: 601–602.

Follet, W. I. 1957. Bronzed Grackles feeding on emerald shiners. *Auk* 74: 263.

Fraga, R. 1972. Cooperative breeding and a case of successive polyandry in the Bay-winged Cowbird. *Auk* 89:447–449.

———. 1978. The Rufous-collared Sparrow as a host of the Shiny Cowbird. *Wilson Bull.* 90:271–284.

———. 1985. Host-parasite interactions between Chalk-browed Mockingbirds and Shiny Cowbirds. In *Neotropical ornithology,* ed. P. A. Buckley, M. S. Foster, E. S. Morton, R. S. Ridgely, and F. G. Buckley. Amer. Ornith. Union, Ornith Monogr. 36:829–844.

———. 1987. Vocal mimicry in the Epaulet Oriole. *Condor* 89:133–137.

Friedmann, H. 1929. *The cowbirds: A study in the biology of social parasitism.* Springfield, Ill.: C. C. Thomas.

———. 1963. Host relations of the parasitic cowbirds. *U. S. Natl. Mus. Bull.* 233.

Friedmann, H., and L. F. Kiff. 1985. The parasitic cowbirds and their hosts. *Proc. Western Found. Vert. Zool.* 2:225–302.

Gavin, T. A. 1984. Broodedness in Bobolinks. *Auk* 101:179–181.

Gochfeld, M. 1975. Comparative ecology and behavior of red-breasted meadowlarks *(Aves, Icteridae)* and their interactions in sympatry. Ph.D. diss., City University of New York.

———. 1979. The begging of nestling Shiny Cowbirds: Adaptive or maladaptive? *Living Bird* 17 (for 1978):41–50.

Goodall, J. D., A. W. Johnson, and R. A. Philippi B. 1957. *Las aves de Chile,* vol. 1. Buenos Aires: Platt Establecimientos Gráficos.

Goodfellow, W. 1901. Results of an ornithological journey through Colombia and Ecuador. *Ibis,* ser. 8, 2:300–319.

Gosse, P. H. 1847. *The birds of Jamaica.* London: John van Voorst.

Hamilton, W. J., Jr. 1951. The food of nestling Bronzed Grackles, *Quiscalus quiscula versicolor,* in central New York. *Auk* 68: 213–217.

Hamilton, W. J., III, and G. H. Orians. 1965. Evolution of brood parasitism in altricial birds. *Condor* 67:361–382.

Hann, H. W. 1937. Life history of the Oven-bird in southern Michigan. *Wilson Bull.* 49:145–237.

Harrison, C. J. O. 1963. Interspecific preening display of the Rice Grackle, *Psomocolax oryzivorus. Auk* 80:373–374.

Hernandez, M. D. 1986. Brown-headed Cowbirds feeding young in Coffee County, Tennessee. *Migrant* 57:98. Abstract in *Auk* 105 (2, suppl.) 20-B. 1988.

Hilty, S. L., and W. L. Brown. 1986. *A guide to the birds of Colombia.* Princeton, N.J.: Princeton University Press.

Horn, H. S. 1970. Social behavior of nesting Brewer's Blackbirds. *Condor* 72:15–23.

Howe, H. F. 1976. Egg size, hatching asynchrony, sex, and brood reduction in the Common Grackle. *Ecology* 57:1195–1207.

———. 1977. Sex-ratio adjustment in the Common Grackle. *Science* 198:744–746.

———. 1978. Initial investment, clutch size, and brood reduction in the Common Grackle (*Quiscalus quiscula* L.) *Ecology* 59:1109–1122.

———. 1979. Evolutionary aspects of parental care in the Common Grackle, *Quiscalus quiscula* L. *Evolution* 33:41–51.

Howell, S.N.G., S. Webb, and B. M. de Montes. 1992. Colonial nesting of the Orange Oriole. *Wilson Bull.* 104:189–190.

Howell, T. R. 1964. Mating behavior of the Montezuma Oropendola. *Condor* 66:511.

Hoy, G., and J. Ottow. 1964. Biological and oological studies of molothrine cowbirds (Icteridae) in Argentina. *Auk* 81:186–203.

Hudson, W. H. 1920. *Birds of La Plata,* 2 vols. London: J. M. Dent and Sons.

Hutcheson, W. H., and W. Post. 1990. Shiny Cowbird collected in South Carolina: First North American specimen, *Wilson Bull.* 102:561.

Jackson, N. H., and D. D. Roby. 1992. Fecundity and egg-laying patterns of captive yearling Brown-headed Cowbirds. *Condor* 94:585–589.

King, J. R. 1973. Reproductive relationships of the Rufous-collared Sparrow and the Shiny Cowbird. *Auk* 90:19–34.

Lack, D., and J. T. Emlen, Jr. 1939. Observations on breeding behavior in Tricolored Red-wings. *Condor* 41:225–230.

Lamb, C. C. 1944. Grackle kills warbler. *Condor* 46:245.

Lanyon, W. E. 1956a. The comparative ethology and ecology of sympatric meadowlarks in Wisconsin and other north-central states. In *Summaries of doctoral dissertations,* vol. 16, pp. 173–174. Madison: University of Wisconsin Press.

———. 1956b. Territory in meadowlarks, genus *Sturnella. Ibis* 98:485–489.

———. 1962. Specific limits and distribution of meadowlarks of the desert grassland. *Auk* 79:183–207.

———. 1966. Hybridization in meadowlarks. *Amer. Mus. Nat. Hist. Bull.* 134:1–25.

———. 1979. Hybrid sterility in meadowlarks. *Nature* 279:557–558.

Laskey, A. R. 1950. Cowbird behavior. *Wilson Bull.* 62:157–174.

Leak, J., and S. K. Robinson. 1989. Notes on the social behavior and mating system of the Casqued Oropendola. *Wilson Bull.* 101:134–137.

Leonard, M. L., and J. Picman. 1986. Why are nesting Marsh Wrens and Yellow-headed Blackbirds spatially segregated? *Auk* 103:135–140.

Lincoln, F. C. 1950. *Migration of birds.* U. S. Fish and Wildlife Service, Circular 16.

Lowther, P. E., and S. I. Rothstein. 1980. Head-down or "preening invitation" displays involving juvenile Brown-headed Cowbirds. *Condor* 82:459–460.

Maxwell, G.R. II, and L. S. Putnam. 1972. Incubation, care of young, and nest success of the Common Grackle *(Quiscalus quiscula)* in northern Ohio. *Auk* 89:349–359.

Mayfield, H. F. 1965. The Brown-headed Cowbird with old and new hosts. *Living Bird* 4:13–28.

McIlhenny, E. A. 1937. Life history of the Boat-tailed Grackle in Louisiana. *Auk* 54:274–295.

Meyer de Schauensee, R., and W. H. Phelps, Jr. 1978. *A guide to the birds of Venezuela.* Princeton, N.J.: Princeton University Press.

Nero, R. W. 1956. A behavior study of the Red-winged Blackbird. *Wilson Bull.* 68:5–37 (part 1, Mating and nesting activities), 129–150 (part 2, Territoriality).

———. 1963. Comparative behavior of the Yellow-headed Blackbird, Red-winged Blackbird, and other icterids. *Wilson Bull.* 75:376–413.

Nice, M. M. 1937. *Studies in the life history of the Song Sparrow,* I. Trans. Linnaean Soc. New York 4.

Nolan, V., Jr. 1978. *The ecology and behavior of the Prairie Warbler Dendroica discolor.* Amer. Ornith. Union, Ornith. Monogr. 26.

Norman, R. F., and R. J. Robertson. 1975. Nest-searching behavior in the Brown-headed Cowbird. *Auk* 92:610–611.

Norris, R. T. 1947. The cowbirds of Preston Frith. *Wilson Bull.* 59: 83–103.

Orians, G. H. 1960. Autumnal breeding in the Tricolored Blackbird. *Auk* 77:379–398.

———. 1961a. The ecology of blackbird *(Agelaius)* social systems. *Ecol. Monogr.* 31:285–312.

———. 1961b. Social stimulation within blackbird colonies. *Condor* 63: 330–337.

———. 1973. The Red-winged Blackbird in tropical marshes. *Condor* 75:28–42.

———. 1980. *Some adaptations of marsh-nesting blackbirds.* Princeton, N.J.: Princeton University Press.

———. 1983a. Notes on the behavior of the Melodious Blackbird. *Condor* 85:453–460.

———. 1983b. *Agelaius phoeniceus* (Tordo Sargento, Red-winged Blackbird). In *Costa Rican Natural History,* ed. D. H. Janzen, pp. 544–545. Chicago: University of Chicago Press.

———. 1985. *Blackbirds of the Americas.* Seattle: University of Washington Press.

Orians, G. H., and G. M. Christman. 1968. A comparative study of the behavior of Red-winged, Tricolored, and Yellow-headed blackbirds. *Univ. Calif. Publ. Zool.* 84:1–85.

Orians, G. H., and G. Collier. 1963. Competition and blackbird social systems. *Evolution* 17:449–459.

Orians, G. H., M. L. Erckmann, and J. C. Schultz. 1977. Nesting and other habits of the Bolivian Blackbird *(Oreopsar bolivianus). Condor* 79:250–255.

Orians, G. H., C. E. Orians, and K. J. Orians. 1977. Helpers at the nest in some Argentine blackbirds. In *Evolutionary Ecology,* ed. B. Stonehouse and C. Perrins, pp. 138–151. London and New York: Macmillan Press.

Payne, R. B. 1969a. Breeding seasons and reproductive physiology of Tricolored Blackbirds and Red-winged Blackbirds. *Univ. Calif. Publ. Zool.* 90:1–137.

———. 1969b. Giant Cowbird solicits preening from man. *Auk* 86:751–752.

Pearson, D. L. 1974. The use of abandoned cacique nests by nesting Troupials *(Icterus icterus):* Precursor to parasitism? *Wilson Bull.* 86: 290–291.

Petersen, A., and H. Young. 1950. A nesting study of the Bronzed Grackle. *Auk* 67:466–476.

Picman, J. 1984. Mechanism of increased puncture resistance of eggs of Brown-headed Cowbirds. *Auk* 106:577–583.

Pleasants, B. Y. 1981. Aspects of the breeding biology of a subtropical oriole, *Icterus gularis. Wilson Bull.* 93:531–537.

Post, W. 1981. Biology of the Yellow-shouldered Blackbird *Agelaius* on a tropical island. *Florida State Mus. Biol. Sci. Bull.* 26(3):125–202.

———. 1992. Dominance and mating success in male Boat-tailed Grackles. *Anim. Behav.* 44:917–929.

Post, W., and M. M. Browne. 1982. Active anting by the Yellow-shouldered Blackbird. *Wilson Bull.* 94:89–90.

Post, W., and K. W. Post. 1987. Roosting behavior of the Yellow-shouldered Blackbird. *Florida Field Naturalist* 15:93–105.

Post, W., and J. W. Wiley. 1976. The Yellow-shouldered Blackbird—present and future. *Amer. Birds* 30:13–20.

———. 1977a. The Shiny Cowbird in the West Indies. *Condor* 79:119–121.

———. 1977b. Reproductive interactions of the Shiny Cowbird and the Yellow-shouldered Blackbird. *Condor* 79:176–184.

———. 1992. The head-down display in Shiny Cowbirds and its relation to dominance behavior. *Condor* 94:999–1002.

Post, W., A. Cruz, and D. B. McNair. 1993. The North American invasion pattern of the Shiny Cowbird. *Journ Field Ornith.* 64:32–41.

Prescott, K. W. 1947. Unusual behavior of a cowbird and Scarlet Tanager at a Red-eyed Vireo nest. *Wilson Bull.* 59:210.

Robertson, R. J., and R. F. Norman. 1976. Behavioral defenses to brood parasitism by potential hosts of the Brown-headed Cowbird. *Condor* 78:166–173.

Robinson, S. K. 1985a. The Yellow-rumped Cacique and its associated nest pirates. In *Neotropical Ornithology,* ed. P. A. Buckley, M. S. Foster, E. S. Morton, R. S. Ridgely, and F. G. Buckley. Amer. Ornith. Union, Ornith. Monogr. 36:898–907.

———. 1985b. Coloniality in the Yellow-rumped Cacique as a defense against nest predators. *Auk* 102:506–519.

———. 1986. Competitive and mutualistic interactions among females in a neotropical oriole. *Anim. Behav.* 34:113–122.

———. 1988. Ecology and host relationships of Giant Cowbirds in southeastern Peru. *Wilson Bull.* 100:224–235.

Rohwer, S., C. D. Spaw, and E. Røskaft. 1989. Costs to Northern Orioles of puncture-ejecting parasitic cowbird eggs from their nests. *Auk* 106: 734–738.

Roseberry, J. L., and W. D. Klimstra. 1970. Nesting ecology and reproductive performance of the Eastern Meadowlark. *Wilson Bull.* 82: 243–267.

Rothstein, S. L. 1974. Mechanisms of avian egg recognition: Possible learned and innate factors. *Auk* 91:796–807.

———. 1975. An experimental and teleonomic investigation of avian brood parasitism. *Condor* 77:250–271.

———. 1976a. Experiments on defenses Cedar Waxwings use against cowbird parasitism. *Auk* 93:675–691.

———. 1976b. Cowbird parasitism of the Cedar Waxwing and its evolutionary implications. *Auk* 93:498–509.

———. 1977a. The preening invitation or head-down display of cowbirds, I: Evidence for intraspecific occurrence. *Condor* 79:13–23.

———. 1977b. Cowbird parasitism and egg recognition of the Northern Oriole. *Wilson Bull.* 89:21–32.

Rowley, J. S. 1984. Breeding records of land birds in Oaxaca, Mexico. *Proc. Western Found. Vert. Zool.* 2:76–221.

Schaefer, V. H. 1980. Geographical variation in the insulative qualites of nests of the Northern Oriole. *Wilson Bull.* 92:466–474.

Scott, D. M. 1991. The time of day of egg laying by the Brown-headed Cowbird and other icterines. *Canadian Journ. Zool.* 69:2093–2099.

Scott, D. M., and C. D. Ankey. 1980. Fecundity of the Brown-headed Cowbird in southern Ontario. *Auk* 97: 677–683.

Scott, D. M., P. J. Weatherhead, and C. D. Ankey. 1992. Egg-eating by female Brown-headed Cowbirds. *Condor* 94:579–584.

Scott, T. W., and J. M. Grumstrup-Scott. 1983. Why do Brown-headed Cowbirds perform the head-down display? *Auk* 100:139–148.

Sealey, S. G. 1980. Breeding biology of Orchard Orioles in a new population in Manitoba. *Can. Field-Nat.* 94:154–158.

Selander, R. K. 1958. Age determination and molt in the Boat-tailed Grackle. *Condor* 60:355–376.

———. 1960. Sex ratio of nestlings and clutch size in the Boat-tailed Grackle. *Condor* 62:34–44.

———. 1961. Supplementary data on the sex ratio in nestling Boat-tailed Grackles. *Condor* 63:504.

———. 1964. Behavior of captive South American cowbirds. *Auk* 81: 394–402.

———. 1970. Parental feeding in a male Great-tailed Grackle. *Condor* 72:238.

Selander, R. K., and D. R. Giller. 1961. Analysis of sympatry of Great-tailed and Boat-tailed grackles. *Condor* 63:29–86.

Selander, R. K., and C. J. La Rue, Jr. 1961. Interspecific preening invitation display of parasitic cowbirds. *Auk* 78:473–504.

Short, L. L., Jr. 1968. Sympatry of red-breasted meadowlarks in Argentina, and the taxonomy of meadowlarks (Aves: *Leistes, Pezites,* and *Sturnella*). *Amer. Mus. Novitates* 2349:1–30.

Sick, H. 1984. *Ornitologia brasileira,* 2d. ed., 2 vols. Brasilia: Editora Universidade de Brasilia.

———. 1993. *Birds in Brazil,* trans. W. Belton. Princeton, N.J.: Princeton University Press.

Skutch, A. F. 1954. *Life histories of Central American birds,* vol. 1. Pacific Coast Avifauna 31. Berkeley, Calif.: Cooper Ornithological Society.

———. 1967. Life history notes on the Oriole-Blackbird *("Gymnomystax mexicanus")* in Venezuela. *Hornero* 10:379–388.

———. 1969. A study of the Rufous-fronted Thornbird and associated birds. *Wilson Bull.* 81:5–43, 123–139.

———. 1972. *Studies of tropical American birds.* Publ. Nuttall Ornith. Club 10. Cambridge, Mass.

———. 1977. *A bird watcher's adventures in tropical America.* Austin: University of Texas Press.

Smith, N. G. 1968. The advantage of being parasitized. *Nature* 219:690–694.

———. 1983. *Zarhynchus wagleri* (Oropéndola Cabecicastaña, Oropendola, Chestnut-headed Oropendola). In *Costa Rican natural history,* ed. D. H. Janzen, pp. 614–616. Chicago: University of Chicago Press.

Stiles, F. G. and A. F. Skutch. 1989. *A guide to the birds of Costa Rica.* Ithaca, N.Y.: Cornell University Press.

Strosnider, R. 1960. Polygyny and other notes on the Red-winged Blackbird. *Wilson Bull.* 72:200.

Studer, A., and J. Vielliard. 1988. Premières données étho-ecologiques sur l'Ictéridé brésilien *Curaeus forbesi* (Sclater 1886) (Aves, Passeriformes). *Revue suisse Zool.* 95:1063–1077.

Tashian, R. E. 1957. Nesting behavior of the Crested Oropendola (*Psarocolius decumanus*) in northern Trinidad. *Zoologica* (N.Y. Zool. Soc.) 42:87–98.

Teather, K. L. 1987. Intersexual differences in food consumption by hand-reared Great-tailed grackle *(Quiscalus mexicanus)* nestlings. *Auk* 104:635–639.

Tinbergen, N. 1951. *The study of instinct.* Oxford: Clarendon Press.

Walkinshaw, L. H. 1949. Twenty-five eggs apparently laid by a cowbird. *Wilson Bull.* 61:82–85.

Webster, M. S. 1994. Interspecific brood parasitism of Montezuma Oropendolas by Giant Cowbirds: Parasitism or mutualism? *Condor* 96: 794–798.

Whittingham, L. A. 1989. An experimental study of paternal behavior in Red-winged Blackbirds. *Behav. Ecol. Sociobiol.* 25:73–80.

Wiens, J. A. 1965. Behavioral interactions of Red-winged Blackbirds and Common Grackles on a common breeding ground. *Auk* 82:356–374.

Wiley, R. H. 1976a. Affiliation between the sexes in Common Grackles. *Z. Tierpsychol.* 40:59–79 (part 1, Specificity and seasonal progression), 244–264 (part 2, Spatial and vocal coordination).

———. 1976b. Communication and spatial relationships in a colony of common Grackles. *Anim. Behav.* 24:570–584.

Wiley, R. H., and A. Cruz. 1980. The Jamaican Blackbird: A "natural experiment" for hypotheses in socioecology. In *Evolutionary biology,* vol. 13, ed. M. K. Hecht, W. C. Steere, and B. Wallace, pp. 261–293. New York: Plenum Press.

Wiley, R. H., and M. S. Wiley. 1980. Spacing and timing in the nesting ecology of a tropical blackbird: Comparison of populations in different environments. *Ecol. Monogr.* 50(2):153–178.

Williams, L. 1952. Breeding behavior of the Brewer Blackbird. *Condor* 54:3–47.

Wittenberger, J. F. 1978. The breeding biology of an isolated Bobolink population in Oregon. *Condor* 80:355–371.

———. 1980. Feeding of secondary nestlings by polygynous male Bobolinks in Oregon. *Wilson Bull.* 92:330–340.

———. 1982. Factors affecting how male and female Bobolinks apportion parental investments. *Condor* 84:22–39.

Young, H. 1963. Breeding success of the cowbird. *Wilson Bull.* 75:115–122.

Index

About the Author

Alexander Skutch was born in Baltimore, Maryland, in 1904 and grew up in the neighboring country, early developing a love of nature. He was educated in private schools and Johns Hopkins University, from which he received, in 1928, a Ph.D. in botany. That same year, he went to Panama to continue studies of the anatomy and physiology of the banana plant. There he became deeply interested in Neotropical birds and, after learning how little was known about their habits, resolved to spend years studying them. Paying his way by collecting botanical specimens for museums in the United States and Europe, he watched birds from Guatemala to Panama and in Ecuador. After a tour for the U.S. Department of Agriculture in Peru, Ecuador, and Colombia, he bought, in the then wild and isolated valley of El General in southern Costa Rica, the farm where he has made his home since 1941. In addition to land cleared for pastures and crops, it retains a large tract of unspoiled rain forest, rich in birds, which he preserved while the surrounding woods were replaced by coffee and sugarcane. Here, at Los Cusingos, he has lived by subsistence farming, writing, spells of teaching ornithology for the University of Costa Rica, and leading birding tours in Central and South America. From this base he has made prolonged visits to other parts of Costa Rica and to Venezuela, to study different birds. He has published many articles in magazines and scientific journals and twenty-six books on natural history, mainly birds, his travels, and philosophy, his main interest after ornithology.

Alexander Skutch is an honorary member of the American Ornithologists' Union, the British Ornithologists' Union, and the Cooper Ornithological Society, and honorary president of the recently formed Asociación Ornitológica de Costa Rica. Among his awards are the William Brewster Medal from the American Ornithologists' Union, the Arthur A. Allen

Medal from the Cornell Laboratory of Ornithology, and the John Burroughs Medal.

In 1950 he married Pamela Lankester. In 1993 their land was acquired by the Tropical Science Center of San José, Costa Rica, which has undertaken to preserve Los Cusingos indefinitely as a nature reserve.

The New American Expat

The New American Expat

Thriving and Surviving Overseas in the Post-9/11 World

William Russell Melton

INTERCULTURAL PRESS
A Nicholas Brealey Publishing Company

YARMOUTH, ME • BOSTON • LONDON

First published by Intercultural Press, a Nicholas Brealey Publishing Company, in 2005.

Intercultural Press, Inc.,
a Nicholas Brealey Publishing Company
PO Box 700
Yarmouth, Maine 04096 USA
Information: 1-888-BREALEY
Orders: 207-846-5168
Fax: 207-846-5181
www.interculturalpress.com

Nicholas Brealey Publishing
3-5 Spafield Street
London, EC1R 4QB, UK
Tel: +44-(0)-207-239-0360
Fax: +44-(0)-207-239-0370
www.nbrealey-books.com

Printed in the United States of America

09 08 07 06 05 1 2 3 4 5

ISBN: 1-931930-24-4

Library of Congress Cataloging-in-Publication Data

Melton, William Russell.
The new American expat : thriving and surviving overseas in the post-9/11 world / William Russell Melton.
p. cm.
Includes index.
ISBN 1-931930-24-4
1. Americans--Foreign countries--Handbooks, manuals, etc. I. Title.
E184.2.M45 2005
909'.0413083--dc22
2005001856

To Siân, Erik, and James,

who have patiently endured my quest for new adventures,

and Victoria,

who has inspired it.

Table of Contents

Acknowledgments

I would like to express my sincere appreciation to the following people:

Ted McNamara, Martha Childers, and Greg Raver-Lampman, for their contributions to the content of this book. Ted, Martha, and Greg have a combined experience of more than 30 years of living and working overseas. They have added their insight and experience gained from living and working in Japan, Austria, Switzerland, Germany, India, Kenya, Iran, Jamaica, Ecuador, France, Egypt, and the Czech Republic. I appreciate their willingness and enthusiasm in reviewing this book and contributing their thoughts to its pages.

Patricia O'Hare, Publisher, Erika Heilman, Executive Editor, and Judy Carl-Hendrick, Managing Editor, of Intercultural Press, for their invaluable comments and suggestions in editing, improving, and finalizing the content of this book.

Victoria Moran, author and speaker, my personal editor (and my wife), for her inspiration, patience, and love that I feel every day of our lives.

Preface

My first foreign assignment was in 1982 in Saudi Arabia. I was raised in Kansas, and my exposure to foreign countries, like that of many Americans, was limited to a couple of short trips to Europe.

Still, I was intrigued by the phone call from the headhunter searching for candidates for a Saudi-based legal position with a major U.S. corporation. There was something attractive and exotic about the unknown, pulling on me to break away from my Midwestern bonds.

Within a few weeks of receiving and accepting an offer, I had closed out my life in the United States and was on a plane to Jeddah.

I landed there at the end of the haj, the annual Muslim pilgrimage to the holy city of Mecca. As I passed through immigration and customs into the airport concourse, I quickly realized that I was not in Kansas anymore. Tens of thousands of Muslim pilgrims were standing, sitting, and lying on the floor, trying to stay comfortable while awaiting their flights home. Surrounded by representatives of all the races and nationalities of the Muslim world, I was overwhelmed by the palette of colors and the multitude of different fabrics and styles of clothing they were wearing. The sights and sounds of the airport hall were like nothing I had experienced or even imagined before that day. Nevertheless, I felt strangely comfortable, as though I had been there many times before.

I made a mental note to remember this experience (the first of many thousands of times I would say this to myself) and smiled inwardly as I stepped into the crowd. My great adventure as an expatriate—one that would last the rest of my life and that would profoundly change my personality, values, and attitudes—had just begun.

—William Russell Melton
March 2005

Introduction

The New American Expat in the Twenty-First Century

The world has experienced an enormous change in the last several years, brought on not only by the events of September 11, 2001, but by the revolution in technology, communications, and transportation during the last decade. Advances in these areas, combined with political and economic restructuring in Eastern Europe and China, have converged to make the world a much smaller place. For better or worse, everyone is now a real or potential neighbor.

These trends are often talked about in the context of globalization, but to me, *globalization* is just a politically correct term for Americanization.[1] The process is fueled by U.S. economic, political, and military power. I don't see the world becoming more French,

[1] I would like to clarify my usage of the term American. The reader will note that commencing with the title and continuing on throughout this book, I use this term to refer exclusively to citizens and residents of the United States of America. With apologies to the millions of other Americans inhabiting the continents of North and South America, I can think of no other term for U.S. citizens and residents that is not linguistically cumbersome. In order to retain a conversational and easy-to-read style throughout the book, I also refer to Americans as "we" and "us," thus including myself within the group even when discussing a characteristic or viewpoint that I do not personally possess or hold.

English, Chinese, or Japanese. Rather, I see the world becoming more American almost every day. The influence of American technology, pop culture, corporate power, and, in some cases raw military might, is so pervasive that it has been called a *new American empire*. While the merits of Americanization can be debated by politicians and economists, its impact upon the world community is undeniable. What we are experiencing is a massive (and surprisingly voluntary, in many cases) Americanization of all aspects of global life, culture, business, and society.

In this setting, no one can continue to be isolated (or insulated) from the influences of the rest of the world. Like it or not, we all are now forced to deal with people on the other side of the world as neighbors, thus creating the opportunity to make both new friends and new enemies. Our new world neighborhood presents us with two divergent paths from which to choose: one based on mutual understanding and tolerance of our differences and a contrary path characterized by cultural, religious, and ethnic intolerance and the enmity that invariably follows. It is up to us which path we choose. Unfortunately, as the world has indeed gotten smaller, many of these neighbors have not become closer, but rather have been pushed (or pulled themselves) farther away.

These tensions were heightened by the events of September 11, 2001, a day that shattered our assumptions about the security of our social, economic, and political institutions and sent us hurtling down a path of uncertainty.

It is against this background of expanding American influence and the threat of terrorism that I write this book. These factors have had a profound impact on the American expatriate ("expat")—who is living in an always challenging, and sometimes hostile, foreign environment. Exciting new markets and overseas opportunities are now easily within reach of an increasing number of Americans. On the other hand, the new American expat is confronted with an entirely new set of challenges in living and working in foreign countries—a difficult task to begin with.

Although Americans abroad today may choose from endless new opportunities, they must also face risks very different from those experienced by the expats of previous eras. In some ways, today's challenges are easier to deal with because of improved communication and transportation, increased development, and the accessibility of American conveniences and pop culture in most countries. At the same time, the world of today's expats is much more competitive. The competition in foreign countries comes from not only a larger pool of able and willing expats, but also an increasing number of local employees with the skills and knowledge required to perform jobs previously reserved for expats. As a result, financial rewards for today's expat have decreased dramatically—the numerous allowances and luxurious lifestyles enjoyed by our predecessors are pretty much a thing of the past.

The New Anti-Americanism

The new safety and security issues raised by 9/11 and other recent events further complicate the situation of Americans abroad. Current attitudes toward the United States are radically different from those of just a few years ago. While the Cold War divisions have disappeared, a new polarization along religious and cultural lines has emerged as the United States began exerting more global influence. The response of local cultures has sometimes been positive — but too often was not only negative, but hostile, a trend that continues to spiral downward. The events of 9/11 and our response to them have further polarized the view of the U.S. held by numbers of people around the world *and* Americans' perception of other peoples and places. As a result, the mutual hostility between the U.S. and some other—primarily Muslim—countries seems likely only to increase, fueled as much by our mutually erroneous and stereotyped perceptions of each other as by political and religious differences.

Stereotyping is one of the most common, and harmful, of all human traits. *The Microsoft Encarta Dictionary* (http://encarta.com) defines a stereotype as an "oversimplified standardized image or idea held by one group of another." Unlike a generalization, which is a statistically based viewpoint that acknowledges the possibility (even the likelihood) of individual exceptions, a stereotype is based on emotion. As such, it is an inherently biased response that presupposes no exceptions. We all make generalizations about groups of people, and they are rarely harmful. Unfortunately, most of us also fall prey to stereotyping other people as well, and such stereotypes are never benign.

It has long been common among all ethnic, racial, national, tribal, and religious groups to develop stereotyped perceptions of other groups. This tendency occurs within countries, cities, and even neighborhoods, and is in no way limited to cross-border relations. However, it is in the international setting that stereotypes are perhaps most dangerous. Often in such cases, stereotypes are based on highly inaccurate information derived from movies, biased news sources, and government propaganda, rather than from personal experience. These stereotypes are thus the most difficult to overcome, and the most likely to serve as tinder for international crises. At their worst, they can contribute significantly to hostile behavior, aggression, and even war.

There are four important points to remember about stereotypes. First, they are often based on an exaggerated version of some real trait. Second, they focus on traits that a group perceives as different from its own. Third, they are driven by insufficient or incorrect knowledge of the group being stereotyped. Lastly, stereotyping is a two-way street—Americans are as guilty as anyone of developing stereotypical views of everyone from local taxi drivers to entire countries.

American expatriates are no more immune to stereotyping than Americans at home. I will not dwell on the most negative of the stereotypical views of foreigners commonly held by American expats, because you will hear them soon enough (if you haven't already) when you move overseas. I do not want to give credibility by repeating them here. However, it is probably a good idea for you to know the stereotypes, both good and bad, that the rest of the world has historically applied to Americans, since you may soon to be subject to them. First, the good news: As Americans, we are likely to be welcomed sight unseen as having a good work ethic, a can-do attitude, confidence, powers of innovation, high energy, good organizational skills, decisive leadership, focus, generosity, kindness, and a sense of fair play.

Unfortunately, we are also likely to be "pre-viewed" as being loud, obnoxious know-it-alls; high-minded and overly moralistic; too rich and obsessed with making money rather than friends; soft, wasteful, and way too impatient; arrogant in our attitudes toward others, yet ignorant of the rest of the world; overweight and sartorially challenged; and, last, possessed of an unnatural fixation with our teeth. Recently added to these are the more ominous stereotypes of Americans as aggressive, imperialistic, and militaristic. As with many stereotypes, both the positive and negative, American stereotypes are based on exaggerated versions of real traits—with the degree of exaggeration being in the eye of the beholder.

My hope is that this book will help the new American expat demonstrate to the world that these latter negative stereotypes cannot be fairly applied to all, or even most, of us. Just as important, I hope you will take from this book an attitude that rejects the negative stereotyping of others who are different from us.

Do They Hate Us?

The level of global anti-Americanism has escalated in recent years to a popular perception that the rest of the world hates Americans. This perception is understandable in view of the recent wars, conflicts, and tensions in which America has become involved. The frequency with which the question "Why do they hate us?" appears as the subject of books, media pieces, and editorials has raised the idea that they *do* hate us to a truism. Nevertheless, in more than 25 years of living and traveling extensively in other countries, I have never experienced a single hateful or hostile encounter with any individual or group that was based on my being an American.

True, there are many people in the world who hate Americans. Terrorists and criminals hate us because we are a threat to them. Certain religious extremists hate us because they hate everyone who does not adhere to their belief system. Many ordinary (and innocent) people hate us because they or their family members have been injured or killed in wars or military conflicts with the United States, or because they have other real or perceived grievances against our government. There are also countless others who know little about Americans but hate us solely because they are taught to do so. However, in relative terms, the people who hate Americans are largely limited to a small group of people who are mostly centered in easily identifiable regions.

For the most part, you will encounter people who—with justification or without—strongly disapprove of the policies and actions of the U.S. government. They express such disapproval through both peaceful and violent means. However, they do not necessarily translate this disapproval into a general hatred of Americans as *individuals*. Instead, they cling to an almost universally held notion that Americans at their core are an innately good, fair, and generous people.

I have encountered this attitude many times in my discussions with people from foreign countries. I have lived in and traveled to countries in which anti-Americanism was perceived to be at a high level. I have had numerous business and personal contacts with people from countries where everyone supposedly hates us. I have encountered many people who at various times are frustrated with, insulted by, or enraged at America as a country. But they did not transfer or apply these feelings to me as an individual American. In their minds, they disassociated individual Americans from the monolithic and impersonal face of the U.S. government, as though we have nothing to do with its policies and actions.

No, with few exceptions, they do not hate us.

Why is this point important or relevant to the subject matter of this book? First, it is important for you to understand that, should you embark on a new career as an expat, you will not be greeted with hostility at every turn. Unless you venture into known high-risk areas or bring it upon yourself by displaying an arrogant attitude, you will probably be as untroubled by anti-American sentiment as I have been in my time abroad.

Furthermore, whether speaking of benign stereotypes or more hostile foreign attitudes, every American who moves to (or even visits) a foreign country has an opportunity to influence the global view of America and Americans. We are all ambassadors for our government, our culture, and the American people. We *do* have the ability to influence how others see us as individuals and as a country, even through seemingly trivial, everyday actions. We also have the opportunity to return home with a better understanding of, and tolerance for, the viewpoints and attitudes of those who are different from us. This is something that I hope to help you accomplish.

Succeeding Where Others Have Failed

Another goal of this book is to assist you in becoming resoundingly successful in every stage of your expat experience—from finding your first foreign position to coming home upon completion of your last foreign assignment. Historically, Americans aren't very good at working abroad. A high percentage of Americans fail to successfully complete their foreign assignments—much higher than necessary. To most Americans, moving to another country is a big deal. To most foreigners, it is more like moving from California to New York is to us. In other countries, people don't buy a book like this one or attend seminars on living and working abroad. Living and working in other countries is part of their normal lives. For us, it tends to be considered something exotic, and often we don't do it particularly well. Still, I believe that most Americans have the potential to be successful expats. The same traits that have enabled Americans to create the most dynamic and successful country in history can help us to succeed in any foreign setting.

The dichotomy between the potential of Americans to succeed as expats and their actual track record has contributed to my own conflicting views on hiring Americans for foreign positions. On the one hand, I know from experience that if I want an employee who is driven to succeed, who will most closely match my own work ethic, who will have a can-do attitude about the job, and who will be results-oriented (and all those other "good" American stereotypes), then I want an American. On the other hand, I also know from experience that far too many Americans are inflexible. We have trouble adjusting to different cultures and practices. Our personalities don't always travel well. We complain too much about local conditions and "inconveniences," and, as a result, fail to complete our foreign assignments *successfully*. (We may be able to hang around until the end of our contract but we are not successful by traditional measurements of job performance.)

My goal is to get you to the point at which no one would attribute any of these negative traits to you.

It's Not Just a Job—It's an Adventure

I also hope to instill a little excitement in you, so that your foreign experience is truly an adventure and doesn't end up like a temporary transfer to Wichita. Don't just "get through it." Don't hole up in an American compound or neighborhood, focusing only on your job. Experience the local scene and taste the local pleasures. Really try to understand the local way of thinking and doing things. Re-invent and extend yourself. Do things you didn't, couldn't, or wouldn't do back home. Explore your new world and explore your inner self at the same time. You can turn this into your greatest adventure and learn things about yourself that would have remained hidden all your life back in the good old USA.

I don't just mean that you should visit museums and historic buildings (although this can be good, too). In addition you might pursue some more adventurous activities:

- Climb a mountain. I managed a high hill in Switzerland.
- Stay out all night and see the sun come up. Sunrise from the Sacre Coeur in Paris is quite stunning. My friend Martha preferred to go to bed, but then get up early and watch Paris awaken while enjoying coffee at her favorite sidewalk cafe.
- If you are musically inclined, play in a local band or orchestra or sing in a local chorus or choir. I played French horn in the Jeddah Expat Orchestra, and harmonica with a blues band in Amsterdam.
- Take those off-the-beaten-track side trips. I rode a steamer from Mandalay to Pagan, Myanmar (then known as Burma), and met people who had never seen a Westerner. One of my friends took a similar trip on a paddle steamer down the Sudd river in Sudan.
- Take part in local events. I ran the Riyadh marathon (although this was definitely my last 26-mile run in the Saudi desert).
- Take bicycle trips or long walks. I walked across the island of Singapore one sunny day.
- Do not pass on opportunities that may lead to an interesting or bizarre experience. I once accepted a dinner invitation and wound up being entertained by Idi Amin Dada, self-proclaimed Conqueror of the British Empire, who enthusiastically played polkas on his accordion after dinner (no, he was not any good, but none of us felt like telling this alleged Butcher of Africa and murderer of 300,000 people our opinion of his musical talent).

Immersing yourself in the local scene is how you really start to understand the personality of a country and its people. By all means, come back with some good stories to tell.

Content and Focus of This Book

In one sense, it is unrealistic to even attempt to write a book for *all* American expats. The issues and challenges facing you in a foreign environment will differ considerably depending on the stage of your career and your life. You may be young, single, and unencumbered with children and financial responsibilities, or you may have a spouse and children whose welfare you must consider and provide for. You might be a more mature candidate for a foreign assignment (perhaps an empty nester) who is seeking a career change or a new adventure in your life. While these situations have much in common, each presents its own unique concerns and challenges. In writing this book, I have tried to draw on my own experience as an expat at different stages of my life—a young, single American on my first foreign assignment, a husband and father relocating my entire family to a foreign country, and an old-timer working and living in various foreign countries while my wife pursued her own career back home in New York City.

In the following chapters, I will be addressing the full range of issues and challenges confronting today's expats:

- Factors to consider in making your expat decision
- Potential benefits and detriments of the expat experience
- The personality traits you will need to possess (or develop) in order to succeed
- Finding your foreign position
- Negotiating your expatriate compensation package
- Moving you, your family, and your possessions to your foreign home
- Living and adjusting to life in your new country

- Working successfully as a U.S. expat in another country
- Staying safe, secure, and out of trouble in today's foreign environment
- Returning to your U.S. home and job when it's all over

There are numerous other books, as well as websites, dealing with these issues. Many of these are excellent resources that I have used often and will recommend later in this book. However, I will be taking a somewhat different approach from what I found in most of them. I come from the vantage point of a lawyer and business person who has been vested with the responsibility of managing foreign operations profitably, achieving targets, and growing revenues. Rather than address the issues associated with living and working overseas from an academic or human resources standpoint, I will approach them as a person who had to meet tough business goals every quarter while dealing with personal issues on my own time.

On each foreign assignment, I knew that my employer's goal was very simple—to successfully create and/or manage a business operation in a foreign country. My success or failure at this task was how I was judged, and generally I was not given a lot of time to get settled in or to adjust to a new environment. Most of you who choose to embark on a foreign assignment will likely encounter the same pressure and sense of urgency in dealing with the issues discussed in this book.

An American World Citizen

My final goal for the new American expat of the twenty-first century is to create as many American world citizens out of this group as possible. This concept is easily misunderstood. When an American speaks of being a world citizen, there is a group of other Americans who immediately accuse him or her of being anti-American or unpatriotic. Unfortunately, I have rarely met these other Americans among my expatriate colleagues because they usually choose not to see much of the rest of the world. I say *unfortunately,* because I believe that the only hope they might have for developing a different attitude about the world is to start by experiencing more of it.

When I speak of being an American world citizen, I am referring to a set of experiences and attitudes that reflect a true knowledge and understanding of the world and our place in it. An American world citizen's *experiences* include foreign travel, knowledge of world events, language study, working and living in other countries, and a wealth of friends from different nations and cultures. All of these experiences ultimately spring from a simple desire to learn about the rest of the world.

An American world citizen's *attitudes* include an appreciation for the differences in people and an understanding that not everyone in the world wants (or needs) to be exactly like us. An American world citizen realizes that, whether in business, diplomacy, or personal relations, buy-in by others is always more profitable than surrender. An American world citizen knows that the respect of our foreign colleagues will always bring us nearer our goals than will their fear. Lastly, an American world citizen, when departing from a country, leaves everyone there with a better impression of America and Americans than was commonly held before he or she arrived.

I promise that if you take advantage of the opportunities that are offered in your life as an expat, you will find yourself becoming an American world citizen. And in so doing, you will find that the people of a foreign country—any foreign country—are far more like us than they are different.

In sum, every American has the opportunity and potential to be successful in a foreign assignment, enjoy the experience, make a positive impact on the world, and be a great ambassador for our country. We all have it within our control to create the right environment and attitude for success overseas, whether in our personal or business life. My hope is that this book will help you ask the right questions and make the right decisions so that you can do exactly that.

Chapter One

The World at Your Doorstep

Americans are working overseas in increasing numbers. It is estimated that up to 8 million Americans (excluding military personnel) are currently living and working in foreign countries, and this number continues to grow at an unprecedented rate. For the majority of these Americans, their foreign experience lasts from three months to two years, so there is a constant turnover of Americans moving overseas, returning home, and being replaced by new expats.

Perhaps you're one of them already. Or you've received a tempting job offer in a country you hadn't thought about since your high school geography class. Or you simply have a touch of wanderlust your friends and family may not understand. This book is for you.

Chapter 1 lays the groundwork for your expatriate experience, explores the pros and cons of working overseas, and discusses the "right stuff" needed to be successful in your international experience.

Although Americans experience a unique set of challenges in today's world, it would be a mistake to suggest that this significant upsurge in expats is a purely American trend. It is rather an essential component of life and business for individuals and companies around the world—a direct result of the internationalization of businesses from every continent on the globe. Furthermore, internationalization is no longer the province of the huge multinational corporations. Now even small companies are testing international opportunities at very early stages in their business development, and they need experienced and knowledgeable people to build and manage their international commerce and operations.

A significant part of this growth has resulted from the opening and development of former nonmarket economies that previously had been more or less closed to American interests. These huge markets, ranging from Russia and the Eastern European Soviet bloc countries to China, are now most definitely open for business and welcoming Americans and other expats into significant areas of their economy and their commercial enterprises. I believe it is accurate to say that, while there may be countries to which you personally don't want to move or that are viewed as dangerous for political reasons, there are fewer countries today that are closed to American expats than at any time in history.

Although it is true that more expat positions are available than ever, the nature of foreign assignments has changed over the years. In the past, there was often a general goal of developing the international knowledge base of U.S. employees so that they could come back and use that knowledge for the benefit of the company and share it with other employees. Also, there was frequently a perceived need to fill key foreign positions with familiar faces—long-term employees who were loyal to the parent company, who understood the corporate culture and procedures, and who could keep an eye on things for the folks back home. This is increasingly rare in today's world of budget restraints combined with twenty-first century communication and ease of travel. It is simply no longer necessary to develop a cadre of internationalists or to permanently move key U.S.

managers to a foreign office. Everyone can, to a certain extent, become internationally savvy simply though the Internet, videoconferencing, and relatively short trips.

In today's world, foreign assignments are usually driven by a defined tactical objective that only a U.S. expat can achieve. Assignments now tend to be transition positions, with a clear short- to mid-term goal of replacing the expat with a local manager or other employee. U.S. employers today send expats for specialized foreign assignments: to open a new territory, train local employees, fill a void in technical expertise, or facilitate liaison with (or control by) corporate headquarters during the start-up stage. In all these cases, there is a value-added proposition provided by the expat for the short to mid-term, and usually a plan to transfer the expat's knowledge and skills to a local employee.

This strategy is reflected in the most recent *GMAC Relocation Trends Survey Report* (GMAC Global Relocation Services, 2004), which found a pronounced trend in U.S. corporate practices toward shorter foreign assignments, less costly compensation packages for overseas employees, and a transition to either local employees or local-compensation standards for their U.S. expat employees.

In my case, my foreign *legal* assignments were driven by the requirement for a U.S.-trained lawyer to be based at the company's international headquarters. The need was for a lawyer familiar with the legal issues encountered by U.S. companies in their international business. My foreign *management* assignments were generally driven by the need for someone familiar with the company's policies, strategies, and products, and who was known to the U.S. senior management, to establish a new foreign operation or turn around an underperforming one. In almost all cases, I provided knowledge and skills that were unavailable in the foreign country and allowed my employer to achieve its tactical objectives. In addition, even when my assignment was of indefinite duration, it was never viewed as permanent by my employer or me.

Why Do You Want to Do This?

This is a simple question—deceptively simple because the answer is usually very complicated. However, a clear understanding of your answer to the question of why you want to live and work in a foreign country is crucial to your ultimate success or failure. If you don't know what your goals are, you almost certainly will not achieve them. Judging from the high percentage of failures, an unfortunately high percentage of American expats don't have a clear answer to this question.

The reasons why a person is motivated to take a job in a foreign country almost always fall into one of three categories:

1. **Professional**. Career advancement, career expansion, job promotion, more money, increased knowledge of international markets, professional development, résumé enhancement, and building an international network are the most common reasons.
2. **Inquisitive**. Personal (or family) growth, interest in expanding your personal experiences, interest in expanding your knowledge of the world, adventure, and desire to experience new and different challenges are the reasons that most often lead to success.
3. **Remedial**. Running away from a difficult personal situation or financial problems, a recent unpleasant change in life (divorce, for example), and a perceived boring and pointless life are also fairly common reasons—and they aren't always bad. Sometimes a change in environment and new start will have a positive effect. Generally, however, the problems you are running away from just follow you, and they will be even more difficult to deal with in an unfamiliar setting.

You need to be able to articulate to yourself and to any prospective employer the exact reasons you want to work overseas, what your personal and professional goals are, and why you want to work in a particular country or for a particular company. You then need to carefully evaluate whether you are likely to achieve these goals. In other words, are you stepping into a situation in which you have a reasonable chance of success? If not, you need to rethink your thinking on this.

I must admit that I probably had no clearly defined goals for my first foreign assignment other than curiosity, a sense of adventure, and a desire to see and experience another part of the world. Not thinking much about career or financial issues, I fit into category two—inquisitive. I wanted to expand my knowledge of the world. While I probably didn't articulate this goal at the time, I did achieve it. However, the reasons behind my subsequent overseas postings have been much more calculated. For each one, I could clearly articulate to my employer and myself exactly what my goals were, and each one was part of a strategic personal and professional plan.

In addition to considering why you want to work abroad, you must also ask a related and equally important question: *Why does your company want to send you overseas?* You need to understand the company's goals in the foreign or international market, its current stage of international development, and its growth and expansion plans.

What role does your employer expect you to play in achieving these goals? Are you getting in on the ground floor of something new, with all the potential for growth that comes with success? Or are you intended to play a specific role in a mature international organization? Both of these environments can be personally and professionally rewarding, but they are completely different. Some personalities fit well with one but not the other.

How do the employer's international goals and structure fit with your own personality, strengths, working style, and personal goals. Bottom line: Are they a good fit for you? Are your goals and your employer's goals compatible and equally achievable? If not, neither you nor your employer will end up getting what you want.

At the same time, consider whether your employer is willing to provide you the support and opportunity to achieve these goals. How committed is your employer to the foreign operation where you are to be based, and to you personally? How important is the success of this assignment to the success of the employer's operations overall? How much assistance will your employer provide to you in terms of preparation, logistics, and financial support? How committed is your employer to making you and your family comfortable? What resources will your employer commit to your success?

Remember, it is not possible for your employer to fail in your foreign assignment. All responsibility for failure will rest with, and be attributed to, you (whether you believe this is fair or not). This is simply the nature of the beast, which is why it is so important to deal with all these issues before you go rather than after you get there.

Lastly, under no circumstances should you accept an expat assignment only because your employer wants you to go. This can be an instigating factor, and often is, but there must be positive reasons personal to you. This move is going to affect *your* life, *your* career, and *your* family as much as or more than it affects your employer. If there are not sufficient personal reasons for you to go, you will fail in any event—and you and your family, as well as your employer, will suffer the consequences.

Pros and Cons

There are four positive outcomes that may result from moving to a foreign country:

1. Career progression
2. Financial gain
3. Strengthened family relationships
4. Positive personal development

There are also four negative outcomes that may make you wish you had just stayed home:

1. Career regression
2. Financial loss
3. Damaged family relationships
4. Negative personal development

As you will notice, this proposition can go either way for you. The simple act of moving overseas and working in a foreign assignment is in no way guaranteed to have positive results for any aspect of your life:

Career

Expatriate positions tend to provide you with an opportunity to exercise responsibility, get noticed, and make an impression. A foreign assignment can thus be a stepping-stone to a promotion or a new position with higher compensation. It can increase your marketability and competitiveness in the job market. Do not assume, however, that such good results will automatically occur. Being in a high-profile position is not always a good thing for your career. It depends on what you do with it. A foreign assignment can also have negative implications for your career. This is particularly true if you are perceived as having failed, but it can happen even if you are successful. You might miss out on promotions, be passed by your colleagues on the climb up the company ladder, lose touch with your network of supporters, and/or fall behind in learning new skills.

Finances

If well planned, an expatriate position can be very good for you financially. You may be able to save a significant amount of money and return from your foreign assignment with a nice nest egg. On the other hand, you might not even make enough to cover your costs of living in the foreign country. Or you might blow everything on luxuries and living beyond your means (or needs). Financially, being an expat is very tricky. There often are hidden costs and taxes that you did not incur in the United States. You'll have to deal with cost-of-living variables, inflation in the foreign country, and exchange rate fluctuations. You may be reduced from a two-income family to a one-income family. The compensation package that works for you at home may not work for you in a foreign country. In fact, it almost never does. Poorly planned, a foreign assignment can have a devastating effect on your financial health.

Family

A foreign posting with family members can provide an opportunity for your entire family to develop closer bonds. You will be able to share experiences that will make your family stronger. You will of necessity get out of the rut of watching television and playing video games, and actually spend time together. All members of your family will have an opportunity to learn, grow, and develop to an extent they never would have back home. Often, however, this opportunity isn't realized. A large percentage of expat failures are due to family issues—unhappy family members and stresses on family relationships. Spouses have to quit jobs and put careers on hold. Children have to attend new schools and make new friends. Boredom and feelings of isolation can set in. Cultural and language differences can make life frustrating, especially for family members who weren't enthusiastic about the move in the first place. If your relationship with your spouse is on the edge, a foreign assignment could send it reeling over the cliff.

Personal Development

From a personal development standpoint, what could possibly offer you more opportunity to expand yourself than living in a foreign country? There is so much to experience and learn. You will have the opportunity to return to the United States at the end of this assignment with a new outlook on the world, the ability to speak a foreign language, a new set of friends who could remain close for the rest of your life, knowledge and understanding of a foreign culture, and memories of experiences that few Americans have the opportunity to enjoy. Even so, I am amazed at the number of U.S. expats I have encountered during my career who seem to have passed on all of this. They stay in their American compounds, eat their American food, socialize with their American friends, complain about almost everything in the foreign country, and return home without really ever having left it.

At this stage of planning, however, neither a positive nor negative outcome is preordained. As you read this book and contemplate pursuing this opportunity for a change in your life, you have it totally within your power to determine which of these outcomes will be yours. It is all a matter of your personal preparation and attitude.

Expat Success Plan

Your expat assignment can be successful for you and your family if you want it to be. In addition to understanding your and your employer's goals, as discussed earlier in this chapter, you need to have an Expat Success Plan that covers all four issues highlighted in the previous section: career, finances, family, and personal development. Don't bother formulating a plan right now because you aren't ready to do so—it probably would be a short exercise. At the end of your reading, however, sit down (with your family if you have one) and develop an Expat Success Plan for these four aspects of your foreign assignment:

1. What are you going to accomplish in this overseas position? How will you ensure that this position actually enhances your career? At the end of the assignment, what will you have added to your skills and knowledge base? How will you leverage it for success when you return?

2. Discuss and plan for the financial implications and effects of the foreign assignment on you and your family. Understand all the financial variables (compensation, costs, taxes, exchange rate, and inflation issues) and create your plan for financial success.

3. What are the issues that concern your family members? What are the challenges that might make life difficult or frustrating for them? Try to anticipate these and have a plan for dealing with them. What are you and your family going to do while living overseas to strengthen the relationships between family members?

4. Develop a plan for what you want to accomplish in terms of your own personal development. Think about what you want to learn and experience, places you want to go, people you want to meet. Elevate the process from the theoretical to a solid plan with a realistic timetable. Otherwise, you will end up getting to the end of your assignment with many of your goals still unfulfilled. Remember, the idea is to come back with some good stories to tell.

The Right Stuff

There are two kinds of "right stuff" required for succeeding in an expat position. Evaluate yourself honestly with respect to both of these. First, you need the right stuff from a job standpoint. You need to evaluate whether you have the skills and knowledge to be successful in this specific position. To do this, you must understand not only the range of your own skills and knowledge base, but also the expectations and needs of the employer with respect to this position. One aspect of having the right stuff is the ability to do a realistic appraisal of your suitability for the foreign job.

If you get an overseas assignment, you are likely to be out on a limb more than at any time in your previous professional career. You will have much less support and relatively few lifelines from the United States. You must be prepared to operate and make decisions on your own more than ever before. This will be difficult if you don't have the qualifications for this particular position in the first place. In almost every case, you are being sent abroad to *do*, not to learn.

Your employer may see you primarily in terms of these job-related skills—possibly to the exclusion of personality factors. Don't make the same mistake yourself in evaluating whether you are right for a foreign position. Success in international business often depends more on personal relationships and interactions than on technical or other job skills.

This brings us to the second category of right stuff required for success as an expat: your personality. Many Americans do not exhibit the very specific personality traits that are mandatory for living or working in a foreign environment. This isn't so much because we can't, but because we have never had to. It is well known that most Americans travel outside of the United States very little (if at all), so how can we be expected to know how to act when we do?

I found myself in this exact position when I landed in Saudi Arabia on my first expat assignment in 1982. In the hope that you might avoid many of my missteps in this area, I will share the lessons I've learned (often the hard way) in my subsequent 20-plus years of international experience. In order to be successful in overseas endeavors, the new American expat must either have or develop the following 10 expat personality traits:

1. Good Communication Skills

The parody of an American or English visitor to a foreign country is that we believe that if we simply get in the face of non-English speakers and shout loudly and slowly enough, everyone will understand us (whether they understand a single word of English or not). I wish life were that easy. Even when dealing with a foreign national who speaks fluent English, good communication can be difficult. This is complicated even more by the fact that many people you meet will indeed speak and understand English reasonably well, but not nearly as well as you imagine. It is thus imperative that you as an expat develop good communication skills.

You will have to focus on these skills much more than was ever required in the United States—it must become almost an obsession for you. You must listen closely and ask questions. When you speak, speak slowly. Repeat what people say to you to confirm your understanding. Patiently allow people to complete their sentences without interrupting or attempting to complete them yourself (even though you might think you are just being helpful). You must allow intervals of silence between statements to give your listeners time to fully understand your meaning. When necessary, ask people to repeat themselves or to explain what they mean. Do not use idioms or overly complex sentences. (My friend Ted remembers the shocked look on the faces of his Swiss colleagues when an American spoke of "beating a dead horse" during their meeting.) Follow up your verbal communications with e-mails or other written confirmation. Check with people after a conversation to confirm that they truly understood what you said and are taking action accordingly. A simple *yes* or nod of the head does not mean that they understood your point.

In doing all this, you must be able to build personal relationships through your communication skills, and be respectful and polite when you speak to people (no matter what their social or job level may be).

Speaking of communication, can you be a good communicator if you don't speak the language of your adopted country? Yes, you definitely can. Being a good communicator has more to do with style and attitude than with foreign language skills. Nevertheless, don't discount the value and desirability of being able to communicate in the local language—both in personal and business settings. Fluency in, or even a strong working knowledge of, the language of the country is a tremendous asset in getting a foreign assignment, doing well professionally, and thriving personally. More important, it is an invaluable tool for opening a window on the culture and truly understanding the people of that country.

Fluency (or even competency) in a foreign language is not, however, required for you to be successful in your foreign assignment. Great linguists aren't necessarily good managers or communicators, and they are not necessarily the highest performers. Being able to speak a foreign language will never be the number one prerequisite for a foreign position (unless it is a job as a translator or linguist), but it will give you a leg up. It can separate you from the other applicants and, together with your other qualifications, make you uniquely qualified for the position. It will earn you respect from your foreign colleagues and enhance your ability to develop the close personal relationships that are so important for success in international business.

I have attempted several times to learn a foreign language, and have at times become proficient at a "street level." I have also found a way to get along in countries where I didn't speak the language at all. I was able to do this partly because English is now so widely spoken (particularly in my chosen industry—computers) and partly because of my own attitude and style of dealing with people. Still, as a person who has never mastered a foreign language, I admit that I wish that I had. You will be more effective in your job and get more out of your foreign experience if you have, or develop, an ability to communicate in the local language of your foreign assignment.

2. Adaptability/Flexibility

This may be the most obvious of personality traits required in a foreign assignment, and having lived in seven different countries, it is one with which I have had a lot of practice. Every time that I've moved to a new country, I have encountered new behaviors, ways of thinking, perceptions, social customs, foods, styles of dress, housing standards, weather, business practices—the list is practically endless. Every country is different in these respects, and you will often find significant differences even between regions or cities in the same country (for example, I found Frankfurt and Munich to be different in many significant respects, even though they are only a few hours apart by train).

My first assignment was in Saudi Arabia, a country traditionally viewed as difficult for Westerners and certainly not a place someone would choose for relocation in today's world without considerable thought. Nevertheless, I had a relatively easy time of it there because I started out living in a walled compound occupied primarily by Westerners. I thus could adapt at my own pace, ultimately moving into an apartment in downtown Jeddah. By the time I moved to Singapore, my seventh country, I was so used to adapting that I didn't even notice that it was different. True, many Asians describe Singapore as "Asia Light," in the sense that it has become very Westernized, so this was undoubtedly a factor in my ease of adjustment as well.

Surprisingly, Germany was the country in which I had the most difficult adjustment. I finally realized that this was because Germans generally tend to be very rule-oriented (this is a generalization, not a stereotype). They obey laws and regulations to the extreme, even in the small details of their personal lives. For example, I found it amazing that most Germans actually obey Walk/Don't Walk signs, whether or not a vehicle is approaching. As a New Yorker, I have always viewed these signs as merely advisory. In all aspects of business and personal life, we Americans generally want to find a reason that a law doesn't apply or that we are an exception to it. We are willing to push the envelope and see what happens. Germans, on the other hand, tend to give the strictest interpretation possible to laws and regulations, and are loath to do anything that even *might* constitute a violation.

One of my friends had a very different experience with German rules. She and her husband had stopped at a nice spot along the autobahn and were sunbathing on the side of the road (in their bathing suits). A policeman came by and was aghast that they would do anything so rude and vulgar. Yet I distinctly remember innocently walking through a park one nice day in Munich and stumbling across a German couple sunbathing *totally naked*! (Remember, I grew up in

Kansas, and we don't do that sort of thing in Kansas.) Apparently, this was perfectly acceptable behavior although sunbathing in bathing suits alongside the autobahn was not.

As the managing director for our German operation, I had to make a significant adjustment in my own style of doing business to adapt to this German adherence to rules (although I never did quite figure out where one was allowed to take off one's clothes and where one was not).

Another example of having to adjust to cultural differences is my experience in France—a favorite whipping boy for American business people long before the U.S. government invented freedom fries. The book on the French is that they don't come into the office until 10:00 A.M., and then take a two-hour lunch every day, leave the office by 5:00 P.M., and have more combined vacation and holidays each year than total workdays. Of course, this is one of those stereotypes that is based on an exaggerated version of the truth. There is no doubt that the French have a different work ethic—and view of the importance of their jobs relative to the rest of their lives—than do most Americans.

The typical American tendency upon arriving for an assignment in France is to immediately try to transform the French employees into pseudo-Americans. I have learned that this transformation just isn't going to happen: I have to adjust myself to their norms. Once I relaxed and accepted this state of affairs as the way it is, everything was okay and I got along fine with the French employees. Besides, the French think we are all crazy with our workaholic hours, dedication to our jobs, and a mere two weeks' vacation per year. If you really think about it, maybe they're right.

Throughout my years of experience as an American expat, I have developed the following attitudes about being flexible and adaptable when I am living in another country:

- It is their country, not mine.
- They have the right to run it the way they want to.

- Things aren't *always* better in the United States.
- Most people in the world don't actually want to be just like Americans.
- If I try their approach, I might like it.

I recommend that new American expats adopt these attitudes to help them adjust to a new country and get through the rough spots of acclimation. Sometimes I wonder if this list shouldn't be printed on a card and carried by every expat, to be pulled out in those times of stress and frustration that we all experience.

3. Openness

By this, I mean *not* being arrogant but instead being open to new and different ideas—the willingness to listen to your foreign colleagues and consider what they have to say. After all, they have lived in the country all their lives, and someone in your company made a decision to hire them. So why not listen to them? Unfortunately, this isn't always the slam-dunk that it should be. When I was newly installed as the managing director for the U.K. subsidiary of a U.S. software company, I received an e-mail directive from the global vice president of marketing stipulating that all advertising campaigns would be developed by the corporate headquarters' marketing group, and that my U.K. marketing team should merely change the spellings where necessary and roll them out in the United Kingdom.

I knew, and my U.K. marketing director certainly knew, that this was a recipe for disaster. U.S.-style, in-your-face advertising campaigns have never been known for their subtlety and dry humor. They tend to be viewed as over-the-top by your average Brit, who appreciates a more restrained message. It took much more debate than it should have, and ultimately required going over the head of the marketing VP, to win the day on this point. The outcome was an agreement that what we really needed in our marketing campaigns was a balance between global consistency and locally effective content.

However, the process for reaching this conclusion was a time-consuming distraction that we did not need.

4. Tolerance/Patience

You will encounter in every country a variety of different practices and customs that are different from those you are used to in the United States. People in other countries have different religious rituals and practices, different standards of cleanliness and hygiene, different eating styles and views of what is considered edible, and different bathroom and grooming habits. You must refrain from denigrating or looking down on these customs and habits merely because they are different. Our customs and habits seem natural only because they are what we have always done. To people from other countries, a lot of the things that we regularly do seem disgusting—for example, blowing our noses into a dirty handkerchief and then putting it back into our pocket, and sitting in dirty bath water. If you really think about these and other habits of Americans, they're not all exemplary. The point is that as an expat you have to accept these differences and allow people to observe and practice their own customs, rituals, and habits without prejudice from you.

A related personality trait that is particularly difficult for Americans to embrace is that of patience. I was fortunate in having Saudi Arabia as my first foreign assignment, because there is no better place in the world to learn patience. There is an Arabic phrase, *Bukrah Insh'allah*, which means "tomorrow, God willing." This phrase seems to be the answer to most questions asked in Saudi Arabia. Saudis (and people in most other countries) simply do not have the American sense of urgency about getting things done *right now*. Getting angry or upset doesn't change this, so I had to learn to be patient and work within the system. After four years in Saudi Arabia, I am proud to say that patience has never been an issue for me again, which is good because I've found that it is a necessary trait in all my foreign assignments.

5. Sense of Humor

A lot of things are going to go wrong in your life as an expat, perhaps starting with your first flight to your new country and probably continuing until you get back home. Flights will be delayed. Hotel reservations will be wrong. Important items will go missing. The wrong items will be delivered. Deadlines will be missed. People will be at the wrong place at the wrong time. You will misunderstand what people say to you. You will get lost. You will make mistakes. You will feel silly. You will feel frustrated. You may feel like packing up and going home.

It is the nature of life as an expat that more things will go wrong and fewer things will go right than you were used to back home. You have basically two choices. You can allow these to build up your stress levels to the point at which you are no longer enjoying your expat experience and you have become an ineffective employee. Or you can see the humor in all of this and become like a duck with water rolling off its back.

Since you can never change anything that has already happened, it is far better to just laugh at the situation and move on. Share your frustrations and problems with other expats. They will have their own to share with you, and you can all have a good laugh together.

6. Humility

It is also important that you be willing to admit your mistakes, and to laugh at yourself as well as the situations in which you find yourself. You will make many mistakes as you learn about your new country. No matter how hard you try do things right, you will commit numerous faux pas. You will find yourself in moments of embarrassing silence while everyone around you gets that open-mouthed look of puzzlement at something you have just done or said. I have found that the best recovery from these situations is often to openly laugh at myself and attribute my actions to being just another ignorant American. This seems to be something that all foreigners can relate to, so they laugh with me, and the moment of embarrassment passes. People the world over appreciate a person who has humility and is willing to laugh at his or her own mistakes.

7. Creativity/Resourcefulness

You will find in your foreign assignment that you usually don't have the support structure and backup resources available to you that you were used to in the United States. Many U.S. companies operate their foreign operations on a tight budget with funding and resources allocated on the basis of the amount of revenue that the foreign office is generating. Foreign operations often take a while to get started and to generate stable revenue flows and profits for the parent company. Therefore, you might find yourself working in an office that is operating on a limited budget. Your budgets for marketing, personnel, travel, office rent, and other important items may be a fraction of what you believe they should be. In these situations, it helps to be creative and resourceful, and it helps to be a good negotiator. You will learn to "borrow" and leverage resources from your parent company and from your colleagues in other foreign offices.

When I was managing director for the new Singapore office of a U.S. company, I went out of my way to develop a good relationship with the vice president of marketing and did not hesitate to ask him to help me out by funding needed Asia-Pacific marketing campaigns from his budget. I also borrowed service consultants from the vice president of consulting services for short-term customer projects, so that I could perform these projects without adding to my permanent local personnel costs. I leveraged the marketing campaigns developed by our U.K. office because, with minor tweaking, they were appropriate for Singapore and Australia as well. I thus didn't have to pay creative and setup costs out of my budget.

In every country in which I have been based, I have also paid particular attention to the ideas of local employees. They know how to get things done in their country, how to save money, how to source locally, and how to negotiate with the local vendors. They will be a great source of creative thinking for you if you let them.

8. Decisiveness

It is human nature to become indecisive in unfamiliar situations, and I have seen many expats back off from making decisions that they were being paid to make. They became uncertain and hesitant, sought too many backup opinions, and tried to defer responsibility for decision making to someone else (usually someone back in the corporate headquarters). The problem with this is that the people back in the corporate headquarters usually aren't in the same time zone, they don't know the local situation, and they aren't as qualified as the expat to make the right decision. It is very important as an expat that you be willing to take responsibility for making decisions, enjoy the rewards of making good decisions, and live with the consequences of making bad ones. That is why you have been placed in the position you're in and been given the opportunity you've been given. Seize it while you can.

9. Commitment/Perseverance

Sometime very early in your first expat experience, you will likely become so frustrated that you want to go home. There is nothing wrong with this feeling. It happens to most expats on their first assignment, and it will happen periodically throughout the ebbs and flows of your expat career. It is a natural human reaction and an easy answer to the often overwhelming challenges and problems that you will encounter in adjusting to your life in a new country.

However, it would be wrong to actually do it—go home, that is. When this feeling occurs you need to revisit that plan that you developed for yourself, see if you are on track, and figure out how to make adjustments if you aren't. In the cold light of rational analysis, you will almost certainly realize that things are not nearly as bleak as they seem. By all means you must stay the course, because this feeling of frustration and homesickness is a temporary phase that you will get through. Do not allow yourself to go home a failure. You must have a commitment to fulfilling your personal plan for yourself, and to fulfilling the expectations of those who are relying on you.

10. Independence/Self-Reliance

When you make a decision to move to a foreign country, either alone or with your family, you are making a decision that will cut you off (at least partially) from your existing personal and professional support network. There are ways to alleviate this problem (many of which are discussed later in this book), and you and your family members will create new support networks in your new country. But they will not provide the same level of support as you enjoyed back home, at least in the beginning. Allow some time for these to develop.

To succeed as an expat, you need to be comfortable with the idea of independence and self-reliance. You will likely be spending more time alone than you are used to. You will not have ready access to those friends and professional advisors to whom you have traditionally turned in times of trouble or crisis. You will find yourself needing to solve problems for yourself, which is something that not everyone is accustomed to doing.

The reverse side of this is that, in time, you will indeed create a new circle of friends and new support networks, both among other expats and local citizens. My experience has been that these relationships can become very strong. Many will remain an important part of your life long after you have completed your foreign assignment and either moved on to the next one or returned home.

Along with considering and evaluating yourself with respect to these 10 personality traits, you also need to evaluate the country you are considering. You can have all the right stuff, but this may do you little good if the nationals of that country will not give you a fair opportunity to succeed. Attitudes are changing for the better throughout most of the world, and I personally do not believe that a person's gender, race, religion, or ethnicity is relevant today to his or her likelihood of success in the vast majority of countries. However, there are still certain countries where one or more of these may impede or prevent you from achieving your employer's goals and your own.

I am all for aggressively and confidently going out into the world and working to change attitudes. Still, you need to understand what you are up against, and be prepared to work within certain cultural rules and customs no matter how distasteful or demeaning they may seem. You won't likely be able to change them to any significant extent while you are there, although I do believe that every small contact can be a positive step in this direction. If your goal is to be a crusader and shake things up, please stay home. Or shake things up on your own time instead of your employer's. You owe it to your company to focus your efforts on its business and not allow this to be disrupted or relegated to a secondary role by your personal agenda.

Evaluate Yourself?

Now would be a good time to take stock of the issues discussed in this chapter, and evaluate yourself and your own situation in light of the various personality traits and other factors discussed above. If you are considering a specific foreign position, evaluate your suitability for this position from the standpoint of your current job skills and knowledge. Then go through each of the 10 personality traits required of the new American expat and honestly evaluate whether this is a strength or weakness for you. Determine whether you indeed have the right stuff already or whether you have some work to do in certain areas.

Many of the books on expat life provide personality tests to assist you in this—career assessment/evaluation tools especially designed for the expat candidate. They have lists of questions, rating systems, and charts to assist you in evaluating your suitability for an expat position. These are very interesting. However, I don't believe it is necessary to take a test to self-evaluate your suitability for a foreign assignment. And more important, I question how accurately any such test can be as a predictor of your likelihood for success or failure. As I was looking through one of these tests, it occurred to me that, assuming I had taken it prior to my first expat position and answered the questions honestly, I almost certainly would have flunked. I thus might have returned to Kansas and spent the rest of my life wondering what I had missed. This is because *at that time* I didn't have the right stuff. I wasn't even close to possessing the personality traits that I describe here as being necessary to succeed as an expat. But I developed them very quickly. So can you.

If you enjoy taking these kinds of tests, by all means take one. You can find several of them in the books that I list in appendix A. However, please do not consider any negative results as a factor in your decision as to whether you can be successful in a foreign assignment. Don't rely on any external opinion—whether it comes from another person or from a test—to tell you that you don't have the personality characteristics required to succeed as an expat. Just take it as an indication of how much work you have to do to get where you need to be. You may not have the right stuff right now, or you may not know whether you have it, but I promise that you can develop it if you are determined to do so.

Chapter 1 **Summary** and **Checklist**

Evaluating a Prospective Foreign Position

1. Understand and articulate your reasons for desiring to move overseas and the goals you want to achieve.
2. Understand your company's reasons for moving you overseas, its goals, and the role it expects you to play in achieving these.
3. Determine whether your goals and your company's goals are compatible and achievable and whether they are supported by sufficient commitment and resources from your company.
4. Develop an Expat Success Plan covering your career, finances, family, and personal development.
5. Evaluate whether you have the requisite job-related skills and knowledge to be successful in the expat position. You are being sent abroad to *do,* not to learn.
6. Evaluate where you currently stand in relation to the 10 expat personality traits required of the new American expat (the "right stuff") to be successful overseas.

Chapter Two

Finding Your Job Overseas

Finding a foreign job can require significant research, networking, searching, and interviewing. This chapter explains how to go about finding your overseas job, either with your current company or with a new employer.

Trends in Expat Hiring

Before embarking on your foreign job search, it is important to understand the recent trends in expatriate hiring practices of U.S. companies. As stated previously, there has been a move toward shorter assignments, with a majority now lasting for one year or less. Many expatriate assignments are now focused on achieving short-term tactical objectives or skill transfer, with an ensuing transition to locally hired employees. Also, there is a higher percentage of single men, and a lower percentage of families with children, among the U.S. expatriate population than ever before. Lastly, compensation packages and benefits made available to expatriate employees have been cut back by many employers. As you would probably surmise, these current trends are attributable to cost-reduction measures by U.S. employers, something to keep in mind as you search for an overseas position.

Under Your Nose: Being Transferred by Your Current Employer

Without a doubt, one of the best places to look for a foreign assignment is with your current employer. If your company has either international operations or plans to develop them, you could have immediate value in this area because of your knowledge regarding the company, your specific job skills, your relationships with other employees, and your ability to transfer this knowledge. You are a known quantity and a known performer.

Staying with your current employer is an ideal situation from your standpoint as well. Many companies will not send a new employee to work in an overseas operation, preferring that credentials be established in a U.S.-based position first. By staying with your current employer and transferring, you can avoid having to go through this start-up period. Your knowledge of the company and your relationships with U.S. personnel will have immediate value to the company's foreign operation, making for a smoother transition. You also are more likely than a new hire to make a successful return to the U.S. company upon completion of your foreign assignment.

Research your company's international business and foreign operations. Be able to discuss them intelligently on an operational basis and to talk current and historical numbers (i.e., revenue, expenses, profit margins, head count, revenue per head count, trends, and comparisons with U.S. operation). Develop an understanding of the company's goals for its international operations and the challenges and problems it faces. Network as much as possible—both with the U.S. staff involved in supporting the international operations and with the foreign staff when they visit the United States. The foregoing will enable you to understand where you can add value to your company's international operations and position yourself for an opportunity when it becomes available. Just as important, it establishes you as a person who is interested in working overseas. Your interest will be remembered by the company's management when there is a need in one of the foreign offices for the expertise or management skills that you are qualified to provide.

Make yourself available to assist on international projects (but only with the approval of your current supervisor). If an opportunity arises to make even a short trip to a foreign office (e.g., for a training session or to work on a specific project), grab it. While you are there, establish relationships with the foreign staff. Always be looking for ways that you can add value and fill a need.

When a position becomes available working in your company's international operations, utilize the knowledge and relationships you have developed to position yourself as the ideal candidate. However, don't be afraid to define and propose such a position yourself. If you see a problem that you can solve or a need that you can fill, don't wait for someone else to see it. Build your proposal and present it.

This approach worked well for me on two of my foreign assignments. In the first, I was sent to our European headquarters in my capacity as international general counsel for one month to help close the quarter. I didn't return home after that, except to sell my house and pack, and I eventually became the managing director for the United Kingdom and the Netherlands offices for this company. All of this was the result of my seizing the one-month opportunity to build relationships with the European staff and to identify a specific longer-term role there in which I could add value.

In the second, I had a U.S.-based position in a company that was acquired by a larger company. I was probably headed for redundancy in that position. The acquiring company had mature international operations, with foreign offices in seven countries. I researched the company's international operations, identified areas in which I could add value, and presented my proposal to the vice president of international operations. I was transferred to a new position overseas one month later.

My friend Ted had a similar path to his first foreign assignment. He transferred into the international division of his employer, but was initially based in the United States. After working there for several years, the entire division was transferred to Switzerland with Ted as a top officer. More than 10 years later, he is still living in Switzerland as a permanent resident (although he has moved on to a new U.S employer).

Student Options for Foreign Jobs

Students and recent graduates have a wealth of opportunities for working overseas through internships with private companies, government agencies, and educational institutions. These positions are short term (generally from two months and up) and ordinarily don't pay well, if at all. However, they can be invaluable in planting the seeds of your international network. They also provide an opportunity to improve your foreign language skills and knowledge of another country, and to establish your credentials for future positions. In numerous countries, a specific short-term student work permit is available for U.S. students who meet the requirements.

Not all internships related to international issues or business involve working overseas. Even if they are U.S.-based, however, they can be a valuable way to open doors and establish your credentials for a subsequent foreign position.

The placement or career office of your own college or university is a good place to start in your search for an internship. There you'll find information on the various independent intern programs described in the following section and also on intern or exchange programs that the school sponsors or is formally affiliated with.

International Intern and Exchange Agencies

What follows is a list of several independent intern, training, and exchange agencies that specialize in assisting students and graduates in finding short-term foreign positions. These agencies are a good gateway to a foreign internship because they have their own network of potential employers, as well as databases of current work opportunities, to assist in your search. Just as important, they handle your work permit documentation, thus providing an easy way for you to legally enter and work in a foreign country. They also will assist in your preparation for departure, help you find housing in the foreign country, and provide in-country orientation and support services. However, you must start your search process with these agencies well in advance of the time you want to work overseas:

- Council on International Education Exchange (www.ciee.org). This organization has been in existence since 1947 and can assist you in finding internships in Australia, New Zealand, Canada, France, Ireland, and Germany.
- Association for International Practical Training (www.aipt.org). AIPT focuses on technical fields such as engineering, computer science, mathematics, the natural and physical sciences, architecture, and agricultural science. It operates in more than 70 countries.
- AIESEC (www.aiesec.org). This organization has a network that covers more than 80 countries. It focuses on three different types of internships: business management, technical (primarily information technology), and community development.
- BUNAC (www.bunac.org). BUNAC can assist you in finding internships and obtaining work authorization in the United Kingdom, Ireland, Australia, New Zealand, and Canada.
- International Association for the Exchange of Students for Technical Experience (www.iaeste.org). IAESTE was founded in 1948 and operates in more than 80 countries. It focuses on assisting students and recent graduates in finding internships in engineering, the natural sciences, computer science, mathematics, and many other technical disciplines.

- International Cooperative Education Program (www.icemenlo.com). The ICE Program is not restricted to any specific disciplines or countries. It operates globally and in a variety of fields and has placed students in more than 15,000 internships since its founding in 1971.

Information on these organizations' programs, application processes, and fees can obtained by visiting their respective websites.

Internships at Government Agencies and International Organizations

High-quality internships relating to international issues or business are also available at many government agencies and international organizations. The competition for these positions is intense, and applications must be submitted early. Some of the more interesting international-related internships are available at the following agencies:

- U.S. State Department (www.careers.state.gov)
- Central Intelligence Agency (www.cia.gov)
- National Security Agency (www.nsa.gov)
- International Finance Corporation (www.ifc.org)
- International Monetary Fund (www.imf.org)
- World Bank (www.worldbank.org)
- Agency for International Development (www.usaid.gov)
- U.S. Department of Commerce (www.commerce.gov)
- Peace Corps (www.peacecorps.gov). Although not a traditional internship, a two-year position with the Peace Corps is an excellent way to perfect language skills, become knowledgeable about a country, and give yourself international credibility for future job applications. In addition, you will be doing something good for the country in which you are based and for the United States.

- United Nations. Each organization within the United Nations has its own website and many of them offer internship opportunities. You can access all UN organizations through www.unsystem.org.

Information on internship programs for these agencies and organizations can be obtained directly from their websites or through the portal www.studentjobs.gov cosponsored by the U.S. Office of Personnel Management and the U.S. Department of Education. The portal contains links to the relevant student-intern pages of the websites for all of the previously mentioned internships, as well as many other federal government agencies and international organizations.

Internships with Companies

Internships with companies are also available. These opportunities are not limited to large multinational companies. I hired (and paid) a summer intern when I was the vice president of international sales at a company with less than US$50 million in revenue per annum. The intern was a business major who spoke fluent German. My company was in the process of establishing a network of resellers throughout Europe, including in Germany. The intern was thus able to achieve his goal of learning about international business while at the same time playing a significant role in helping us achieve our business goals.

Information on internships with selected companies will be available at your college or university placement or careers office. If you don't find one there that interests you, don't hesitate to contact a company directly and promote yourself for an intern position. Use your network (or your parents' network) to open doors. Search for intern programs for specific companies or industries on the Internet. Check with professional or trade organizations in your field to find out if they sponsor or have information on internships. For example, the American Bar Association has on its website a 42-page listing of foreign summer programs for law students, many of which include internship opportunities.

In positioning yourself for an internship with a company, it is important to understand as much as possible about the company and its international business and to establish that you are a person who will be able to help the company achieve its international goals (if not immediately, then in the future). Even though you are currently seeking only a short-term internship, you may well be interviewing for a full-time position upon graduation that will hopefully result in your being transferred overseas. You thus should follow the steps described under "Application Process" on page 46 and attempt to position yourself as a qualified candidate for a future foreign assignment.

You might not work in a foreign country during the internship itself (unless you are already there in connection with classes, or you possess needed foreign language skills). Nevertheless, your intern experience will be a valuable stepping-stone to your first foreign assignment.

Additional Information on International Internships

Two websites that contain good information on student positions, internships, and other short-term foreign jobs are www.backdoorjobs.com and www.transitionsabroad.com. Also, the *Directory of International Internships* (Michigan State University, 2004) is a good resource. It lists internships sponsored by educational institutions, government agencies, and private organizations and is updated regularly. It can be ordered at www.isp.msu.edu (click on *Study Abroad—Internships—Internship Websites and Resources*). Several good books on intern opportunities in foreign countries are listed in appendix C, and many others can be found through online booksellers.

Opportunities for Teaching Overseas

There are two basic types of opportunities for teaching in foreign countries: teaching at international schools and teaching English as a second language. Both are fulfilling in and of themselves. If desired, they can also be a gateway or stepping-stone to another position in the foreign country. Many Americans take an initial job teaching English as a second language to students in a foreign country and (through networking) eventually obtain a job in their chosen field.

Teaching in International Schools

There are nearly 1,000 private schools throughout the world that provide elementary and secondary education based on the U.S. educational system. These cater to the children of expats from the United States and other countries. Some of them are sponsored and supported by the U.S. government. The U.S. Department of Defense also operates approximately 150 primary and secondary schools at military bases around the world. Many of the teachers who work at these schools are certified teachers from the U.S.

One of the best ways to find a teaching position at one of these schools is through organizations that support and staff these schools. These include the Association of American Schools in South America (www.aassa.com), the European Council of International Schools (www.ecis.org), International Schools Services (www.iss.edu), and Search Associates (www.search-associates.com). Several of them sponsor recruitment fairs for teachers and administrators in various locations throughout the United States.

Two valuable websites to visit for information on international schools and teaching overseas are: www.overseasdigest.com (which has a comprehensive list of international schools) and www.transitionsabroad.com. Information regarding international schools that are sponsored by the U.S. government is available at www.state.gov (click on *History, Education and Culture—Overseas Schools*).

Teaching English as a Second Language

Teaching English as a second language is a booming business as a result of the global interest in learning English, which has become the default language for international commerce, technology, aviation, education, and diplomacy. The market for English teachers has expanded even more with the many countries that opened their doors to Western commerce in the latter part of the twentieth century.

The two requirements that you must meet in order to teach English as a second language are to be a native-English speaker and to have a college degree. However, you can also obtain training and a certification from schools and agencies that specialize in this area. Doing so is increasingly becoming a good idea in view of the intense competition for positions in the most desirable countries. These programs are variously called TESL Certification (Teaching English as a Second Language), TEFL Certification (Teaching English as a Foreign Language), and other similar names. A basic certificate can be obtained through a one-month program at a specialized TESL training agency. A master's degree can be obtained through a one-year college or university program.

Three websites for information on TESL jobs and certification are www.eslcafe.com, www.english-international.com, and www.tefl.com. These will get you started and provide links to the websites of many other organizations specializing in TESL. Graduate programs where master's degrees in TESL can be obtained are listed at www.gradschools.com.

Exchange Programs for Scholars, Teachers, and Administrators

Short-term teaching assignments and exchange programs are also available for U.S. scholars, teachers, and administrators. Educational institutions in other countries often seek out professors, teachers, and administrators from the United States. Sometimes hiring is done as part of an official exchange program with a specific U.S. educational institution, or to bring in a specific expertise that a U.S.-educated individual possesses. These types of positions can last for a semester or for several years—or they can be for an indefinite duration. Even a job of short duration can be a good way to see how you like living in a particular country, and then serve as a stepping-stone to a more permanent position. Information on these programs can be obtained from the Fulbright Teacher and Administrator Exchange Program (www.fulbrightexchanges.org) and the Council for International Exchange of Scholars (www.cies.org), as well as at the teaching-exchange portal www.theeducationportal.com (click on *Employment—Exchange Programs*).

Finding a New Job Overseas

Who Hires American Expats?

Obviously, U.S. companies that have (or wish to develop) foreign operations or international business are your most likely source for a foreign-based job. However, foreign companies that have operations in, or do business with, the United States may also have attractive positions outside the U.S that are open to American expats. Indeed, they may have some positions for which an American is even preferred (e.g., because of a perceived need for someone having specific U.S. market knowledge or a U.S.-based experience or certification). Many departments and agencies within the U.S. federal government, as well as international organizations such as the United Nations, hire U.S. citizens for overseas positions. Nonprofit and public organizations, particularly those involved in providing services or aid to people in other countries, also are a possible source for a foreign assignment.

Where Do You Want to Go?

Before you start your search, you need to determine whether there is any specific country or region to which you want to move. Are you simply interested in a foreign assignment because of the various potential benefits discussed in Chapter1? If so, the entire world is a possibility for your expat job. However, you might have a specific country or countries in which you want to work because of your foreign language skills (or lack thereof), family background or connections, love of a particular country, or even a particular interest of yours (such as skiing or scuba diving). These considerations will affect the way you approach finding your foreign position and will make your search much more focused.

How Do You Find a Foreign Job?

There are numerous sources today for finding a position in a foreign country. Many of them are available to you through the Internet.

Internet Job Sites

A good starting point is the wide range of Internet job sites that are available—specifically those that specialize in international positions. These include such well-known sites as www.monster.com (click on *Overseas Jobs* from Site Map), www.quintcareers.com (click on *Geographic Specific Jobs* from Detailed Site Map), www.escapeartist.com, www. rileyguide.com (click on *Job Listings—International Job Resources*), and www.overseasjobs.com. You can create search parameters on some of these sites that define the countries, industries, types of positions, titles, compensation range, and/or other variables to limit your search results to jobs that will be of interest to you. These sites also contain links to numerous other foreign job websites, many of them specific to a particular country or region. You also can find Internet job sites that specialize in specific types of positions or industries, thus helping you to focus your search.

Listings of international jobs are also found in international job bulletins, such as the *International Employment Gazette* (www.intemployment .com), and the *International Employment Hotline* and *International Career Employment Weekly* (both available from the International Jobs Center at www.internationaljobs.org). These job bulletins specialize in international positions and are updated regularly. A subscription fee is charged for these, so you will want to check them carefully to make sure that they are of value to you.

Search Firms and Recruiting Agencies

These can be a fertile source for international positions. However, it can be difficult to get the attention of these firms because of the thousands of résumés that they receive. They typically take assignments only from employers seeking to fill a particular position, not the other way around. As part of their search process, however, available positions are often described on their websites and interested applicants are invited to submit résumés.

There are literally thousands of search firms in the United States, ranging from small niche firms specializing in specific industries or geographic areas to multinational firms with offices in numerous countries. Many recruiting and search firms focus only a particular industry or type of position: if you can find a firm that specializes in your field and/or industry, this will assist you in limiting your search to positions that will interest you. Do an Internet search for a search firm that specializes in international business or in a specific industry or field. You can conduct such a search using "search firm," followed by "international," and then by a word or combination of words describing the field, industry, and/or country or region in which you are interested. For example, when I conduct a search using the string: "*search firm*" "*international*" "*lawyer,*" I find 1,539 possible entries for search firms that have at least a partial focus on international legal positions. When I add the word "Europe" to this string, the number of entries is reduced to a more focused 550. I could further limit the search results by limiting it to specific countries or to a specific industry or legal specialization (such as "*technology*").

The Kennedy Directory of Executive Recruiters (www.kennedyinfo.com) is another resource for identifying search firms and recruiting agencies. This directory lists more than 5,600 search firms in the United States, Canada, and Mexico, categorizing them by industry, geographic, and job-function specializations (including contact information for foreign offices). The international version of the directory lists 2,500 firms in more than 80 countries. These are somewhat pricey for individuals, but are available at some public and university libraries. A "personal edition" is available for US$50, which provides scaled-down contact and categorization information for firms in the U.S., Canada, and Mexico. The Kennedy website also allows you to conduct an online search for recruiters (categorized by function, industry, and country) and to purchase the results of such a search for a per-contact fee.

Listings of and links to search firms specializing in international positions also are available at the Internet job sites listed previously.

When approaching a search firm or recruiting agency, remember that they generally work only for companies that engage them to fill specific positions. They won't spend time on you unless you fit the criteria of a specific position they have been engaged to fill. However, they will never contact you about one of these positions if they don't know you exist. It is therefore worthwhile to submit your résumé to selected international recruiters in your industry and field. Most of these firms will allow you to submit speculative résumés (i.e., not relating to any specific position) through their websites. However, you are more likely to get their attention if you submit your résumé in response to a specific position that they are attempting to fill for an employer. In either case, it is critical that you design your résumé carefully. Résumés are electronically loaded into candidate databases and then matched with the criteria of positions being filled on the basis of keyword searches. It is possible that no human being at any of these firms will see your résumé until it is pulled as a result of such a search. You must therefore make sure that all of the key words that are relevant to your desired position, industry, qualifications, and any other important criteria are included in your résumé.

You also can find agencies that will market you to potential employers for a fee. Often, these agencies merely do a large e-mail blast to a specific industry group or other database of companies that might be interested in someone with your qualifications. Others will provide a more personalized service and circulate your résumé to named executives at targeted companies. They will then follow up and attempt to create an interest in you, either by trying to slot you into a position that is actively being recruited or by creating a new position that didn't previously exist. In general, I do not recommend paying any agency an up-front fee to market you or find a job for you.

Directories of U.S. and Foreign Companies

Various directories and online information sources are available that provide information on U.S. companies with international operations and on foreign companies with U.S. operations. They don't provide information regarding specific job openings, but they can assist you in targeting potential employers. All of these directories and services are available for purchase in a variety of formats: print, CD, and/or online access. These can be expensive, but are available at many public and university libraries. Some of the better known and useful directories and online information sources are as follows:

- *Directory of American Firms Operating in Foreign Countries,* published by Uniworld Business Publications (www.uniworldbp.com). This directory provides information on branches, affiliates, and subsidiaries of U.S. companies in foreign countries. It is available in print or CD format. You also can conduct an online parameterized search from the Uniworld website and obtain specific information on U.S. companies in your state that have foreign operations or information on U.S. companies that have operations in a specific country or region.
- *Directory of Foreign Firms Operating in the United States*, also published by Uniworld Business Publications. This directory provides the same information as the one just discussed, but in reverse. Online searches can be conducted from the Uniworld website to obtain specific information on foreign companies with operations in your state, or information on foreign companies from specific countries or regions that have U.S. operations.
- *America's Corporate Families and International Affiliates*, published by Dun & Bradstreet (www.dnb.com). This directory provides information on more than 12,000 U.S. companies and lists their branches, divisions, and subsidiaries.
- If you can gain access to *Hoover's, Online*, *Harris InfoSource*, or *Lexis-Nexis*, you can search their databases of company information. These online services are targeted at corporate

customers, but they are also available at many libraries for online research by individuals.

- Another valuable research tool that I have used in the past is the membership directory for my local international or world trade club. Many cities and regions have such organizations, which focus on international trade issues. These organizations attract as members both U.S. and foreign companies that are involved in international business, and they publish membership directories with information on these companies and contact details. The Federation of International Trade Associations (FITA) (www.fita.org) includes more than 180 such international or world trade clubs among its 450 members. Contact details for these clubs and links to their websites are available on the FITA website.
- Local chambers of commerce also have membership directories that often provide information about a company's foreign operations and, of course, bilateral chambers of commerce (such as the German American Business Association and British American Chamber of Commerce) can provide information on companies from each country that have operations in the other.
- Lastly, the U.S. International Trade Administration (a division of the U.S. Department of Commerce) operates U.S. and Foreign Commercial Services offices in more than 100 U.S. cities and at more than 80 U.S. embassies and consulates overseas. These offices can provide information about U.S. companies with operations in specific countries.

As you are conducting your search and creating a target list, at least consider starting out in a U.S.-based position. It is sometimes difficult to get hired for an immediate foreign assignment. Companies prefer to send employees overseas who have knowledge regarding the home office procedures, policies, and relationships. Such employees can more effectively serve as the representative of the U.S. home office and as the conduit for the flow of information. Don't discount the possibility of starting with an American company in its U.S. headquarters operation (or in the U.S. office of a foreign company). You can then establish yourself and transfer overseas when a suitable position opens up. I used this approach for one of my expat positions, and I found that the time spent in the U.S. office greatly enhanced my value when I transferred overseas.

Government Agencies, International Organizations, and Nonprofit Organizations

There are numerous organizations that are heavily focused on international issues and/or business. They hire many Americans for foreign assignments. A position with one of these organizations can be both an extremely rewarding career in its own right and a networking tool for a future job with a company. A list of major U.S. federal agencies and international organizations involved in international issues or business, and contact information for each, was provided earlier in "Student Options for Foreign Jobs."

It is also possible to get started on your expat career with a volunteer position in your field. Many nonprofit organizations are focused on providing economic and business expertise in countries where it is desperately needed. For example, the Citizens Democracy Corps/MBA Enterprise Corps (www.cdc.org) seeks experienced business people (or MBA students/graduates) to mentor small-to-medium-sized businesses in a variety of countries that are classified as emerging markets. Volunteers for the International Executive Service Corps (www.iesc.org) have assisted businesses in more than 120 countries. The United Nations Volunteer Program (www.unv.org) provides similar assistance globally through volunteers. These programs utilize volunteers, so you won't be paid much other than expenses. However, if this is not a problem for you, it can provide a good way to learn about a country, to network, and to pass on your knowledge and experience where they're needed most.

Government agencies, international organizations, and nonprofit organizations often have political and social agendas. You need to understand these and make sure they are not inconsistent with your own.

Professional and Trade Associations

These associations can be a good resource both for networking and for information on possible foreign jobs. Their journals include articles about overseas activities and projects of members and may have references to specific opportunities for accredited professionals in your field. The associations themselves often have international committees or sections that focus on the special issues relevant to international business. You should become involved in selected associations that are relevant to your field. Don't just join and pay your annual dues: get actively involved in the work of the international committee and establish a reputation for yourself. This is a great way to learn about (and create) opportunities for an expat assignment.

Networking and Informational Interviews

As I have said, there is no more valuable way to find any job, including a foreign one, than by networking. Network within your own personal contacts. Network through professional organizations. Network at your current employer. Network with people you have worked with in previous jobs. Notwithstanding all of the Internet job sites, recruiting agencies, and other resources available to assist you in finding a job, many desirable positions are still filled the old-fashioned way—through personal contacts.

One specific kind of networking that can be particularly valuable is the informational interview. This involves identifying and setting up an appointment with a knowledgeable person who can answer questions for you, share ideas and point you toward other contacts and possible opportunities. Ideally, this person would be someone working in your field who has lived and worked overseas, perhaps in a specific country that interests you. Or it might be a person who is a native of that country and currently lives in the United States.

My experience has been that most people are willing to take time to talk to someone who approaches them politely and professionally with a request for an informational interview. The right person can be a veritable fountain of knowledge regarding living overseas, specific foreign countries, informational resources, contacts, and possible opportunities.

You should prepare carefully for your informational interview: Learn as much as you can about the person's background and company. Research any specific industries or countries in which you are interested so that you can discuss them intelligently. Prepare your questions ahead of time and stick to your agreed upon schedule and time limits. Most important, remember that this is an informational interview and not a job interview. Attempting to turn it into a job interview can have the result of making the person uncomfortable and thus less open in his or her discussions with you.

Go There Without a Job?

What about going to a foreign country to look for a job? This is actually not uncommon and can sometimes be successful. I have known several people who have used this approach, and some of them are still living overseas and enjoying it very much. If you are in the foreign country, it is easier to present yourself as a candidate and you have the ability to make a more convincing and powerful case for yourself. However, keep in mind two potential problems with the approach. First, this means you are entering the country without the necessary work visa or permit. In many countries, there is an unfortunate axiom that "you cannot get a work permit unless you have a job, but you cannot get a job unless you have a work permit." The challenges of getting documented are discussed in more detail later in this book, but for now I'll simply say that most countries (including the United States) do not have a policy of making it easy for foreigners to work within their borders. To do so potentially takes jobs away from their own citizens.

Research this issue carefully before you travel to a foreign country with the idea of finding a job after you get there. Every country has different rules and procedures regarding work permits for foreigners. You might get lucky and fall into a category in which foreign workers are actually sought due to a shortage of local personnel, or you might receive favorable treatment on account of your nationality or ancestry. You might also discover that it is impossible to obtain a work permit, or that it might take months or years to get one. Before you go, make sure you understand exactly how you will get work authorization, the procedures you will have to follow, and how long it will take.

The second potential problem with this approach relates to your compensation. If you are hired by a U.S. company, you are more likely to benefit from various special allowances and payments that are designed to assist you in your relocation and make up for any shortfall in local salaries in the foreign country. These are discussed in detail in the following chapter on expatriate compensation packages. Conversely, if you are hired by a local company in the foreign country, you will almost certainly be viewed as a "local hire" and will be compensated at the same level as the local employees in that position. You are also not likely to receive any financial assistance in your relocation to the country or upon your return to the United States. Depending on the country and your personal financial situation, this may not be a problem for you. However, you should consider the financial impact before you spend the time and money to travel to a foreign country for a job search. Research the standard compensation levels for your position and costs of living before you go.

Beware of Scams

There is no end to the scams that are perpetrated on unsuspecting job seekers, particularly on the Internet. People searching for jobs are hopeful and enthusiastic, and as a result they may become vulnerable and naïve. Unfortunately, scams are especially perilous in the case of foreign job opportunities, which involve unfamiliar and remote settings.

Be on the lookout for obvious red flags: demands for up-front money, the use of P.O. boxes instead of addresses, suspiciously attractive offers or promises, and/or a reluctance to meet with you in person. Do not pay money up-front, or provide credit card details, Social Security numbers, or any other personal information, to any company or person unless you are absolutely certain of their identity. Ask for and check references (but keep in mind that references also can be faked). Contact a Better Business Bureau or equivalent agency in the foreign country if there is one. You can also contact the U.S. Consulate for that country or that country's embassy in the United States. Inquire whether there are complaints about the company or agency and whether the story you are being told is possibly true or obviously suspect.

A damaging job scam that is making the rounds at the time of writing this book involves hiring people located in the United States to process payments via PayPal accounts for a foreign company. The unsuspecting victims responded to listings on well-known Internet job sites believing that they were getting in on the ground floor of a foreign company's expansion into the U.S. market. They thought that they were accepting payment for U.S. software sales and then wiring the funds to a foreign software vendor after deducting their commission. In fact, they were receiving money from stolen credit cards and were on the hook to repay PayPal for the fraudulently transferred funds. They were also potentially subject to criminal prosecution for their involvement in credit card theft. The victims had conducted all discussions with the foreign company via e-mail. In this way, they agreed to a scheme that involved using their own PayPal accounts to receive the funds and then wiring the funds to the foreign company. On reflection, this clearly was not a wise thing to do. However, they had relied on the fact these jobs were obtained through listings on well-known Internet job sites.

The Application Process—Getting Their Attention and Getting the Job

In order to position yourself for a foreign-based job, you must be able to articulate why you are an excellent fit for that employer, that job, and that country—and you must be able to do so more effectively than the other candidates for the job.

- What is it about you that makes you, rather than someone else, the ideal candidate?
- Do you have specific country or market knowledge?
- Do you have specific skills that are necessary for the job?
- Do you have specific international experience that is relevant for the position?
- Do you possess a relevant foreign language ability?
- Do you exhibit the personality characteristics required for success in an expatriate position?

In sum, how do you set yourself apart from the other candidates and present yourself as the perfect candidate for a specific expatriate position?

For my first foreign assignment in Saudi Arabia, I had no relevant language skills and only a very general knowledge about Saudi Arabia. I had never even visited the Middle East. However, I had a specific, specialized job skill—international business and commercial law—that was required for that particular job (plus I was single, had no children, and was very mobile). I was an attractive candidate for a position for which I had a specialized expertise. For a different position 15 years later, I could present myself as a person who had years of experience living and working in foreign countries and specific experience managing foreign subsidiaries of U.S. technology companies. Even though I had a wife and three children by this time, I was considered the most qualified person for the job. In each case, I was able to define and present something about my background, experience, or personal situation that convinced the company that I was the person they needed.

Start as early as possible to internationalize your pedigree and establish yourself as an excellent candidate for an expatriate position. Take classes on international studies or business. Study and learn about any countries in which you are interested—and visit them so that you know them personally. Develop a proficiency in the language spoken there. Try to obtain an internship or other short-term job overseas. Do anything you can to separate yourself from the masses of people who have no special knowledge or experience to offer a potential employer.

First Impressions: Internationalizing Your Résumé

Tailor your résumé and introductory letters to highlight your international qualifications. You want to catch the attention of a potential employer and be immediately perceived as a potential candidate for a foreign position. Emphasize the specific skills and knowledge that establish your suitability for a foreign position:

- Knowledge of the country or region
- Any relevant education
- Foreign language skills
- Prior international work experience
- Your experience traveling to or living in other countries
- Awards or commendations relating to your international experience

- Your successes in prior overseas positions
- Anything that demonstrates that you've got the right stuff for the job (as discussed in Chapter 1)

Interviewing: Establishing Yourself as the Most Qualified Candidate

Research the opportunity before your interview. Obviously, you want to be knowledgeable about the company and, in particular, its international operations. You also want to be informed about the specific country in which the position is to be based—the political situation, economy, market, people, culture, and so on. This is why an internship or other prior experience in the country is so valuable (even a few weeks spent in a foreign country can give you an advantage).

During your interview, you must once again establish your particular credentials for the foreign position. Discuss your special qualifications and your specific knowledge about the job, the country, and so forth. However, it is just as important to demonstrate that you understand the challenges of living and working overseas—and why you will be successful. Give examples of your personal strengths, emphasizing those that are important to international success. Ask questions that show you are tuned in to the specific challenges related to the foreign assignment. You will likely find that the interviewer won't be familiar with these and will instead focus primarily on the technical or job-related skills necessary for the position. By taking advantage of the opportunity to show a deeper understanding of the issues facing Americans living and working overseas, you can set yourself apart from other candidates.

Chapter 2 **Summary** and **Checklist**

Finding and Getting Your Foreign Job

1. Explore opportunities with your current employer. The best, fastest, and most likely path to an expat position is often with your current employer. Research your company's international business. Network with the international staff. Identify and propose a position in the company's foreign operations in which you can add significant value.
2. Consider a student internship. A student internship or similar short-term position can be a good way to start your expat career; these types of positions are available through numerous agencies and companies.
3. Look into teaching. The teaching profession provides a number of ways to obtain an expat position: teaching in one of the 1,000+ overseas elementary and secondary schools catering to children of U.S. expats, teaching English as a second language in a foreign country, and scholar and teacher exchange programs.
4. Use the full range of resources that are available to assist you in finding an overseas job with a new employer: Internet job sites; international job bulletins; international search firms; directories of U.S. and foreign companies involved in international business; membership directories of local international trade clubs and chambers of commerce; professional and trade associations in your field; and the employment opportunities pages on websites

of government agencies, international organizations, and companies you have targeted.

5. Emphasize networking and informational interviews as part of your search process.
6. Be prepared to start in a U.S.-based position with the goal of transferring overseas after you have established yourself with the company, the job, and the people.
7. Remember that moving to a foreign country with the intent to find a job after you get there can be successful, but it raises potential problems regarding work permits and compensation that you need to be aware of before you go.
8. Be on the lookout for scams related to foreign job opportunities, particularly on the Internet.
9. Keep in mind that in the application and interviewing process, your most important goals are to:
   a. emphasize the international aspects of your education, experience, and skills,
   b. establish that you understand and possess the requisite personality traits to be successful overseas, and
   c. generally set yourself apart from the other candidates as not only the most qualified person, but also the one most likely to succeed in an expat position.

Chapter Three

Closing the Deal: Negotiating Your Compensation Package

Expatriate compensation packages are usually very different from a typical U.S-based compensation package. There is no standard approach, and the structure and elements of the package largely depend on the value of the employee and his or her negotiating skills. However, all employers have one common (and increasingly emphasized) concern: their desire to control costs. If allowed to get out of hand, an expatriate compensation package can quickly destroy a company's budget models. The expat also has his or her own concerns, because there are a multitude of unusual and hidden costs associated with moving to and living in a foreign country. This chapter explains the issues related to expatriate compensation packages and the common elements of an expatriate contract, and it offers suggestions as to how to structure such a contract.

The focus of this chapter is on U.S. expats who relocate to another country as an employee of a U.S. company or one of its foreign operations.

If you accept a position abroad as an employee of a company that is based in the country in which you are working, much of the information in this chapter (other than the discussion titled "Additional or Duplicate Expat Costs") may not apply to your situation. Rather, you are likely to be compensated on the same basis as the local employees in that country—unless the foreign employer has specifically recruited you or otherwise requested that you move to that country.

Establishing Your Value and Expectations

As a starting point, it is important that (1) you understand your real value and (2) have realistic expectations. *Value* is always a difficult concept to define, and it is certainly no less difficult when trying to negotiate an expatriate compensation package. Of course, value, in its simplest form, has two major determinants: the cost of a local employee in the foreign country for the same position and your compensation level in the United States. The company will strive to keep its compensation costs for an expat as close as possible to the compensation that it would need to pay an employee hired locally in the foreign country for the same position. Conversely, from your standpoint, you will not want to suffer any reduction in net compensation from what you are paid in the U.S. However, while these concepts provide a starting point for determining your value, they do not provide a final answer. Many additional factors must be added to the value equation, such as the following:

- The availability and qualifications of local employees to do your job efficiently and correctly—if there are none, then the possible compensation level for this position in the foreign country is irrelevant
- The added value derived from your familiarity with the company, its operations, and the U.S. staff
- The difference between a standard compensation level for this position in the foreign country and your previous

compensation in the United States—it could be much lower or, possibly, slightly higher

- The difference, if any, between the relative levels of the U.S. and foreign positions (e.g., is the foreign position a promotion, thus justifying increased compensation?)
- The difference between your current cost of living and the cost of living where you're going

Understanding these factors means understanding your value to the company and the real value of an expatriate compensation package to you in comparison to your current U.S. package.

By the way, any potential employer will probably have reliable information regarding the standard compensation package paid to local employees for the same job in the country to which you are considering relocating. However, it is good idea to supplement and confirm this information with your own research. You can do so through the Internet job sites and recruiting agency websites described in chapter 2, or at specialist websites such as www.mercerhr.com and www.salaryexpert.com, or by conducting an Internet search using the string "*salary survey*" "[insert specific country or city]."

Regarding *expectations*, there was a time when expatriate compensation packages greatly increased your real income during a foreign assignment. The extra compensation allowed expats to either live a more luxurious lifestyle than they had at home, save significant amounts of money, or both. You may have heard tales of these expat packages from old-timers, but they are no longer the norm except in countries of unusual hardship or danger. The compensation of expats has greatly changed—in part because most foreign assignments are no longer viewed as a burden on the employee, and in part because locally hired employees are now available and qualified to do many jobs that formerly were the exclusive province of expats. Both of these factors have combined to reduce the comparative value of expatriate compensation packages.

I have known many American expats who, 10 years ago, could routinely expect a wide range of special premiums and allowances like those discussed below, even for assignments in countries like the United Kingdom. Today, employers approach expatriate compensation with a much sharper pencil. American expats have thus been forced to become more realistic in their expectations.

There Is No "Standard Deal"

Unfortunately, there is no standard approach that can be offered as a cookie-cutter expatriate-compensation model. There are many different approaches used by companies, and even different approaches within the same company for different countries (these are discussed in upcoming sections).

Adjusting Over the Long Term

Regardless of the expatriate compensation package you negotiate, you should be prepared to adjust its terms if your overseas assignment starts to take on the flavor of a permanent relocation. It is unlikely that you can continue with the additional compensation and/or allowances of an expat package indefinitely. There is an expectation in today's world that these additional benefits are a cushion either to cover you during a temporary assignment or to help you adjust to local conditions for a longer assignment. (The more "undesirable" a foreign post is considered, the longer you can expect this additional financial compensation.) In most cases, however, your company will at some point probably want to cut back on your perks and make your compensation package simpler and more in line with salary standards for that operation. This might involve a combination of eliminating certain special allowances or incorporating them into your regular salary. If you are unwilling to make these adjustments, you likely will find yourself replaced by a local employee or a more flexible expat.

There's More to The Expat Life Than Money

While all of this is important, don't make the mistake of focusing exclusively on financial remuneration, while discounting the nonfinancial benefits that you and your family can derive from a foreign assignment. Regardless of compensation, this experience can have a decidedly positive effect on your career—both at your current company and future ones. It can also have a positive impact your personal life and development and that of your family. Factor these into your evaluation process so you won't ignore the bigger picture.

Get Some Help

In addition to the research and other self-help recommended in this chapter, you should seriously consider getting professional assistance with the negotiation of your expatriate compensation package and contract. This is an important document that will be binding on you, possibly for the duration of your foreign assignment. If you don't cover all issues thoroughly, you may find yourself in the awkward position *after* your relocation of requesting a renegotiation to deal with issues of which you were unaware. This is a bad way to start a foreign assignment. The general advice provided in this chapter should not be the sole basis for decisions regarding your specific situation. Money spent on a lawyer and/or international tax expert to get your deal right is money well spent.

Your employer may provide you with cost-free access to such experts to assist in your compensation planning and negotiations. This is fine, and it can be a real help to you in framing and understanding the issues. However, even the best-intentioned employer has its own goals and agenda that may differ from yours. Furthermore, it is you and your family, not your employer, who must live within the terms of your compensation package. Therefore, when I say "money spent on experts is money well spent," I am referring to *your* experts and *your* money.

Understanding the Issues

You must understand your financial issues—income, additional/hidden costs, U.S. and foreign taxes, currency exchange fluctuations, and inflation factors—thoroughly. Research these. Understand what your employer will do for you regarding each of them and what you must cover yourself. Prepare a detailed budget and cash flow plan (both start-up and ongoing) based on the issues raised in this chapter and the information derived from your own research. Treat this exercise as you would any critical business project and use appropriate financial-planning software or spreadsheets.

Don't be surprised if your cost structure turns out to be totally different from what you are used to in the United States—expats usually incur many new costs that are not commonly incurred in the U.S. Furthermore, the income that your spouse has been earning in the U.S. may be reduced or eliminated in the foreign country, thus squeezing your budget even more.

Additional or Duplicate Expat Costs

Expats always incur additional costs, or in some cases they pay duplicate costs for something overseas that they continue to pay for in the United States (e.g., rent for a foreign apartment while continuing to pay rent or mortgage costs for a home in the U.S.).

What follows is a list of the major additional or duplicate costs that you are likely to incur as an expat. Some of these you have been incurring in the United States all along, but the cost of such items or services in the foreign country might be higher. Not all of these expenses are obvious to a person considering his or her first expat position, and not all of them will apply to every expat. However, it is important that you be able to quantify all of those that apply to your situation and that you account for them in your financial planning. You should not assume that your company should pay, or reimburse you for, *all* of these costs. The level of employer responsibility for such expenditures is a matter of company policy and negotiation. However, the starting point for such negotiations, and for your expat financial planning, is to understand which of these costs you will incur and their financial impact on your situation:

- Packing, shipment, and/or storage of furniture and other personal items, and the costs of return shipments
- Storage of U.S. automobiles
- Housing-related costs in the United States: sale of house (closing costs, fees, and/or commissions), buyout of your existing lease, or fees paid to a management firm to manage the rental of your house while you are overseas
- Medical and dental checkups, vaccinations, and documentation for family and pets
- Airfare for relocation, home-leave, and your ultimate return to the U.S.
- Transporting of pets
- Temporary hotel or other accommodation upon arrival
- Temporary car rental until permanent transportation is arranged
- Housing costs in the foreign country: deposits, rent, and utilities (may be a duplicate expense or higher than previously paid in the U.S.)
- Automobile rental and fuel costs (may be a duplicate expense; likely to be higher than previously paid in the U.S.)
- School tuition costs for children (public schools may not be suitable)
- Food, clothing, entertainment, and other daily expenses (may be higher than previously paid in the U.S.)
- Additional insurance coverage: health, dental, personal property (previous U.S. policies may not provide adequate coverage for foreign claims and/or losses)

- Communication costs (e.g., telephone calls to family and friends in the U.S.)
- Safety and security measures (depending on the country, protective measures for family, home, and/or yourself may be required)

Special Factors That Can Change Quickly

You will encounter several compensation issues that are unique to expatriate contracts, and that are subject to wide fluctuations that can make your financial situation uncomfortable (or, in the extreme, untenable) if not factored into your compensation package:

Cost-of-Living Differences and Inflation

You will find varying degrees of uniformity (or lack thereof) in the cost of living for various countries around the world. This is not dissimilar to the cost-of-living variables found in assorted U.S. cities. For example, a move from a mid-size town in the Midwest to Los Angeles can provide a major shock when it comes to housing costs, as well as a severe diminution in the size and availability of housing. On the other hand, a reverse move can put the relocating individual or family into the lap of (relative) luxury.

The major cost-of living variables to be considered usually are as follows:

- Housing and related expenses—such as utility costs.
- Auto-related costs—particularly gasoline, but often the cost of purchasing or leasing an automobile (plus associated import, use, and license fees) can also be more expensive than the United States.
- Food and costs of eating out—this is somewhat dependent on whether you seek out and purchase expensive U.S.-source groceries and eat at American-style restaurants or take a liking to the more reasonably priced local cuisine.

The difficulty in calculating cost differentials is threefold: First, costs of living tend to be specific to regions and cities rather than countries. Relocating from New York City to a small town in the English Midlands can possibly result in a reduction in your cost of living. Conversely, a move from a small town in the U.S. South to London can result in a significant increase. You will find these variances in most countries, so your evaluation of cost-of-living differences must be based on a very specific comparison between the particular U.S. city or region in which you are currently living and the particular foreign city or region to which you moving.

Second, it can be difficult to get a handle on the true cost-of-living differential even for a specific city, if you attempt to calculate all of the potential factors individually. For example, housing costs may be less expensive in a particular city, but automobile costs may be significantly more expensive (particularly if fuel costs are three to four times higher than in the United States and/or the government imposes high taxes on the purchase or rental of automobiles, such as in Singapore).

Lastly, your well-researched cost-of-living calculations can be nullified almost overnight by sudden inflation or swings in exchange rates, which often arrive together. (The specific impact of these is discussed below.)

As a result, the best approach to calculating the difference between the cost of living in your home city and a foreign city is to use one of the many indexes that can be found by conducting an Internet search on the string "*cost of living index*" "*global.*" Mercer Human Resource Consulting (www.mercerhr.com) is an example of a major consulting firm that makes such cost-of-living indexes available for a fee. Alternatively, do a search on "*cost of living*" "[*insert specific or city or country*]" to find comparative cost-of-living information for a specific city or country. The information available through these websites can be quite comprehensive. For example, the Mercer reports take into account more than 200 cost variables in each locale, and cover more than 250 cities worldwide (more than 30 U.S. cities). Individual country reports are available, as well as a comparative index that ranks all of the covered cities by a total cost-of-living rating. You can get an idea of how the comparative index works by going to www.finfacts.ie, which reproduces a portion (50 cities) of the 2003 Mercer comparative index. There you can see that a relocation from New York City to Zurich as of 2003 was almost a total wash, with both ratings being within 0.3 percentage points of each other. However, moving from Chicago to Zurich involved a 19 percent increase in total cost of living (from an 83.9 rating to a 100.3 rating).

Alternatively, my friend Ted recommends the "Big Mac Combo" test; in other words, what does a Big Mac Combo cost in the foreign country compared to at home (and he reminds us that ketchup costs extra in most foreign countries). As a vegetarian, this test is not of much value to me, but I understand his point.

Some of these cost-of-living indexes (including Ted's) are free, but the more accurate and up-to-date indexes charge a subscription or other fee to access the cost-of-living comparative data. Access fees can be expensive for an individual, but you should not hesitate to request that any company considering you for a foreign assignment purchase such data for you.

Once you have determined your initial cost-of-living differential, you need to stay up to date with changes in these costs. Inflation can sneak up on you and then run away at unbelievably high percentage rates. It can affect an entire economy or only certain sectors. Inflation can be caused by a variety of related factors: loss of international confidence in a country's economy or currency, exchange rate fluctuations, excess demand for housing or goods, new taxes, and/or labor agreements affecting major industries. It is not necessary that you understand the underlying economic reasons for inflation. However, you do need to understand the impact it has on your compensation package and cost structure and be prepared to protect yourself against its deleterious effects.

Exchange Rate Fluctuations

The effect of exchange rates on the real value of your income and expenses in the foreign country cannot be overemphasized. Exchange rates can change quickly, and you should regularly check the current exchange rate between U.S. dollars and the currency of your new country at websites such as www.oanda.com and www.xe.net.

I have lived in the United Kingdom when the exchange rate was approximately US$1 = £1, when it was approximately US$2 = £1, and when it was somewhere in between. This means that the cost of a monthly apartment rental fee of £1,000 fluctuated in U.S. currency from $1,000 to $2,000. Similarly, I lived in Europe as the U.S. dollar/euro exchange rate quickly eroded from approximately US$.90 = 1 euro to a rate close to US$1.30 = 1 euro. All of my euro expenses (for example, rent, food, entertainment, transportation, and so on) increased in terms of their U.S. dollar cost by almost 40 percent. If my compensation had been paid solely in U.S. dollars, I would have quickly found myself in a financial dilemma.

Protecting Yourself

There are several steps you can take to protect yourself against the potentially devastating effects of major changes in your foreign cost of living or in exchange rates:

- First (and most important), stay up to date on cost-of-living and inflation indexes and on current exchange rates. Don't merely rely on anecdotal evidence that "things seem more expensive." Collect the requisite data confirming and supporting your belief.
- When negotiating your expatriate contract, consider the possibility of splitting your compensation between U.S. dollars and the currency of the country in which you will be living. Being paid in sufficient local currency to cover your living costs in the foreign country will help protect you

against currency fluctuations like those discussed above. However, you often will prefer to receive the remainder of your compensation in U.S. dollars, so that it will be available to pay U.S. expenses and for your use upon return to the United States at the end of the assignment. Many U.S. companies will be willing to work with you to create a reasonable split between foreign currency and U.S. dollar compensation that protects you from severe exchange rate swings.

- You also can agree with your company that your compensation package will be reviewed either at specific intervals or in the event of a specified (and documentable) percentage change in cost of living, inflation, or exchange rates in the foreign country. Upon the basis of such review, your compensation would be adjusted to get you back to the "real value" originally intended by you and your employer. Of course, you should keep in mind that this could cut both ways. It is not a good idea to suggest that your compensation package can only be *increased* to reflect such changes. You must be willing to accept an adjustment in the other direction if the economic facts so justify.

The Importance of Where You Get Paid

In addition to being paid in dual currencies, you might consider requesting that a portion of your compensation be paid in the foreign country and the remainder in the United States. For example, your company might pay your foreign currency compensation to you in the foreign country but pay your U.S. dollar compensation directly into your U.S. bank account. Dividing the payment this way has several potential advantages. It may

- minimize fees and delays associated with transferring funds between your home bank account and your bank in the foreign country,

- avoid potential currency controls in your adopted country that might deny you access to your funds in that country, or the ability to repatriate such funds to the U.S., and
- reduce the foreign taxes assessed against your income, if place of payment is relevant to the determination of your tax liability under the foreign country's tax regulations.

Splitting payment of your compensation can be tricky. It might require dual employment contracts (one for each country of payment) and supporting documentation for tax authorities to justify the basis for splitting your compensation between two countries. You should not attempt to do this without professional advice, particularly regarding the tax implications.

Different Approaches to Expatriate Compensation Plans

Companies take different approaches to expatriate compensation packages. An employer's philosophy and the policies that it adopts for your expat compensation package will depend on a variety of factors:

- How many expats it already has overseas and what kind of packages they have
- Whether it has a defined expat compensation policy or negotiates all expat contracts on an individual basis
- The "difficulty" of living in the country to which you are moving
- The additional cost of living in that country compared to the United States
- The local tax structure in the country
- Any special factors that are specific to you (i.e., relating to your personal circumstances, such as your home and family situation)

- How much the employer needs you overseas
- How well informed you are and how well you can negotiate

Although there is no standard deal, three basic approaches are most commonly used. Under all three, the employer usually has the same (albeit sometimes conflicting) goals: hire the most qualified person, control costs as much as possible, and "be fair" with respect to the compensation paid to the expat. However, as with all things, there can much difference of opinion as to how best to achieve these goals, particularly with respect to the issue of fairness.

The three most common approaches for structuring an expatriate compensation package in today's world are as follows:

Local Hire

This approach involves treating expats like local employees hired in the foreign country, with the expatriate compensation package equalized to the standard compensation level for the same position in that country. Basically, in such cases, the company treats the expat as a locally hired employee, with a flat relocation allowance and perhaps a few start-up allowances that are transitioned out over a relatively short period of time. If at all possible, companies prefer to fit an expat's compensation into the budgets and compensation models of the foreign operation in which they are working.

This approach is increasingly common for expats being transferred to so-called industrialized or "Westernized" countries—namely, Western Europe, Australia, Singapore, and other destinations that have economies and cost-of-living structures similar to those in the United States. This approach can work well in countries such as these, as long as the relocation allowance is sufficient to cover reasonable costs associated with moving, storage, and house-related expenses in your home country. (It works better for expats without children, since public schools in the foreign country may not be an option, and tuition fees for U.S.-curriculum private schools can be expensive.)

However, this approach does not work well at all if the compensation levels and/or standard of living where you're going are significantly lower than in the United States, or if the living conditions are such that special allowances are required to make you comfortable (and/or secure).

U.S. Compensation Plus Allowances and Benefits

This approach involves maintaining your U.S. compensation as is and adding specific allowances to cover extra costs that you will incur as an expat. It tends to result in numerous home and foreign allowances, and in-kind benefits that are designed to compensate you for extra costs incurred and to provide a style of living (e.g., housing) more or less equivalent to what you were used to in the United States. In the past, this was the standard expatriate compensation package. Today, its use is largely confined to: locales with a significantly higher cost of living than in the expat's home city, emerging-market or developing countries, and countries that are considered difficult or dangerous areas in which to live. In the latter situations, it might be combined with a hardship or foreign service premium that is offered as an enticement to attract expats with needed expertise and experience to such countries.

Equivalent Standard of Living

This option involves equalizing your expatriate financial situation and standard of living to their U.S. equivalent (often using the cost-of-living indexes discussed above). The goal is to provide you with a salary sufficient to allow *you* to pay for all the additional costs that you will incur as an expat (with the exception of relocation costs, which are often handled as a separate allowance).

The end result under this approach (i.e., the total value of your compensation package) is often the same as with the previous approach of retaining the original U.S. compensation and adding allowances. In both cases, there is a general goal of providing a compensation package that covers your additional expat costs and allows you to live at a standard equal to that you enjoyed in the United States. Both of these approaches take into account the various additional costs that you will incur as an expat. The primary difference is that the first covers these with numerous separately defined allowances, while the other throws them all into a basket, comes up with a single number for your total salary figure, and then leaves it up to you to cover such costs yourself. Employers increasingly prefer the latter approach, if for no other reason than ease of administration. This can be workable from your standpoint, but it places the burden squarely on you to ensure that you understand and cover all the issues discussed in this chapter within that "single number."

Fitting Your Situation into the Company's Model

You may find that your company does not have any previously defined structure or model for expatriate compensation packages. In this case you are free to negotiate a mutually agreeable model and terms using those concepts discussed in this chapter that are applicable to your situation. However, larger companies (and employers with numerous expatriate employees) may have already put into place a compensation program that applies to all employees transferred to a different country.

These programs are often developed and administered by accounting or human resource consulting firms with an expertise in expatriate compensation, and they are designed to cover the various issues discussed in this chapter. If such a program exists, you will need to find a way to work within its limits. Nevertheless, no standardized program can be appropriate for every person nor can it take into account every variable that might arise in an expat's personal situation.

Do not hesitate to raise issues of concern or to request adjustments to the company's program if necessary. Demonstrate to your employer that you are familiar with the issues discussed in this chapter. Best case, you will be able to negotiate a favorable modification to the company's standard policy. If nothing else, you will show that you are knowledgeable and that you are paying attention. Both of these traits will serve you and your employer well in your foreign assignment.

Possible Expatriate Allowances and Benefits

To the extent that you and your company adopt an approach for your compensation package that includes mutually agreed-upon expatriate allowances and/or benefits, there are a variety of possible options for you to consider. It is difficult to say which of these, if any, you should expect to receive—every expat's situation is different and each company has its own compensation policies and philosophy. In addition, there is the previously noted and ongoing trend for U.S. employers to attempt to reduce the costs of foreign assignments, thus putting many of these previously common allowances and benefits to the knife. However, certain of them are still more common (and supportable) than others, and I will attempt to point these out and provide some guidance on which of them could apply in your situation.

As you consider the individual allowances discussed below, keep in mind that employers will often combine several of these into a single lump sum payment. In this case, you might have money left over but you will also be responsible for any costs in excess of the payment. Alternatively, the company may reimburse certain expenses only up to a specified cap. This is particularly true with respect to costs incurred in the United States prior to, or in connection with, your move (see "Allowances on the Home Side" in the next section). It is very common for a company to simply pay a flat allowance to cover some or all of these expenses and to expect the expat to figure out how to use this money in the most efficient manner.

You should also remember that your likelihood of obtaining these allowances and benefits, in every case, hinges to a great extent on whether you or your employer initiated the request for the transfer, and on the underlying reasons for the relocation to a foreign office. For example, if you request a transfer to a foreign country because you have relatives there, such transfer is likely to be viewed as an accommodation to you. The company will see little need to pay you any expatriate allowances. On the other hand, if the relocation is based on the company's need to have your knowledge and skills in the foreign office (even if you proposed the transfer), your negotiating position with respect to these allowances is much stronger.

Allowances on the Home Side

You will probably incur significant costs in the United States in preparation for your move overseas, and many employers provide a variety of allowances to help you cover such costs.

Exploratory Trip to Prospective Country

Ideally, you (and at least some of your family members) will be able to visit the country for an exploratory visit prior to moving there, and, again ideally, your company will pay the costs of this visit. Such a trip will provide an opportunity to make important preparations for your move—finding housing, visiting schools, and so forth. (See "Do an Exploratory Visit" in chapter 4). Although it is reasonable to request this perk, your actually getting it will depend on the company's policy and possibly on the country to which you are relocating. For example, moving to the United Kingdom is relatively easy compared to moving to China. Therefore, an exploratory trip to the U.K. might be considered an unnecessary expense. You are more likely to receive a positive response if you couch your request in terms of the necessity of a visit to determine *if* you are willing to transfer to the country, rather than merely a desire for an exploratory visit to help prepare for the move.

Relocation

This is perhaps the most easily justifiable allowance, because you are clearly going to incur costs in moving to your new country, and it is difficult to fashion an argument that you should be responsible for the costs of the move unless you are relocating solely at your own initiative. Your relocation costs will include: packing and shipping a certain amount of your furniture and/or personal effects by sea and air, airfare for you and your family, costs of transporting any pets overseas, your costs for a hotel or other temporary housing, and food and transportation during the period when you're first settling in (often covered by a per diem allowance).

U.S. Home-Related Costs

You will probably need to do something with your U.S. home—you can't just leave it sitting there empty and uncared for. If you own your home, your options include selling it or putting it up for rental while you are overseas. Conversely, if you are currently renting your home, you may need to buy out your lease or sublease.

Selling your home will involve costs that you would not have incurred if not for your move overseas: advertising costs, real estate fees and commissions, closing costs, legal fees, and so on. You may also incur monthly carrying costs (i.e., mortgage, taxes, fees, etc.) for a significant period while your house is on the market.

Renting your home to someone else will almost certainly require that you engage a property management agency to handle the lease and the property in your absence. Doing so will require the payment of monthly management fees. You may also suffer a shortfall between your monthly rental income and the carrying costs for your home. Of course, buying your way out of a lease will require payment of either an amount stipulated in your lease or negotiated by you.

It is not uncommon for an employer to cover some or all of these expenses in combination with your relocation allowance, often up to a stipulated cap.

Storage of Furniture and Personal Items

It usually doesn't make sense to move all your furniture to a foreign country, except for a lengthy or permanent assignment. The more economical and practical alternative is to place your furniture in storage until you return. Employers often reimburse expats for the costs of storage.

U.S. Automobile-Related Costs

Although I personally prefer to sell my automobiles when I move overseas, this is not a practical or economically sound tactic in all situations. An automobile can be stored for extended periods, but doing so requires professional storage in a climate-controlled environment with periodic maintenance. An employer occasionally will cover the cost of doing this. It is reasonable to raise the issue, learn what your company's policy is, and, if this is important to you, try to make a compelling case for why it should be covered.

Foreign Allowances and Benefits

You will encounter more disparity with respect to the foreign allowances and benefits that are offered to expats than those on the U.S. side. This disparity is due in part to the previously noted trend of companies' avoiding complex contracts with numerous special allowances in favor of simpler contracts providing for a single all-inclusive salary. However, it is more attributable to the many differences that still exist between standards of living and compensation levels around the world.

In some countries, it is difficult to justify many special allowances at all, and in these countries you find more expats working under the same compensation packages as locally hired employees. Other countries, however, still present the kinds of challenges and living conditions that merit many of the following types of allowances and benefits.

In any event, you should not expect to receive all of these allowances and benefits no matter where you are based, because there is some overlap in the costs they are intended to address. Do not get greedy or have unrealistic expectations; focus your negotiations on real issues that are demonstrably applicable to your situation.

Hardship/Hazardous-Duty Premium

There are fewer countries in today's world that merit any kind of hardship or hazardous duty pay. However, they do still exist, and well-paying expat positions are always available for Americans who are willing to endure extreme hardships or to put their health (and sometimes their lives) at risk. Special skills are not always required to obtain high-paying positions in these countries. Relatively unskilled jobs can merit six-figure incomes plus numerous in-kind benefits. Needless to say, you need to do a careful risk/reward analysis before accepting any such position and have a clear understanding of what you are getting yourself into. Make sure that your benefits include appropriate security and safety measures as discussed in chapter 6.

Cost-of-Living Allowance

It is possible that your compensation level will not significantly change in your expat position, but that the cost of living in your new city will significantly exceed that of your home city. Many companies track the disparity through cost-of-living indexes and pay a supplemental cost-of-living allowance. The company may expect such an allowance to cover most of the other possible foreign allowances discussed in the sections that follow.

Housing

You might find yourself paying for housing costs (including furniture rental) in the foreign country while still paying a mortgage and other costs associated with your home in the United States, particularly if some of your family members remain in the U.S. and continue to occupy your U.S. home. Even if you sell or rent out your U.S. home, your total housing costs in the foreign country may still significantly exceed what you were paying. In these situations, it is not uncommon for a company to assist the expat with some form of housing allowance designed to cover these duplicate or additional costs.

However, I again advise you to keep your expectations within reason. It is unreasonable to expect your employer to cover all costs associated with selling or renting your U.S. home *and* all costs incurred for your foreign home. You should limit what you're asking for to your legitimate duplicate or additional costs over your previous housing costs that were incurred in the United States. Be willing to meet your company halfway on this issue.

Automobile and Transportation Costs

It would be nice to have a car during your foreign assignment, but you really don't want to purchase an automobile in a foreign country if you can avoid it. Bringing it home will be a problem, and selling it there will be a hassle. Even if you expect to be abroad for several years, it is best to stay out of the automobile market. Your company is in a far better position to deal with an automobile at the end of your assignment than you are. It can turn the car over to another employee, sell it, or return it if leased.

If your company will not provide a car, try to at least get an automobile allowance so that you do not incur the cost of a purchase or lease. Such a request is reasonable, particularly if you need the automobile to get to work and/or to perform your job responsibilities.

You also should be aware of fuel costs in the country to which you are moving. In most European countries the price of fuel can be three to four times higher than it is in the United States. If you drive a lot, gasoline can be a significant additional expense (although automobiles in Europe do tend to be more fuel-efficient than in the U.S.). This expense might be covered within the context of a cost-of-living allowance as discussed above. Alternatively, you can request that the company reimburse you for work-related fuel expenses or provide you with a fuel credit card that you can use for work-related driving.

Home Leave

You probably will want to return to your U.S. home at least once a year and, depending on the size of your family, this can be expensive. Many companies provide a home-leave allowance that will cover the airfare for an annual trip home for you and your family. Some employers require that the flights be arranged through the company's travel agent and be paid for by the company. Others will simply pay you the allowance and allow you to use it as you desire. The latter option opens up the possibility of using the allowance to visit a more exotic locale if you don't need, or are not in the mood, to go home.

Education and School Tuition

If you are moving to a country in which English is not the principal language, your children will be unable to attend the local schools unless they can communicate in that language at least well enough to understand what is being said in the classroom. (However, do not underestimate their ability to learn the language quickly. I recall a Mexican family with five children who relocated to the French-speaking part of Switzerland, and all of the children were fluent in French within six months of arrival.) However, even in countries where language is not an issue, the standards or curricula of local schools may not suitable for your children. In such cases you're left with the choice of either paying the high tuition fees charged by U.S.-curriculum private schools or home schooling your children. The latter choice can have many positive benefits, but may not be feasible in your situation. You should definitely explore with your company the possibility of receiving an education allowance to help defray your children's tuition costs (even a partial allowance will help).

Health and Dental Insurance

Your employer's standard health and dental insurance plans may not provide adequate (or any) coverage outside of the United States. Research these plans to understand what coverage is available to you while overseas, and inquire whether the company is willing to purchase supplemental policies that provide you and your family the same level of coverage you would have in the U.S. However, keep in mind that health care often is less expensive in foreign countries (in some, it's free) so this should be factored into your consideration of the necessary coverage and policy limits.

Vacation Allowance

In many countries, it is not uncommon for a local employee to have up to six weeks' vacation a year compared to the two or three weeks common in the United States. It is not unreasonable to request that you receive the same amount of vacation allowance as the local employees in the country in which you are working. You need more vacation time than you had in the U.S so that you can carry out your annual home leave and still have vacation available to travel within and outside your new country. You don't want to miss the opportunity to explore and learn about the rest of the world.

Participation in Corporate Benefits Programs

Your employer will probably have several different types of benefit programs available for U.S. employees—401K, stock option plans, stock purchase plans, and so forth. These may or may not be available to employees in foreign offices. Find out if you are eligible as an expat to participate in these plans. If you find yourself excluded from one of these programs in which you wish to participate, it may be possible to restructure your employment in a way that will make you eligible. Discuss this with the appropriate person in your human resources department.

Costs of Preparing Tax Returns

As an expat, your U.S. tax return will be more complicated than it was before. You will also probably have to prepare and file a tax return in the country in which you are working. Many companies will pay the costs of preparing U.S. and foreign tax returns for their expat employees, particularly if the company operates a tax-protection or tax-equalization plan as discussed in the next section.

Repatriation Allowance

Unless the initial relocation was entirely at your initiative, there is no reason why the company should not pay your costs of returning home (airfare, packing and shipping, transporting pets, temporary housing, etc.). As with the original relocation allowance, these costs are often handled either by a predetermined lump-sum payment or as a cap-limited reimbursement.

Taxation of Expatriate Income

The two most important tax facts that you must understand as an American expat are: (1) you not only continue to be subject to U.S. taxes while living overseas, but also may be subject to taxation in the country in which you are based, and (2) you must file a U.S. tax return every year, even if you don't owe any U.S. taxes.

It is also critical that you understand the overall tax burden you will incur as an expat—taking into account U.S. taxes, foreign taxes, credits, and exemptions. I have known many expats who were surprised at the total taxes assessed against their expatriate compensation. Unfortunately, their surprise occurred at the time of filing their tax returns, instead of while structuring their compensation package in the first place.

Your tax liability is thus one area in which you absolutely must obtain professional guidance *before* you move overseas. (You should do this even before you finalize the terms of your expatriate compensation package.) With proper advice and planning, you can structure your compensation and employment in a way that reduces your total tax burden. Many employers will pay for this consultation, as well as for the preparation of your U.S. and foreign tax returns during your overseas assignment. Expatriate tax liability is a specialized field of tax law, and you should engage someone with specific expertise in this area. Major national and global CPA firms, and most medium-sized firms in large cities, will have such expertise. In addition, certain firms have developed a specialized focus in the area of expatriate taxation. Contact details for several of these firms are provided in appendix D.

U.S. Taxation of Expatriate Income

Unlike many other countries, the U.S. taxes all worldwide income of its citizens no matter where they live. However, current U.S. tax law also provides rules under which you can significantly reduce the amount of U.S. taxes owed on your expatriate income. These include the foreign earned-income exclusion, foreign housing exclusion, and credits for foreign taxes paid. The regulations governing these concepts are complex and require adherence to guidelines regarding such issues as the establishment of a bona fide foreign residence, the number of days you are inside and outside of the United States, and other detailed requirements. You must understand these guidelines before moving to the foreign country and maintain the required records throughout your overseas assignment.

Your specialist tax advisor will be able to help you navigate these regulations so you can take advantage of the tax reduction that is available if you fall within the applicable guidelines. The advisor also will be able to help you determine whether you should stop or reduce your federal and/or state tax withholding (assuming you continue to be paid by a U.S. employer while overseas).

Foreign Taxation of Expatriate Income

Your income earned while living in a foreign country will most likely be subject to some degree of taxation in that country. However, you sometimes have considerable scope for reducing the amount of tax assessed against your income by a foreign government. Other countries apply many different concepts for determining whether you are subject to taxation, the amount of your income that is subject to tax, and the rate at which taxes will be levied. For example, Switzerland has a special six-month work visa under which you can spend up to six months per year working in Switzerland without incurring any liability for Swiss income taxes. Other countries sometimes do not tax income that is paid to you outside of that country (see "The Importance of Where You Get Paid" on page 62). More than 50 countries have also entered into tax treaties with the United States that contain specific provisions for reducing taxes and eliminating double taxation on the citizens of both countries.

The tax rules in each country are different, but in all countries (including our own) you must understand these rules up front, adhere to the applicable guidelines, and compile the documentation required to support your tax returns. Your tax advisor can assist you in understanding and taking advantage of tax rules such as those I've just described and also in exploring other tax reduction strategies that may be available to you.

Tax Schemes

As an expatriate, you will encounter many tax reduction schemes offered by supposed experts that may appear to significantly reduce (or sometimes entirely eliminate) the taxation of your expatriate income. Be very skeptical of these schemes. Some are legitimate for citizens of foreign countries but do not work for Americans. Others are just plain dicey for everyone.

Company-Sponsored Tax Programs

Many large companies, and smaller ones that have numerous expatriate employees, implement a company-sponsored tax program that is designed to (1) assist employees in the calculation of U.S. and foreign taxes and the filing of returns, (2) ease employees' foreign tax burden by reimbursing for foreign taxes in excess of U.S. tax liability, and (3) put all expatriate employees on a fair and equal footing, regardless of the country in which they are based.

There are two fundamental approaches to these programs: The first is "foreign tax protection," under which the company reimburses an expat for any foreign taxes paid in excess of his or her U.S. tax liability. The other approach is "tax equalization," under which the company calculates a hypothetical U.S. tax liability that is equal to the federal and state tax that would have been assessed on the employee's income if the employee were still living and working in the United States. The employer then withholds this hypothetical tax from the expat's income and uses it to pay the expat's actual U.S. and foreign taxes.

Both of these programs may sound simple. In fact, they are not only complex but also subject to many variations in terms of how they are administered. Many companies turn these tax programs over to outside accounting or human resource firms to manage. These firms then assist the expat employees with their tax planning under the programs and prepare all required U.S. and foreign tax returns.

Tax-protection and tax-equalization programs are laborious to read, and few people (other than accountants) enjoy dealing with them. Nevertheless, you must take the time to understand how your company's program will affect your expatriate compensation. Otherwise, you may be setting yourself up for a major surprise. When I lived in Saudi Arabia, there was no Saudi income tax on expatriate income. In combination with the U.S. foreign-earned-income exclusion, this lack of Saudi tax meant that many of the U.S. expats incurred no income tax liability whatsoever on their expatriate income. However, my company had implemented a tax-equalization program, which meant that it withheld a hypothetical tax from the expatriate employees' earnings that was equal to the U.S. income tax that would have been imposed had we been living and working in the United States. (This hypothetical withholding was retained by the company.) One of the expats did not understand this when he transferred to Saudi Arabia. He was understandably surprised and upset when he received his payroll stubs and realized that the company was withholding an amount from each check to cover the hypothetical U.S. income tax.

In fairness to the company, the tax-equalization program had been covered in training seminars and was described in the expatriate employment contract (but its import and effect had somehow been missed by this employee). What's more, the program made perfect sense when viewed from a global perspective. For example, if this same employee had been based in a country that imposed a higher tax rate on his expatriate income than the U.S. tax rate, he would have had to pay only the same hypothetical U.S. income tax that was withheld from his Saudi payroll checks. The company would have been responsible for any excess foreign tax liability above the U.S. tax rate. Thus, all employees, regardless of the country in which they were based, were put on a fair and equal footing (i.e., the employees based in both a no-tax country and a high-tax country paid the same hypothetical U.S. tax and thus were treated exactly the same). The real problem in this situation was the individual employee's failure to pay attention to the details of the tax-equalization program.

Chapter 3 **Summary** and **Checklist**

Outline for Expatriate Contract

The following outline sets forth the major issues that may be covered in an expatriate employment contract. There are likely to be additional issues specific to your situation that should be addressed in your contract. This outline provides only general information on which to base your discussions. You should obtain professional advice with respect to your specific situation and in the negotiation of contract terms. Not all of these issues are routinely covered within the terms of an expatriate contract. It is not uncommon for an employer to take the position that certain issues will be handled "in accordance with company policies." If this is the case, make sure that you know what the company policies are with respect to each issue and whether they are subject to change without your approval during your foreign assignment. Of course, if you are working in a foreign country for an employer based in that country, many of the following issues will not be applicable.

1. *Contract approach and structure.* As a preliminary matter, you and the company need to agree on which of the three expatriate contract structures apply—local hire, U.S. compensation plus allowances and benefits, or equivalent standard of living. This will affect how the following issues will be covered in the contract.
2. *Job description.* What will your title be? What responsibilities will the position entail?

3. *Location.* Where will you be based? Will the company be responsible for obtaining and paying for all work permits and other authorizations required for you to enter and work in the foreign country?

4. *Duration.* What is the anticipated length of the expatriate assignment? Is there any penalty or loss of allowances/benefits if you do not complete the full term of the assignment? Is there any bonus if you do complete the full term? What will happen if the company terminates your employment prior to the completion of the full term (e.g., whether for cause or due to redundancy or other staff cutbacks)? How can the contract be extended beyond the original term? Will there be any adjustments to the contract (e.g., the allowances and benefits paid to you) if the contract term is extended or if you remain in the country for more than a specified period of time?

5. *Monetary compensation.* What is your base compensation? Is it possible to earn bonuses and/or commissions? If so, what are the criteria for paying these? In what currency will you be paid? Where will your compensation be paid? When are you eligible for compensation review and increases in compensation?

6. *Local and foreign taxes (nonincome).* Will you continue to pay U.S. Social Security taxes? Who will be responsible for paying social insurance and other nonincome taxes and charges in the foreign country?

7. *Expatriate allowances and benefits.* Will you receive any of the following? Are there any caps on the amounts that will be paid? Are any of them covered by lump-sum payments?
   a. Relocation allowance: reimbursement for travel expenses, packing, shipping furniture, and personal effects (both

ocean freight and air freight), transporting pets, temporary housing, and other costs upon arrival

b. Reimbursement for U.S. home-related costs: costs and fees incurred in selling your house, lease buyout charge, fees paid to a management firm to manage rental of your home
c. Reimbursement for storage of furniture and personal items in the United States
d. Payment of U.S. automobile-related costs such as storage fees
e. Hardship/hazardous duty premium
f. Cost-of-living allowance
g. Foreign housing allowance: payment covering expenses of leasing a home in foreign country, including furniture rental and utilities
h. Company-provided automobile or automobile allowance
i. Fuel allowance, reimbursement, or credit card to assist in covering gasoline costs
j. Home-leave allowance: airfare for scheduled trips home
k. Tuition allowance: full or partial reimbursement for children's tuition fees
l. Health and dental insurance: payment of premiums for coverage in a foreign country
m. Vacation allowance: How many weeks' vacation will you receive while working an overseas assignment?
n. Repatriation allowance: covers the same types of costs as incurred in your initial relocation, but in reverse

8. *Participation in company benefit programs.* Are you eligible to participate in the company benefit programs that are offered to employees either in the United States or in the foreign office?

9. *Taxation.* Does the company have a foreign tax-protection or tax-equalization program in which you will be participating?

Will the company pay for the preparation of your U.S. and foreign tax returns while you're working overseas?

10. *Review and adjustment.* Will there be any periodic review of your economic situation and contract terms based on cost-of-living, inflation, or exchange rate fluctuations? What criteria will be used to evaluate these issues and how will adjustments to your contract terms be determined?

11. *Governing law and jurisdiction.* This issue relates to the laws that will govern your employment contract and the jurisdiction in which any dispute regarding your employment would be heard. Your rights and protections under the labor laws of the foreign country will be drastically different from those commonly found in the United States. While it is somewhat uncomfortable to agree to be bound by laws with which you are unfamiliar, the laws of most foreign countries are more favorable to employees than are U.S. federal and state laws. In addition, many countries require that all employees working there be subject to the employment laws of that country. On the other hand, if you have a dispute with your employer, you may want to return to the U.S. rather than spend time in the foreign country pursuing a claim through that country's legal system.

Chapter Four

Getting There: Moving Family and Belongings to Your New Home

Getting there is usually not *half the fun—it is a lot of work. However, relocation can be an adventure in itself (and considerably less stressful) if you prepare and plan your move as outlined in this chapter.*

Many books and websites dealing with expatriate issues offer a timeline for accomplishing the tasks discussed in this chapter. These proposed timelines variously commence anywhere from five months to six weeks prior to your departure date. This disparity indicates to me that each of these various timelines represents a particular author's personal experiences in relocating to a foreign country, and that such models are thus not particularly valuable to others in different circumstances. For example, my own timelines for moving overseas have ranged from two weeks to two months. In addition, the structure and timing of your preparation for the move will vary greatly depending on your personal circumstances—for example, are you moving alone or with a family? Are you relocating for a three-year period, or a three-week period, or somewhere in between?

I therefore have elected *not* to propose a timeline applicable to all readers, but instead encourage each of you at the earliest possible date to create your own personal timeline based on you and your family's individual circumstances.

What to Do Before Moving There

Training and Support Your U.S. Employer Should Provide

The majority of U.S. companies offer training programs to employees and their families who are relocating overseas. Employers recognize that you will be more successful in your job if you and your family are comfortable in your new environment. Such training is useful for any destination and is critical if moving to a country significantly different from any place you've experienced before. This training will assist you with learning what you need to know about, and in preparing you for the cultural and emotional adjustment, customs and taboos, work-related practices, living conditions, and other aspects of life in the country you're moving to.

Training programs of this nature are usually provided by professional trainers and consultants who specialize in international and cross-cultural training. They can be easily located in any major metropolitan center in the telephone book, in your local International or World Trade Club directory, or through an Internet search using the string "*international cross cultural training,*" or "*cross cultural training*" "*international.*" Add "[*name of your city or region*]" if you want to locate a training resource in your immediate vicinity.

Many larger companies, and companies with numerous expats in a variety of countries, now outsource not only training and cultural orientation, but also the responsibility for managing the entire spectrum of expatriate policies: compensation plans, benefits and insurance administration, international taxation programs, relocation and repatriation policies, and "in-country" assistance in the foreign country. These are now commonly managed by international human resource and relocation specialist firms, which provide a single point of contact for the expatriate employee to obtain the full range of available support services. Examples of such firms are listed in appendix D.

Do an Exploratory Visit

An exploratory visit to your new country is extremely useful. It affords an opportunity to see the country firsthand, learn about housing and living conditions, and check out potential schools. Such a visit generally serves to ease any tension or uncertainty you and your family might be feeling. However, such a trip can be expensive, particularly if the whole family goes. Some employers will cover the expenses of such a trip, but coverage is not always guaranteed (the more "difficult" the living conditions in a country are considered to be, the more likely your company will be willing to foot the bill). Don't push the issue with your employer if an exploratory visit is not possible. There are numerous alternative ways to learn about the country. You will simply need to be especially thorough in your research and preparation. I have never actually visited a country before moving there, including when I moved with my entire family. We always adjusted fine.

If you do make an exploratory trip, make the most of it. Have a plan for what you want to do, see, and learn. If possible, arrange for someone from your company to spend time with you. Visit various neighborhoods and explore your housing options. Make appointments to visit schools. Stop in at the office where you will be working and introduce yourself. Become familiar with the transportation system. Get a feel for your likely day-to-day living conditions and the cost of basic goods and services.

Finding a Place to Live

It is best if you make progress toward locating housing in your new country before leaving the United States. Upon your arrival, you will want to get out of a hotel and into a "home" as soon as possible. The easiest way to do this is in connection with an exploratory visit as just discussed. However, even without such a visit, your work colleagues can put you in contact with local property agents who can provide information on housing alternatives, prices, and so forth. Factors to investigate are largely the same ones that you would look at in the U.S.:

- Neighborhood characteristics and safety (check this both in the day and at night—the personality of an area sometimes changes dramatically when the sun goes down)
- Distance to work, schools, stores, and conveniences
- Availability of hospitals in the area
- Accessibility to, and dependability of, public transportation
- Condition of utilities (don't expect as much water pressure or hot water as you're used to)
- Condition of appliances (forget super-sized stoves and refrigerators)
- Availability of other expats in the area who can help you get settled in

Obtain copies of leases, and have them translated and reviewed by someone who is familiar with leasing in that country. Before signing, make sure you understand your obligations and liabilities under both the lease and any applicable landlord/tenant laws. Don't assume that any lease presented to you is "standard"—find out whether there are any unusual terms that should be negotiated. In particular, pay close attention to your liabilities relating to termination of the lease: the notice period required for termination, your obligations regarding the condition of the property, and the return of any deposits. Also, it is not uncommon in some countries for the lessee to be liable for the cost of repairing appliances if they break down or for routine maintenance costs that are traditionally paid for by the landlord in the United States. I have rented property in five different countries and the lease terms and applicable laws in each were different.

Lessors in some countries are reluctant to lease to individual expats (because of their transient nature) and insist that the company cosign any lease. Hopefully, your employer will be willing to assist you in this regard if it becomes an issue.

Don't reject out of hand the idea of living in local-style housing, even if it is very different from what you are used to. One of my friends was posted to Lamu Island, Kenya, and found a very nice traditional Arabic home furnished with Lamu antiques for $3 rental per day (this was admittedly a few years ago). It was not stocked with Western-style conveniences, but she thoroughly enjoyed the experience of living there.

Learn Before You Go

The more you learn about the country to which you are moving, the smoother your adjustment period will be. Any place is more interesting if you know something about it, but this is more than just an exercise in casual education. There are many aspects about a country that you must know before you get there. Otherwise you will be setting yourself up for problems. What follows is a list of the major types of information you should research as part of your preparation for relocating to any foreign country:

- Cultural and religious beliefs and customs
- Cultural taboos and locally sensitive subjects
- Specific gestures that are appropriate or inappropriate
- Body language and eye contact practices

- Laws that are different from those in the United States (something you commonly do at home could be illegal in the foreign country—e.g., drink alcohol, wear a particular religious symbol, dress in short pants in public, chew gum)
- Prevailing attitudes toward Americans (both official and among the general population)
- Current political situation (e.g., stability of government)
- Any safety, security, or health risks that need to be dealt with
- The country's history, style, and structure of government
- Languages spoken
- Greeting customs
- Living and housing conditions
- Availability, structure, and quality of the health system (in particular, how do you obtain medical assistance when necessary?)
- Information regarding the banking system and gaining access to money
- Information regarding American or other expatriate organizations in the area
- Driving rules (i.e., what will you need in order to be able to drive?)
- Information about public transportation
- Traditional food, eating customs, availability of Western food
- Tipping customs
- Gender issues (accepted gender roles, customs regarding separation of men and women, restrictions on activities or conduct of women)
- Views toward tobacco and alcohol
- Entertainment customs

- Acceptable and unacceptable styles of dress
- Climate conditions
- Attitudes toward pets
- Telephone and electrical issues. (There are a wide variety of standards for voltage, electrical/telephone plugs, and televisions throughout the world. This is discussed in more detail in the section titled "Get Turned On" in chapter 5.)

There are numerous sources available for researching these issues and learning about a country. A good starting point is the wealth of information available through the U.S. Government. The State Department's travel website at www.travel.state.gov can serve as a gateway to a variety of information sources regarding a specific country, including: travel, safety, and health information for a region; visa requirements for U.S. citizens traveling to the foreign country; lists of English-speaking doctors and hospitals in the country. The State Department also publishes (at this same website) Background Notes for almost every country, which provide information on many of the topics listed previously. Judging from those about countries in which I have lived, these are informative and accurate. They are not cookie-cutter summaries but instead deal with specific and topical issues for each country.

Most countries have a variety of offices in the United States that can provide information: embassies, consular offices, and chamber of commerce offices. Contact details for foreign embassies and consular offices in the U.S. are available at the State Department's travel website (www.travel.state.gov) and at www.embassyworld.com. Information regarding U.S. offices of foreign chambers of commerce can be found at www.chamberofcommerce.com.

Of course, social and educational clubs focusing on a particular country can be found at universities and in cities where a large number of people from that country live. Either of these can be a source both for information and useful contacts in that country.

Numerous online resources provide information about specific countries and on expatriate issues in general. The four best (free) websites I have found in this area are www.overseasdigest.com, www.transitionsabroad.com, www.expatexchange.com, and www.escapeartist.com.The first three are excellent in their layout, ease of use, and quality of information. The fourth also contains a broad range of accurate and useful information, but I find its layout and sheer mass of information to be somewhat overwhelming. Any person considering a move to a foreign country should spend some time surfing the pages of these websites. Two additional fee-based websites that provide both specific country information and general expatriate information are www.goingglobal .com and www.livingabroad.com.

With so many American expats around the world, you will not be surprised that someone has taken responsibility for getting us organized and focused. (As I said in the Introduction, this is something that Americans generally do very well.) There are several private organizations of American expats that work hard to promote and protect the interests of Americans living overseas. Two of the most active are American Citizens Abroad (www.aca.ch) and the Association of Americans Resident Overseas (www.aaro.org). These organizations have numerous local chapters, which serve both as a meeting place and as a source of information and assistance for Americans living in that area. They also lobby the U.S. federal government and state governments on issues of importance to Americans living overseas (e.g., voting rights and expatriate tax issues). Anyone considering life as an American expat should visit these organizations' websites, and you should certainly take advantage of the knowledge, services, and hospitality afforded by their local chapters when you arrive in your new country.

Talk to Someone Who's Lived There

Talk to U.S. citizens who have lived in the country to which you are moving. Ask questions about all the issues listed above. There is no such thing as a dumb question. Expats like me love to show off our knowledge of foreign countries and will sit for hours answering questions. You might find such people working in your company, or at social and educational clubs that focus on the country, or at local International or World Trade Club events. You can also go online to chat rooms at expatriate websites such as www.expatexchange.com and www.anamericanabroad.com to pose questions directly to expats currently living in the country you are considering.

One word of caution: You undoubtedly will encounter Americans who never made the adjustment or developed the right stuff to be successful as an expat. Unfortunately, these types are still around. Americans of this ilk have made a significant contribution toward creating the environment for anti-American sentiments as discussed in the Introduction. They (and the rest of the world) would have been better off if they had simply stayed home. Should you encounter an American expat who is overtly negative about his or her foreign assignment and who dwells on stereotypes, just thank them for their time and find someone with a more enlightened and useful attitude. It is best for you, as a new American expat of the twenty-first century, to leave these old American expats to their complaining about all things foreign, and move on.

Include Your Family in the Planning

If your family is relocating with you, their preparation, adjustment, and comfort should be your top priority. The well-being of family members will have a profound impact on your job success. Include family members in all aspects of the preparation and decision process. Do not underestimate their difficulties. While you are occupied by new challenges, your family members might be sitting at home in your new country feeling frustrated and isolated.

Children are particularly important from a planning standpoint. Relocation is difficult even for adults and can be especially unsettling for children. They will be uprooted from their school and friends, thrown into an unfamiliar environment, and exposed to potential health and safety risks. Conversely, their time overseas can be a great educational experience and a lot of fun if handled correctly.

The real key to your children's success is planning, preparation and, most important, including them in the process. Do not present them with a *fait accompli*. Do not treat their issues as trivial. You need to sell them on this and get their buy-in, rather than force them to go along with a decision that adults have already made. Spend time with them discussing and researching the country. Find interesting things that they will enjoy doing in their new home. Get them involved in researching potential schools. Help them make contact with children in the new country via e-mail so they'll know someone when they get there. Make the entire planning process a family project. If you get your children excited about the move, it will become an adventure for them and their adjustment will be that much easier.

Most of the expatriate internet resources cited in appendix C include sections, articles, and links on the special issues faced by families, and children in particular, when moving overseas. Another informative and well-written resource on this topic is *Third Culture Kids: The Experience of Growing Up Among Worlds* (Intercultural Press, 2001).

Selecting Schools

One of the choices that should be made prior to leaving the United States is the school your children will attend while overseas. Depending on where you are moving, there may be several alternatives, a few, or none. In most major cities, you choices will include public schools, so-called international schools, parochial schools, and home schooling. The choice you and your family make will significantly affect not only the level of success and enjoyment regarding the time spent overseas, but also the future direction of your children's education.

Local public schools are an option if your children can communicate in the language used in the school. Attendance at such a school can be a great cultural and learning experience and a means to become fluent in the local language. However, aside from possible language issues, the standards and curriculum will probably be significantly different from the U.S. schools they have previously attended (and to which they likely will be returning at the end of the assignment). This may negatively (or positively) affect their educational experience both in the foreign country and upon their return home.

International schools typically focus on the educational standards and curricula of one country (or they may offer an international baccalaureate program that is not tied to the educational system of any particular country). Such schools are accredited by either an educational accreditation board of the country on which their system is based, or the International Baccalaureate Organization in Switzerland, or both. Classes are taught in the language of the focus country or in English in schools with no specific country focus.

For example, Singapore has the following international schools: Australian, Canadian, Dutch, German, American, British, Japanese, Swiss, and French, plus several international schools that are not focused on any particular country but instead offer an international baccalaureate program. The standards of international schools tend to be very high, and a high ratio of graduates are accepted into elite universities and colleges. In addition, there will likely be many children from the United States enrolled in an international school, and most of them will be transient like your children. It therefore might be easier for your children to make friends in this environment, thus easing their adjustment to their new home. However, the potential downside of private international schools is that tuition tends to be expensive. Thus, these schools may not be an option without an employer's education allowance.

The embassy or consular office in the United States for your host country can provide information and contact details for schools in the city to which you are moving—both public and international schools. You can also find international schools by doing an Internet search on the string "*international schools*" "[name of city]," or through many of the country-specific online resources discussed above. Information regarding international schools that are sponsored by the U.S. government is available at www.state.gov (click on *History, Education and Culture —Overseas Schools*).

Private religious schools also may be available in the city to which you are moving. These often offer international educational programs as described above, but with an added religious focus consistent with the belief system of the sponsoring religious organization.

Home schooling might be a viable option for your children's education, particularly if schools in the foreign country are unsuitable for some reason or are too expensive. Home schooling can be a rewarding experience for parent and child and is a growing trend even in the United States. However, it is hard work, and care must be taken to ensure that the cultural and social interaction derived from attending school in a foreign country is provided through other activities. Information and assistance on providing a home-school education for your children, as well as information on home-schooling laws and support groups in more than 30 foreign countries, can be found at the following websites: www.americanhomeschoolassociation.org, www.home-ed-magazine.com, www.homeschooling.gomilpitas.com, and www.holtgws.com.

Spouse and Partner Issues

If your spouse or partner is working in the United States, he or she will likely have to give up this employment in order to accompany you overseas. Removing oneself from a career path, even for a short period, can be detrimental and eventually fuel resentment, particularly after the novelty of the new environment wears off. Even if not working, your spouse or partner will nevertheless be uprooted from friends, activities, and support networks. If your partner is female, she may face the additional challenges of a local culture that limits women's role and status or imposes restrictions on dress and mobility. If male, he will have significantly fewer nonworking colleagues of the same gender for a local support group than a female spouse, and possibly more cultural pressure in the foreign country to be gainfully employed. What will your spouse or partner do to keep occupied, fulfilled, and happy? There are numerous possibilities:

- **Seek employment in the foreign country.** This can be difficult, particularly for a wife in those countries that restrict the ability of women to work. However, it is not impossible. He or she will face the same work permit authorization issues as you. However, if your spouse is working in the United States, he or she will have knowledge, experience, and a set of skills that may be valuable in the new country—perhaps even to his or her current U.S. employer or to your company. There may also be opportunities to teach in international schools or to teach English as a second language. Even a part-time position can be both interesting and professionally rewarding. If appropriate, refer your spouse to chapter 2, "Finding Your Job Overseas."
- **Provide contract or consulting services in his or her profession or field of expertise.** With the Internet and today's ease of communication, such services can be provided from foreign locations even to clients back in the U.S.
- **Focus on personal development and education.** This could take the form of an internship, continuing education in a field of interest, obtaining a new degree or certification, becoming proficient in a foreign language, or personal study of a topic of interest. These can be undertaken through institutions located in the foreign country or via the Internet or correspondence classes.
- **Participate in volunteer or community work.** This is a good way to share skills and knowledge that are needed in the foreign country and to learn about its people, culture, and customs.

- **Focus on family development and growth.** There is no better time to spend with children than while exploring and learning about a foreign country, particularly if you have made a decision to home-school your children while overseas.

You should check with your employer about its policies and programs for assisting the spouses of its expatriate employees. Many companies have spouse-support programs in place or will be willing to provide some form of reasonable assistance if requested. You employer may also be able to provide introductions to potential employers and other networking contacts in the foreign country and/or assist in obtaining a work permit for your spouse.

Remember, a majority of all expatriate failures are attributable to spousal dissatisfaction. These issues thus need to be discussed and planned for early—before your departure.

Preparing For Your Foreign Job

Don't forget to prepare stateside for your foreign job—you do not want to just show up with a smile on your face. Talk to people in your company who have worked in or visited the foreign office. Research the business culture, personality, work ethic, and customs of the office and its personnel. Know who is in charge, who is working there, how successful the office has been, and the problems it has been encountering. Develop a strategy for how you will fit in, have an immediate impact, and make a difference.

Whatever your strategy, don't use it as an excuse to be arrogant or patronizing in your attitude. There is no more certain path to failure as an expat than treating your foreign work colleagues with disrespect.

Develop a Plan For Coming Home

Develop a plan for returning to the United States, particularly in regard to your job expectations and possibilities. Before you transfer, you should have an open and honest discussion with your company regarding mutual employment expectations upon successful completion of your foreign assignment. If your reintegration into the home office isn't discussed before you go, you may find that your employer quite frankly does not know what to do with you when you return. Your old job probably will have been filled, and there may be no available U.S. positions where your international skills and knowledge can be put to good use. If this is the case, you need to be aware of it, and plan on finding a new position when your expat assignment is over.

Getting Your U.S. Affairs in Order

House or Apartment

You have to decide what to do with your U.S. house or apartment. The options are clear: keep it and let it sit empty; sell or rent it out if you own; buy out your lease or sublease if you rent. However, the right decision on which option to take is not always so clear. The best option for you will depend on a number of variables. How long do you expect to be gone? If you allow your residence to sit empty, who will take care of cleaning, routine maintenance, and gardening? Are housing prices inflating or deflating? What is the local rental market like for your home (e.g., renting out a two-bedroom apartment in New York City may be easier than renting out a five-bedroom house in a small Midwestern town)? What are the tax implications of your chosen course of action? Are you willing to allow strangers to live in your home? Are there any zoning, building, or neighborhood rules that restrict your ability to rent out your property? Also, be aware that renting out your home is not necessarily a trouble-free option:

- Tenants require screening.
- Leases need to be negotiated and managed.
- Deposits and rent must be collected.
- Mortgages, taxes, insurance, and housing association fees must be paid.
- Maintenance issues have to be handled.

If you do rent out your home, check to make sure that your homeowner's insurance policy covers rental property and that it provides sufficient personal liability coverage. And although the services of a property management agency can run several thousand dollars a year, depending on the extent of services provided, they are usually worth the money.

Your decision about what to do with your house or apartment will obviously be influenced by the extent to which your employer is willing to assist you with the costs of these various options, as discussed in chapter 3 (see "Possible Expatriate Allowances and Benefits").

Automobiles

Generally, I think it is a bad idea to take your automobile overseas. Shipping an automobile is expensive. You might have to make mechanical modifications to comply with standards in the foreign country, and then you'll have to modify them again when you return. Obtaining parts and service for an American car could be problematic. Some countries impose severe restrictions and heavy taxes on the importation of used automobiles. To me, the real choice is between storing and selling your automobile. This requires a careful analysis of several factors: current age and value of your car, amount of money still owed on your car loan, length of time you expect to be overseas, likely depreciation in your car's value, and costs of storage (and will your company cover such costs?). I have moved overseas several times. Sometimes I have stored my automobile. Sometimes I have sold my automobile. When I stored, I should have sold it. When I sold, I should have stored it. Hopefully, you will make better decisions on this issue than I have.

Legal Matters

Use this opportunity to get your legal affairs in order. Update your will and your spouse's (or create them if you have not previously done so). Review your life, home, and automobile insurance policies—

make sure they are current and arrange for automatic payment of premiums while you're away. In particular, make sure that your automobile insurance continues to provide coverage when you're driving outside the United States. Consider whether you need to grant any powers-of-attorney so that someone can act on your behalf with respect to certain matters—for example, property transactions and renewing automobile licenses.

Place wills, insurance policies, powers-of attorney, deeds, automobile, and other title documents in one (secure) place. Also, make copies of other important documents:

- Front pages of all passports, and front and back pages of any work permits/visas for the foreign country
- Front and back of all credit and debit cards
- Drivers' licenses (check expiration dates)
- Social Security cards
- Notarized copies of birth and marriage certificates and divorce decrees
- Prescriptions for glasses and contact lenses
- Educational diplomas, professional certifications, and school and college transcripts
- Inventories of stored and shipped personal belongings

You will be taking originals or copies of these with you, but you should also maintain copies of everything in a secure place in the United States in case you need to replace them. Make sure that your attorney or a trusted family member or friend knows where these documents and copies are and that he or she can access them if necessary.

Banking and Finance Issues

You will need to have sufficient resources and arrangements in place to pay any continuing financial obligations you have on the U.S. side. Don't rely on bills and statements to be forwarded to you. Forwarding will take too long and you will end up being late with your payments, thus hurting your credit rating. You should arrange for automatic payment by your bank of any periodic bills that are of a constant amount. The easiest thing to do for other types of bill payments is to review and pay them online. Online bill-payment services are available through a variety of online providers (e.g., MSN, PayPal, CheckFree, Quicken) or through any major bank. Start using your online bill-payment system a couple of months before you relocate so you are certain that you know how to use it and that it works like you need it to.

You will also need a bank account in the foreign country for paying bills there and withdrawing cash for daily living expenses. You do not want to rely on ATMs for your cash requirements. Drawing foreign currency out of your U.S. bank account is expensive, is subject to monetary limits, and exposes you to daily fluctuations of exchange rates.

Find out if your U.S. bank has a correspondent bank in the foreign country at which you can establish an account—such an account may result in waived or lower fees for certain transactions between the banks. Determine how much of your pay will be deposited into your U.S. and foreign bank accounts, respectively. Learn the procedures, time frames, and fees involved in transferring funds from your home bank to the foreign bank (and vice versa).

You should also obtain a local currency credit card from your new foreign bank—you don't want to suffer the uncertainties and costs of using a U.S. dollar credit card for foreign currency purchases. Check exchange rates at your bank or online at www.oanda.com or www.xe.net regularly. You might be spending or withdrawing more money (in U.S. dollar terms) than you realize. Carry more than one credit or debit card, particularly if you rely on the card for cash from ATMs. Not all credit or debit cards will work in all foreign ATM machines, so make sure you have at least two sources for withdrawing cash. You can find ATM machines in any area for your particular card at the website for your credit card company and/or bank. If you inquire at your bank, you might also find that it has a reciprocal arrangement with one or more foreign banks that allows you to withdraw cash without having to pay a fee to either bank (e.g., if I use my U.S. bank debit card to withdraw cash from a Barclays Bank ATM in the United Kingdom, I am not charged any fees by either Barclays Bank or my U.S. bank).

Notifying Others That You're Leaving

Think of all the friends, family members, vendors, creditors, business associates, professional organizations, clubs, frequent flyer programs, and other people and companies with whom you regularly and willingly correspond. Now, think of all the absolute junk you receive in the mail that you do not want to deal with overseas. A good way to do this is to simply track (for at least a full month) your incoming and outgoing mail and to record the contact details for all individuals and organizations with whom you need to stay in contact.

Next, decide the best way to continue receiving correspondence from the individuals and organizations you want. Your choices are to provide them with your foreign address (which you may not know before you leave the United States), or to have all or part of your mail directed to a U.S. "forwarder" who will take responsibility for forwarding to you only the mail that you actually want to receive. This forwarder could be a family member, a friend or colleague at work, or one of the numerous companies that specialize in forwarding mail to expats for a fee (you can find many of these companies in your telephone book or through an Internet search).

Ultimately, you will probably give your foreign address to some individuals and companies and rely on forwarding from a U.S. mail forwarder to receive correspondence from others. Note that some companies will not pay the postage to send mail to you in a foreign country, and some credit card companies will cancel your account if you do not have a U.S. address. At the same time that you are arranging for the overseas forwarding of mail that you do want to continue receiving, you should also cancel magazine subscriptions and other mail that you do not want forwarded to you.

Health and Medical Issues

Checkups, Immunizations, and Prescriptions

Every family member should have a complete medical and dental checkup before leaving the United States and get any necessary treatment prior to the trip. Find out if you have any health or dental problems that require attention. Obtain any needed immunizations (including certificates of immunization that may be required to enter certain countries). Ask your doctor to write letters describing any medical conditions you or your family members have or prescription medicines you require, and get these letters translated into the language of the country to which you are moving. Consider wearing a bracelet or carrying card with you at all times that identifies any specific condition that should be known in the event of a medical emergency. Also, get copies of medical records for all family members (you will need these in your new country) and refills of prescription medicines sufficient to last until you can identify a foreign source for these. Keep all prescription medicines in their original containers so that you can prove they were legally obtained.

In addition, all family members should have eye examinations to make sure that the prescriptions for their glasses and contact lenses are still accurate. Take copies of all lens prescriptions with you to your new country, as well as backup glasses and a long-term supply of contact lenses.

Research the Health Situation in the Foreign Country

Your research of the country to which you are relocating should include detailed information regarding that country's general health situation and medical system. Know what diseases are prevalent, what health risks exist, what immunizations and certifications are required for entry, the quality of available health care, and recommended preventive health measures you should take.

Identify hospitals and doctors in the area in which you will be living (hopefully before you leave the United States, but in any event immediately upon your arrival). Don't get into a situation in which a family member has a medical emergency and you have no idea where to take them or how to get there. Also, make sure that the hospitals and doctors you identify are qualified to treat any medical conditions you may have and that either the staff or doctors speak English if you can't communicate with them in the local language. Identify sources for any prescription medicines you require, and find out what documentation you must have in order to get such prescriptions filled.

Learn how the health-care system works (or doesn't work, as the case may be). Many health-care systems are structured quite differently from ours. Health care is much more likely to be available either free or at a much lower cost than in the United States, due to the preponderance of national health systems funded through significant government subsidies. Alternatively, some countries have a mixture of public and private health-care systems, with the private systems structured more like the U.S. model (i.e., funded primarily through private insurance).

Often, public health care is available to noncitizens and/or nonresidents of the country at the same levels and prices as to citizens and residents. However, you should not count on this. Before you move, find out exactly how you will gain access to the local health-care system and the costs you will incur if medical treatment in the foreign country becomes necessary. Also, find out whether your health insurance covers medical treatment outside the United States and whether it is accepted by hospitals and doctors in your new country (health insurance issues are discussed in detail later in this chapter).

You should not reject the possibility of utilizing the public health-care system in the country to which you are moving without at least researching its quality and the competency of its medical professionals. There is much discussion about the relative merits and quality of government-funded public health-care systems versus private health-care systems, such as that in the United States. We Americans tend to criticize the quality of health-care systems in other countries while discounting many of the obvious problems with our own. I have experienced extended hospital stays in Saudi Arabia, the United Kingdom, and Germany, and have had more minor medical treatment in a variety of other countries. Many of my medical encounters were in public health-care systems. I personally have found the quality and competency of medical care provided in these public health-care systems to be quite satisfactory. I also found them to be exceedingly inexpensive (sometimes totally free) and usually available to me without regard to whether I was a citizen or resident of that country. A 10-day stay in a public hospital in Germany, with a variety of daily tests and treatment, cost 3,600 euros (approximately US$4,000 at that time), inclusive of all hospital and doctors' fees. The treatment was very competent, although admittedly Spartan, and I fully recovered from my illness. My point is simply that it is worth learning about a country's health-care system before you automatically assume that it is inadequate to meet your medical needs.

Health Information Resources

A good place to start your research regarding the health situation in the country or region to which you are relocating, and to find answers to many of the questions posed above, is in the *Traveler's Health* section on the U.S. Center for Disease Control website (www.cdc.gov). Alternatively, you can call the CDC hotline (877-FYI-TRIP) or get information that is available on the website via fax at 888-232-3299. The CDC website is easy to use and includes health information and recommendations for all regions of the world: recommended immunizations, types of diseases prevalent for the region, how to avoid getting sick, what you need to take with you from a health standpoint, and what to do after returning home. It also publishes a booklet called *Health Information for International Travel* that can be purchased from the website.

The World Health Organization (WHO) also has a website, similar to that of the CDC, at which you can find much information regarding the health situation in all of the 192 countries that are members of WHO (www.who.int). Some travelers find that WHO information is less alarming than that coming from the CDC. If you tend to hypochondria, you may want to go to the WHO website first.

Perhaps the best source for information on English-speaking doctors, hospitals, and dentists in the country to which you are moving is the local U.S. consular office for that country. You can access this information through the State Department's travel website at www.travel.state.gov. This website also has a list of more than 30 companies that provide air ambulance/medevac services for medical emergencies that require your evacuation to the United States.

The CDC website also links to the website for the International Society of Travel Medicine (www.istm.org) and the American Society of Tropical Medicine and Hygiene (www.astmh.org), both of which list doctors and health clinics throughout the world and include contact information, specialties, languages spoken, and other pertinent information. In addition, the International Association of Medical Assistance to Travelers (www.iamat.org) is a nonprofit membership organization (no fee required to join) that publishes a directory of medical specialists in 125 countries and provides other services and information for people traveling or living outside of the United States. Finally, information and contact details regarding dentists in foreign countries are available at www.smilesite.com.

Your own doctor and dentist may be able to assist you in locating appropriate medical and dental care in a foreign country through their professional affiliations. They have the advantage of understanding your specific medical and dental needs and thus are in a position to assist you in finding the best health care for your situation. Lastly, many credit card companies maintain lists of health-care contacts in foreign countries that are available to their cardholders.

Check Your Insurance Coverage

Check your company's health insurance policy very carefully. The standard policy may not cover medical care outside the United States, or it may cover only narrowly defined emergency situations or impose other limitations that severely restrict coverage. Also keep in mind that the U.S. Social Security Medicare program does not provide any coverage outside the United States.

In all likelihood, you will need either an international rider added to your employer's standard health insurance policy or special international coverage through an independent company. The State Department's travel website (www.travel.state.gov) has a list of insurance companies that specialize in expatriate coverage, as does www.overseasdigest.com (click on *Directory—Expat Insurance Info*). It is also easy to find international health insurance companies by simply doing your own Internet search of "*international health plans.*"

Many different types of international health insurance plans are offered through these companies, with a broad range of coverage and prices. Some of the questions you should ask in evaluating these plans include the following:

- What health conditions are covered—and not covered?
- Is the coverage limited to emergency medical situations or is routine health care also covered? How is "emergency" defined?
- Are there any limitations on coverage with respect to pre-existing conditions? (These insurers are not necessarily subject to the HIPAA regulations regarding preexisting conditions.)
- Is there an annual or lifetime cap on the amount of coverage? What are the applicable deductibles and co-pays?
- Is medical evacuation to the United States covered? Under what circumstances?
- Is preapproval required for treatment or hospital admission?

- Does the health plan provide coverage in all countries in which you will be living and traveling? Does it also provide coverage while you are in the U.S. or only for health care required outside its borders?
- Is there a list of approved health-care providers? Are you required to use these providers, or do you simply receive a higher percentage of coverage if do?
- Do you have to pay medical costs up front and then wait for reimbursement from the insurance company, or does the health-care provider invoice the insurance company directly?

Pet Health Issues

You should carefully consider whether it is a good idea to take pets with you on your foreign assignment. Don't just think about yourself and how much you will miss your pet. The most important issue is the pet's comfort and safety. Consider what life is going to be like for your pet in the foreign country before you lock them up in a kennel and ship them overseas.

Some cultures do not readily accept certain types of pets. Certain kinds of dogs are prohibited in some countries (pit bulls, for example). Or there may be restrictions on the size of dogs that may be brought into a country. There is no uniformity on these restrictions between countries, and they sometimes vary even within different cities or regions of the same country. Lengthy quarantine periods are required in some countries before your pet will be released to you. The climate of your new country may not be comfortable for your particular pet. There may be diseases in the foreign country that will make your pet sick. Veterinary care may be inadequate or unavailable where you will be living.

Even the transport of your pet to the country can be detrimental to its health, particular if it is old or temperamental. Consider all of these issues when determining whether your pet will benefit or potentially be harmed by an "expatriate" experience.

The best place to find answers to these questions, and to learn the specific immunization, certification, and license requirements for your pet, is through the embassy or consular office in the United States for the country to which you are moving. You can also contact professional pet shipping companies through the Independent Pet and Animal Transportation Association (www.ipata.com) for assistance on these issues. You will probably want to work with one of these companies in any event to arrange for the safe and comfortable transport of your pet to the foreign country (discussed later in "Getting Your Pet There Safely and in Good Health").

Have your pet thoroughly examined by your veterinarian and make sure it is healthy enough to make the trip and to survive comfortably in the new country. Get all required vaccinations and health certificates that are required for your pet to enter the country (as well as to transit any countries where you have connecting flights). Obtain a long-term supply of any special medicines, treatments, foods, or dietary supplements that your pet requires or enjoys. If these are not available in your new country, you will have to arrange for regular shipments of them.

Getting Documented

Passports, Visas, and Work Permits

In order to leave the United States and legally enter the country to which you are moving, you and your family members will need U.S. passports and appropriate entry visas and works permits issued by that country. It is rarely difficult for a U.S. citizen to obtain a passport, although the procedures have tightened up since 9/11. The standard processing time is now 25 business days, although you can obtain an expedited turnaround in emergency situations. The best source for information on the procedures, requirements, and fees for obtaining a U.S. passport is the State Department's travel website (www.travel.state.gov). Simply click on *Passports*. Answers to all your questions will likely be found there, as well as telephone and e-mail contact information.

Alternatively, you can call the National Passport Information Center directly at 877-487-2778 for information if you prefer to bypass the website.

The primary source for information on obtaining the visas and work permits required for you and your family to relocate to a foreign country will be the embassy or consular office for that country in the United States. A list of these offices, with contact information, can be found on the same State Department website or at www.embassyworld.com. The State Department's travel website also has a section that lists visa requirements for U.S. citizens to enter foreign countries (click on *International Travel—Documentation Requirements*), but the information in this section covers only short-term tourist and business visas (not work permits or residence visas). However, this can be useful information for any exploratory or other prerelocation trips to your new country. There are also numerous passport and visa expeditors who can assist you in obtaining passports and visas quickly. However, these firms deal primarily in U.S. passports and foreign tourist and business visas rather than foreign work permits and residence visas.

Work permits and residence visas for foreign countries tend to be handled on an individual basis by experts who specialize in such matters. In all likelihood, you will require the assistance of a visa expert in the country to which you are moving in order to obtain the necessary authorizations for you to work in that country and for you and your family to live there for an extended period.

Hopefully, your company will handle all visa and work permit issues for you—it will probably have been through the process before and will have access to the appropriate experts to handle the filing and processing of applications. If you are doing this on your own, information and contact details for immigration lawyers in most countries can be obtained through the U.S. consulate in the country to which you are moving. Websites and contact information for all of these consulates can be accessed through the State Department's travel website at www.travel.state.gov.

Special Rules for Special People

It is possible that you may be able to take advantage of special rules in a foreign country that are designed to either attract people with your specific skills or make it easier for people with your background to obtain the requisite visas and work authorizations. For example, many countries allow people who qualify for a particular status (e.g., bona fide students) to obtain work permits for a limited period. The international intern and exchange agencies discussed in chapter 2 can assist you in obtaining short-term work permits using these laws. In addition, foreigners who have specific skills that are in short supply (e.g., computer programming) are often allowed to enter and work in a country under a special visa classification.

You ancestry might even entitle you to a special status—for example, if certain of your ancestors were citizens of Ireland or Italy, you may qualify for citizenship in that country (something that would open up the entire European Union to you). Other countries sometimes have similar laws under which you might qualify for dual citizenship. The best place to inquire is at the embassy or consular office for that country in the United States. Obtaining a dual citizenship through ancestry usually does not jeopardize your U.S. citizenship. However, be sure to confirm this before you make any applications for foreign citizenship. You don't want to discover after the fact that your act of applying for a dual citizenship was construed by the American government as a renunciation of your U.S. citizenship. Information and rules regarding dual citizenship for U.S. citizens is available on the State Department's travel website (click on *Law and Policy—Citizenship and Nationality*).

The Timing Conundrum

One problem you may encounter is that the process for obtaining a work permit can take quite some time to complete—in some cases, it may be months before you are legally authorized to work in the country. However, you have probably been hired to fill a position that is open right now, and you are needed in the foreign operation as soon as possible. As a practical matter, there is often a lengthy period between the date on which you are hired for an expat job and the date on which you obtain the required authorization to work in the foreign country.

The wait for a work permit is a common problem. I have rarely been in possession of a completed work permit at the time I moved to a new country. However, my experience has been that immigration authorities in most countries are familiar with this issue and will grant some degree of latitude as long as a good faith effort is being made to apply for and receive the required work authorization as expeditiously as possible. On the other hand, don't just show up at the immigration counter and expect to wing it. Check with the immigration expert who is handling your application about the best way to deal with this issue. He or she might advise you as to a specific approach you should take, or provide you with a letter certifying that your application was filed on a specific date.

Entering Your New Country for the First Time

Upon entering the foreign country, you need to be prepared for an interview about the purpose and length of your stay. You might walk right through immigration by just opening your U.S. passport to the photo page, but you are more likely to be stopped and questioned. Many countries have special lines at the airport for foreigners who are entering under a residence visa or work permit. Know the guidelines and limitations of your entry documents and make sure that any answers you provide to the immigration officer's questions are consistent with these documents. Don't lie—even a little. If you are denied entry, this will be a big problem for you in your new job. As stated earlier, it is important to consult with the immigration expert who is handling your visa and work permit application to ensure that you understand what you should say (and not say) in response to the immigration officer's questions.

Other Documents You Might Need

In addition to a passport and a work permit/residence visa (which might be stamped or pasted into your passport), you must carry a number of other documents required for entry into the country. These might include health and immunization certifications for you and your family members, as well as for your pets.

You also should take with you hard copies or electronic files of the other documents you left in safekeeping back home (See "Legal Matters," on page 100 above). You never know when copies of these will be needed—possibly for reference purposes or application forms, to replace lost originals, or to cancel a credit or debit card. You may not have time to wait for someone in the United States to courier or e-mail new copies to you.

Take with you the names, telephone numbers, and other contact details for any U.S.-based companies or individuals you have been contacting regularly and/or whom you may need to contact while overseas. Lastly, carry extra passport photos of everyone, as well as photos of your pets. These will be needed for numerous applications and registrations and will also be of assistance in the event you become separated from a child or lose a pet.

Getting Your Belongings There

International Relocation Specialists: Getting a Good One

The relocation firm that you select should not only be an experienced international mover, but also have experience in moving households to the particular country to which you are relocating. The customs and documentation procedures for every country are different, and a relocation firm without the requisite experience of shipping goods into and clearing customs in your destination country may encounter delays and customs problems.

Obviously, you will want to obtain quotes from several companies, but you also need to ask the right questions regarding both their general international experience and their specific experience shipping to the country to which you are moving. In your questions, focus on the special packing and shipping procedures used to ensure that goods shipped internationally are not damaged. Get references from other customers for whom they have shipped goods to your destination country.

In particular, make sure that your moving company is familiar with the customs regulations, procedures, and documentation for the country. They need to be able to advise you regarding restricted or prohibited goods, possible duties that may be assessed, procedures for clearing customs, any inspections that will be conducted, documentation that must be completed, and time frames for delivering your possessions. Major international relocation firms all have the requisite experience in these areas, and many of them also provide a full suite of expatriate services including compensation management, cultural orientation, visa processing, and other services. Examples of major international relocation firms are listed in appendix D. You also can find international moving specialists in your area at www.1-directory-international-movers.com. There is also a good directory of international movers at www.escapeartist.com (click on *Moving and Living Overseas—Worldwide Movers*).

Storage and Shipping Options

You have five options for dealing with your household items and personal effects:

- Storage: Items that are too bulky or expensive to ship, and/or which you will not need in the foreign country, should be stored in a climate-controlled, secure environment.
- Ocean freight: Larger items, and personal effects that you can live without for several weeks, can be shipped ocean

freight. This is the most inexpensive option for getting your belongings to your new country.

- Air freight: Personal items that are needed right up until departure and immediately after arrival (and that are in excess of checked-baggage limitations) can be shipped air freight. This is expensive, but sometimes is the only option for items that you can't do without for an extended period of time.
- Checked and hand luggage: Items that need to be with you at all times can be carried in your hand luggage or checked. However, most airlines have recently tightened their restrictions (and raised their fees) regarding excess baggage. You should check with your airline on this before showing up at the airport.
- Another man's treasure: You should use this opportunity to get rid of all that junk you have accumulated throughout the years. Expats become very adept at living light after a few relocations.

Cost is obviously a factor in your decision on dividing your personal belongings into these groups. Before making any decisions, you need to understand the costs involved in each—and the extent to which your employer will cover such costs. Once you have a handle on cost issues, the next step is to (1) divide all of your items into the appropriate groups, and (2) stage your packing and shipments in a way that creates the least disruption possible in your and your family's personal lives.

In so doing, keep in mind that you won't need nearly as much *stuff* overseas as you think you will. Be circumspect in your decisions regarding what to ship versus what to store or dispose of. In addition, if you must make a choice between shipping early or just before your departure, you should err on the side of shipping early. It is usually easier and less disruptive to do without your important personal items on the home side than it will be in your new country. You will have enough adjustments and challenges overseas to keep you busy without having to do without needed personal belongings. You should target getting your shipment to the foreign country as soon as possible after your new home is available to you.

My friend Ted took a slightly different approach when he moved overseas. He shipped most of his items to Switzerland, and then he and his family remained in the United States (in temporary housing funded by his employer) until shortly before their belongings arrived in Switzerland. It was much easier for them to adapt to temporary housing in their U.S. home city than in a new city in a foreign country, and it helped the family's adjustment to move directly into their new home overseas the same day they arrived.

Create a clear, detailed inventory of exactly what items are going into storage and into each shipment. You probably will need this for customs clearance anyway. For boxes that are to be shipped, mark at least the room in which each box belongs (e.g., kitchen, children's bedroom). For boxes that go into storage, write or attach a summary of the items included in each box. Eventually, you or someone else will need to get into your storage unit to retrieve something that you forgot or that you need overseas.

Lastly, don't forget to keep records and receipts for all expenses incurred in connection with your relocation (including those relating to the disposition of your U.S. home). You will need them for submission to your employer for those expenses that will be reimbursed, and also for your tax records.

Take It or Leave It?

Generally, I follow the rule that the less you take overseas the better. I try to put as much as possible into either storage or the trash bin. I have done it the other way—moving an entire house full of furniture to a foreign country. I found such a move to be unnecessarily expensive for my employer and very damaging to my furniture. Also, much of my furniture didn't fit very well in the smaller rooms commonly found in foreign housing. Obviously, cost will be a factor in your decision. You need to understand whether your company will assist in covering costs of storage, shipment, and/or furniture rental in the foreign country, and whether limitations or caps are imposed on any of these options.

Items so important that you can't bear (or afford) to lose them should probably be left with a trusted friend or family member or put into a secure storage unit or safe deposit box at home. These would include items such as antiques, expensive jewelry, and family heirlooms.

However, I strongly recommend that you ship a certain number of "home anchors"—personal items that are meaningful to you and your family and that will provide a degree of comfort in your new surroundings. Doing so is particularly important for your children. These items might include family pictures, favorite home decorations, items that have a strong personal or emotional meaning to you, and your children's favorite toys or stuffed animals. Perhaps you can even include holiday decorations in your shipment. It is fun to celebrate your traditional holidays while overseas, and it helps you adjust. It is also a good way to introduce aspects of your home culture to new friends in the foreign country.

Lastly, you should consider including in your shipment a supply of any personal-care items that are particularly important to you—shampoo, deodorant, toothpaste, skin lotions, antacids, cosmetics, and so on. You may not be able to find your preferred type or brand in the foreign country.

Border Crossings

Your shipments of personal belongings will be subject to inspection by customs officials. Be cognizant of the restrictions and prohibitions on what can be legally brought into the country. Not everything that we consider normal in the United States is welcomed or allowed in other countries. Restricted or prohibited items might include the following:

- Weapons of any type
- CDs and tapes that contain obscene materials or sexual content (many countries have a decidedly less permissive attitudes than the United States on this subject)
- Magazines and books with pictures of women in Western-style dress or with advertisements for alcohol
- Religious symbols or artifacts (e.g., Bibles and crosses)
- Books dealing with controversial or political topics or that are critical of the country's government

Even if an item is not restricted or prohibited from being brought into a country, it is possible that a duty will be imposed on its importation. You can obtain information on customs rules, import restrictions, and duties at the country's embassy or consular office in the United States. The right international relocation firm can also provide competent guidance on these issues.

Getting Yourself and Your Family There

Tips for "Stress-Less" Traveling

It is impossible to eliminate all stress associated with relocation to a foreign country. However, this is the beginning of your great adventure, and you want to get off to a positive start. You can make the trip considerably less stressful and more enjoyable for yourself and your family if you observe a few basic guidelines:

Know Exactly Where You're Going and How You'll Get There

Carry a copy of your complete itinerary, with telephone numbers and addresses for hotels and people you may need to contact on arrival. Have all transportation and hotels prebooked. Carry confirmation numbers and copies of any receipts with you.

Spend the Money to Acquire Good-Quality Luggage

You'll want wheels, lightweight material, durability, and good compartmentalization. This is an investment that will pay off during your expat assignment. Place a colored strap around all your checked luggage. This will provide extra support and will help you identify your bags among all the other similar luggage. As you probably know, you should never put anything valuable such as jewelry or expensive cameras in checked luggage, particularly since it must now be kept unlocked for security reasons. Don't forget to include on all ID tags your contact information in your new country.

Anticipate that There May Be Problems and Delays

Keep this probability in mind, particularly if you must change flights. When deciding what to take with you in your carry-on luggage, assume that you may spend lengthy periods (perhaps even overnight) without access to your checked luggage. Depending on the ages and health conditions of family members traveling with you, your carry-on luggage might include the following:

- Prescription medicines, painkillers, antacids, antidiarrhea medicine, cough medicine, throat lozenges, antihistamines, motion sickness medicine, antibiotics, and bandages (my friend Martha carried a spare package of antibiotics with her when she moved to the Sudan, and they ended up saving the life of one of her friends who was stricken with dysentery in a remote location)
- Extra clothes
- Books, games, and other amusements for the children
- Bottled water, juice, and snacks
- Extra pairs of glasses, contact lens fluid, and cases
- Swiss Army knife or other utility knife (although you can no longer carry knives on board an airplane, it is a good idea

to pack one in your checked luggage; it will be useful as you get set up in your new home)

Before Your Departure, Obtain Sufficient Foreign Currency to Get You to Your Final Destination

Exchange rates and fees at airports are usually not favorable, and there is no guarantee that currency exchange facilities will even be open if you arrive during the night.

Arrive at Airports Early

The amount of time required for check-in and security clearance is an unknown factor in today's world. Also, check your connection times at transfer airports. Make sure you have enough time for you and your checked luggage (especially any pet being transported on the same plane) to make all connecting flights. Don't accept without question a travel agent's assurance that the connection is "legal." All that means is that the travel agent is allowed by the airline to book the connection. These legal connection times assume that every flight is on time and that there will be no problems or delays. In addition, travel agents are not familiar with the layout of most airports or the time required to travel between gates—they haven't been to most of the airports for which they are booking flights. Before I understood this, I missed several so-called legal connections (or my luggage was delayed until a later flight). I now allow sufficient time between flights for normal delays commensurate with the season and the layout and congestion of the transfer airport.

Carry All Travel Documents in a Secure Place

These include tickets; passports; visas and work permits; and health and immunization certifications. See "Carrying Money and Valuables" in chapter 6 for suggestions on how to avoid losing these or having them stolen while traveling.

Prepare for Possible Emergencies

All children should be carrying identification, a copy of their itinerary, and information on how you can be contacted. Instruct them on what to do if they become lost or separated from you. Leave a copy of your itinerary and contact details for your destination with someone at home.

Check the Weather at Your Destination and Be Prepared for It

You can do this online at www.accuweather.com or www.weather.com. *USA Today* also provides a daily world weather picture and forecast.

Dress Comfortably, but within the Acceptable Dress Code for Your Destination

You do not want to get off the plane in a Muslim country dressed for a U.S. airport.

Anticipate and Plan for Jet Lag

Determine the time zone difference between your home and destination, and know what time of day or night you will be arriving. Start trying to adapt to the new hours before departure. Upon arrival, force yourself into the schedule for the new time zone as soon as possible. Start eating meals on the new schedule immediately. Drink lots of water and not so much alcohol while traveling.

Be Alert

If traveling through or to a high-risk country, safety and security are critical. Observe the travel security guidelines set forth in chapter 6 under "Living in or Traveling to High-Risk Countries."

Arrive Early

If possible, arrive in your new country one or two weeks before you start work and/or your children start school. This will give you time to get settled in and partially adjusted to your new environment.

Be Aware of Customs Restrictions and Prohibitions

Signs will almost certainly be posted in English in the baggage claim area detailing various items that you aren't allowed to bring into the country (e.g., certain agricultural products and foodstuffs) as well as items on which a duty might have to be paid (e.g., liquor, perfume, and cigarettes). Take the time to read these signs before you walk through the customs zone. (See also the previous discussion under "Border Crossings").

Arrange to Be Picked Up at the Destination Airport for This first Trip

Don't rely on finding a taxi at the airport. Depending on the time of your arrival, taxis may be unavailable or you may have to wait a long time. In addition, you probably won't know the local taxi rules or rates or be able to speak the local language. You will be exhausted and in unfamiliar surroundings. You thus run a significant risk of getting ripped off. If possible, have someone from your company (or arranged by your company) standing at the arrival gate with your name on a sign. This will ease the final leg of your trip and impress your children.

Getting Your Pet There Safely and in Good Health

Before your departure, obtain your veterinarian's advice regarding measures you should take to ensure your pet remains healthy and comfortable during the flight. You will want to consider the pet's age, temperament, and general health condition. Issues such as feeding and watering, travel conditions (e.g., ventilation and sanitation), necessary medications, and the advisability of sedation need to be discussed as well.

The airline's requirements with respect to the transport of your pet must also be considered. Unfortunately, there is little uniformity among airlines regarding pet transport. They variously will or will not accept pets in the airplane cabin or as excess baggage, or they may limit the type, breed, or size of pets that are allowed. Some airlines will not transfer a pet to a connecting flight on a different airline. Airlines' requirements regarding pet health, feeding, watering, and kennel standards also may vary. Whatever requirements your airline may have at the time you schedule your flight, these may change prior to your departure date, particularly if there are seasonal or weather changes. The bottom line is that you will need to research the detailed requirements of your airline regarding pet transport and reconfirm these prior to your departure.

If your pet is traveling on the same plane as you, inform the cabin staff of this if there is a lengthy ground delay or if you are in danger of missing a connecting flight or getting separated from your pet. Include in your air freight shipment any of the following that your pet requires regularly: flea/tick spray, worm tablets, hairball treatment for cats, and any medicines or supplements that your pet requires for good health.

The safest way to transport your pet to a foreign country is to engage a professional pet shipping company. These can be found through the Independent Pet and Animal Transportation Association (www.ipata.com). Your airline may require professional shipping in any event since new security regulations require that cargo on passenger jets originating in the United States only be accepted from known shippers. A professional pet shipper will take responsibility for: pet pickup and delivery; processing health and immunization certificates and other documentation; making flight arrangements and tracking flights; and handling kennel, health, and sanitation issues during the transport.

The First Few Days: Settling In

The best advice I can give you on how to spend your first few days is to chill and relax. There is no need to do too much too soon. Let the folks back home know that you have arrived safely. Start with a little sightseeing and get a feel for the place. Allow yourself to get adjusted to the new time zone. In general, treat these first few days like a vacation rather than a relocation.

Chapter 4 **Summary** and **Checklist**

Getting There

1. *Before You Go:*
   a. Take advantage of training programs provided by your company for you and your family.
   b. Create your own personal timeline for accomplishing the tasks discussed in this chapter.
   c. Do an exploratory visit if possible.
   d. Research housing options, and make as much progress as possible in finalizing your housing arrangements.
   e. Research all the issues regarding the foreign country as identified in this chapter under "Learn Before You Go," using the resources listed.
   f. Talk to people who have lived in the foreign country.
   g. Include all family members (including children) in your research and planning.
   h. Research school options for your children and get them enrolled as soon as possible.
   i. Help your spouse develop a plan for what he or she will do in the foreign country regarding employment or other activities. Inquire whether your company has any spouse-support programs or is willing to assist him or her in some way.
   j. Don't wait until you are there to start preparing for your foreign job. Research your company's foreign operation and be prepared to make an immediate impact.
2. *Getting Your U.S. Affairs in Order:*
   a. Decide what you are going to do with your U.S. house or apartment and start implementing your decision as soon as possible. Find out the extent to which your company

will assist in paying the costs incurred dealing with your U.S. home.

b. Decide what you are going to do with your U.S. car.

c. Get your legal affairs in order: wills, insurance policies, powers-of-attorney, deeds, title documents. Make copies of other important documents as listed in this chapter under "Legal Matters." Place all of these documents in a secure place.

d. Make arrangements to pay any continuing U.S. financial obligations. Research banks where you can establish an account in your new city. Learn procedures, time frames, and fees for transferring funds to and from foreign bank. If possible, set up your foreign bank account before leaving the United States, transfer funds into that account, and apply for a foreign-currency credit or debit card.

e. Notify appropriate individuals and companies of your relocation, and make arrangements for the forwarding of mail that you want to receive in the foreign country.

3. *Health and Medical Issues:*

   a. Get medical, dental, and vision checkups for all family members. Obtain immunizations, required health certifications, letters describing medical conditions, and copies of medical records for all family members. Get sufficient refills on prescription medicines to last until you identify a source in the country where you'll be living.

   b. Research the health situation: prevalent diseases and health risks, structure of the health care system, cost of care, suitable hospitals and doctors in your area, and how you can gain access to the health-care system.

   c. Check your employer's insurance policy to determine if it provides sufficient coverage for you and your family while overseas or whether a supplemental international policy is required.

d. Consider very carefully whether it is a good idea (or even allowable) for pets to accompany you overseas. If you decide to take them, have their health checked and consult with your veterinarian as to how to keep them healthy during the trip and in the foreign country.

4. *Getting Documented:*
   a. Obtain U.S. passports and all required visas, work permits, and other documentation that are required for you and your family members to enter the foreign country. Your employer will most likely handle the application and processing of visas and work permits required by the foreign country.
   b. There may be special rules in a country under which you qualify for a work permit and/or residence visa (and, possibly, citizenship), based on your specific job skills or education, or even based on your ancestry.
   c. When you enter the foreign country for the first time, make sure that (i) you have all required entry documents, (ii) understand the guidelines of such documents, and (iii) do not say anything inconsistent with them. If you have any doubts or uncertainties, consult the immigration expert who is assisting you before you go.
   d. In addition to documents required for entry, also take with you copies of the documents and other items described in this chapter under "Other Documents You May Need."

5. *Getting Your Belongings There:*
   a. Choose an international relocation firm that not only has experience with international relocations but also has experience with the specific country to which you are moving.

b. Divide your personal belongings into: storage, ocean freight shipment, air freight shipment, checked and hand luggage, and "trash." Determine the extent to which your company will assist in paying storage and/or moving costs and whether any limitations or caps are imposed.
c. Stage your packing and shipments so as to create as little disruption as possible.
d. Create detailed inventories of what items are stored and what items are included in each shipment. Mark all boxes as to contents or room.
e. Don't go overboard in shipping furniture and other personal belongings overseas. In particular, don't ship important items that you can't bear (or afford) to lose. However, do take a certain number of "home anchors" that are emotionally or personally meaningful to you. These will help you adjust to your new surroundings.
f. Ship a supply of any personal care items that are important to you. Your preferred types or brands may not be available overseas.
g. Don't include anything in your shipment that is prohibited in your destination country. Your relocation firm should be able to advise you as to restricted or prohibited items and any duties that might be imposed on your shipment.
h. Keep records of all expenses incurred in connection with your relocation—for submission to your company for those that will be reimbursed and also for your tax records.

6. *Getting Yourself and Your Family There:*
   a. Observe the travel tips and suggestions as set forth in this chapter under "Tips for 'Stress-Less' Traveling."

b. Airline rules regarding the transport of pets are not uniform and change suddenly. To ensure your pet's health and safety, it is recommended that you use the services of a professional pet shipper to transport your pet to the foreign country.

c. Use the first few days to get a few immediate tasks accomplished, settle in, and explore. Relax and allow yourself to get adjusted to the new time zone. To the extent possible, treat these days like a vacation rather than a relocation.

Chapter Five

Being There: Living and Working in a Foreign Country

When faced with a totally different environment, many of us have trouble adjusting. On the other hand, if handled properly, living in a foreign country could be the most positive experience of your life. The first part of this chapter explains how the new American expat can adjust to, and get the most out of, the experience of living overseas. The second part of this chapter discusses what to expect, and how to achieve success, in a foreign job—developing the special set of personal skills and the attitude of flexibility and open-mindedness needed to succeed in your new environment.

Getting Set Up in Your New Home

After taking a few days off to recover from the trip, it's time to get set up for life in your new country. While this has always been an important and challenging task for expats, it has taken on a different perspective and a new urgency after the events of 9/11.

In-Country Assistance

Hopefully, your employer will have someone "in country" (either a local employee or an agent) who will meet with you shortly after arrival to assist you in getting acclimated and settled in. This is an invaluable service that your company can and should provide. Such a person can help you

- finalize your home arrangements and get utilities connected,
- rent a car,
- learn how to use local telephones,
- become familiar with local transportation,
- obtain basic food and household items to get you started,
- locate schools, shopping areas, services, medical facilities, and other locales that you will (or may) need to access, and
- assist with the myriad of other questions that will arise.

The list of initial tasks that you should take care of in the short term following your arrival include the following:

Finalize Your Home Arrangements

Hopefully, you started finalizing your home arrangements before you left the United States. If not, review the section entitled "Finding a Place to Live" in chapter 4.

Get Turned On

In conjunction with finalizing your home arrangements, you will need to make it livable and functional—for example, get utilities turned on, buy or rent local appliances and furniture, and set up telephone and Internet service. Don't be surprised if these setups take longer than you are used to.

Make sure you understand how you will be charged for telephone calls for both your land line and mobile telephones, particularly for calls back to the United States. Also, keep in mind that local landline calls are charged per minute in some countries (i.e., they are not free as they are in the U.S.). This can be a shock when your first telephone bill arrives, especially if you have been connecting to the Internet via modem. As for your mobile telephone, check whether you can use it backl in the U.S. (and if it will work in other countries to which you expect to be traveling). The frequencies used in various countries differ. You might need a special multiband chip in order to have coverage in all needed countries. Lastly, be aware that using your mobile telephone outside the country in which your service provider is located can be very expensive. You should inquire about out-of-country roaming charges if you expect to use it regularly in other countries.

You might encounter a variety of challenges as you attempt to get fully connected from an electrical and telephone standpoint. Unfortunately, there is little uniformity in standards between the United States and other countries. Even more unfortunately, the issue is not merely whether or not your electrical devices work in a foreign country—you might destroy or seriously damage an expensive item such as a computer if you do not understand and deal with the matter *before* you plug it in and test it:

- There are numerous types of telephone plug connectors in use around the world. There is not even any uniformity among EU countries. You will very likely need a telephone plug adapter kit in order to connect your computer via modem.
- Telephone lines in foreign countries can be either digital or analog. Your computer modem is analog. Hopefully, you will only get a "no dial tone" message if you attempt to connect through a digital line. However, in the worst case, you might destroy your modem card. It is possible to purchase converters that allow you to connect your analog modem into a digital telephone line.

- The telephone dial tone might be different in your new country, with the result that your modem does not recognize the dial tone. The best solution I have found is to simply adjust my modem settings to ignore the dial tone.
- Electrical plug connectors are very different outside the United States. Again, there is little uniformity. If you expect to be traveling to many countries, you need an electrical plug adapter kit to go along with your telephone plug adapter kit.
- Voltage standards are different. Even if you have the correct plug adapter, you cannot plug a U.S. device into a socket in many foreign counties unless the U.S. device is dual voltage (fortunately, many U.S. electrical devices, such as computers, are now commonly dual voltage). You must check the voltage before you plug it in or you almost certainly will destroy it. Voltage converters can be purchased that allow you to use U.S. electrical devices in countries with different voltage standards.
- Even if you have the correct plug adapter and voltage, the frequency of the electric current might be different from that used by your device (60 hertz versus 50 hertz). This can affect the operation of items such as clocks.
- The quality of the electric power supply in the foreign country might necessitate a more vigilant use of surge protection and/or universal power supply for your important electrical devices than is required in the U.S.
- The standards used for television screens (the number of horizontal and vertical lines of display) are different. There are three basic standards: NTSC (the U.S. standard), SECAM, and PAL. Your U.S. television, VCR and DVD players, and individual tapes and DVDs are NTSC-standard and will not work in many foreign countries. However, it is possible to purchase multistandard systems that recognize

and play all three standards or to convert your tapes and DVDs to the standard used in the foreign country.

The most comprehensive explanation I have found of these issues, and of steps you can take to deal with them on a country-by-country basis, is at www.travel-advisor-online.com. This website also contains information on where to purchase needed adapters, converters, and other items.

Register Yourself

My experience with U.S. consular officers in foreign countries is that they take their responsibilities to American citizens very seriously. They will go the extra mile (within the limits of their authority) to assist Americans in need of help or information. This can include the following:

- Providing names of local doctors and lawyers who speak English
- Providing information on voting in U.S. elections
- Helping you deal with passport and citizenship issues and with Social Security and federal benefits
- Arranging to get funds to you from home
- Trying to find you if your relatives believe you to be lost or missing or if there is an emergency at home
- Providing tax forms and information
- Notarizing/legalizing documents

A complete list of available consular services can be found in the publication *U.S. Consuls Help Americans Abroad* at www.travel.state.gov (click on *International Travel—Travel Brochures*).

Another important service that the U.S. consulate provides is to assist you in the event of a local crisis or emergency—unfortunately, much more of a concern in the post-9/11 world than before. However, the consulate cannot help you if it does not know you are there. One of your initial tasks should be to contact the embassy or consulate to register your presence in the country. Do not wait until you think that an emergency situation might be developing. You may be in danger and not even know it, or a crisis might arise quickly and unexpectedly. If you are registered, the consulate will send you e-mail alerts and try to contact you by telephone in the event of emergency or if important information needs to be disseminated to U.S. citizens. Registration procedures vary from country to country, so call your embassy or consulate directly, or visit its website to learn how to register in your country.

If a baby is born to you while overseas, you should immediately take the local birth documents to your U.S. embassy to obtain a Report of Birth Abroad of a Citizen of the United States and a U.S. passport. (Click on *Children and Family* at www.travel.state.gov for more information on procedures and forms required to complete this process.) You will not be able to travel home or to other countries without this documentation, and it establishes an official record of your baby's claim to U.S. citizenship.

Lastly, you might be required to register yourself and family members with the local police and/or other government departments. The U.S. embassy or consulate will be able to provide information on any such requirements.

Preserve Your Voting Rights

While overseas, you have the right to vote in federal elections via absentee ballot (there are variations among the states as to your right to vote in state and local elections). Information regarding your federal and state voting rights and procedures can be obtained at your local U.S. embassy or consulate or at the Federal Voting Assistance Program website (www.fvap.gov).

You need to plan ahead—you cannot just walk into an embassy or consulate on election day and vote. You must first submit a Federal Post Card Application (which you can obtain from your embassy or consulate or online at the previously mentioned website) and then complete the procedures required for receiving an absentee ballot. This is the sort of thing that is easy to put off. You should complete the process promptly upon your arrival in the foreign country, or come election day you will likely find yourself unable to exercise your right to vote. Even after the controversy in the 2004 U.S. presidential election involving absentee ballots from U.S. citizens living abroad (among other things), many states had difficulty ensuring that overseas voters received their absentee ballots for the 2004 election on a timely basis. Don't wait until the last minute to figure out what you need to do in order to vote.

Locate Medical and Emergency Assistance

Locate and visit the hospitals, doctors' offices, and other medical facilities that you will be using during your overseas stay. Make sure that you know where they are and how to get there. Contact all doctors and make an appointment to introduce yourself, drop off medical records, and get logged into their systems.

Set Up Your Bank Accounts

Visit the local bank that you have selected and finalize setting up your accounts and obtaining necessary credit and debit cards.

Get the Children Enrolled in School

Take your children to their new school and get them enrolled. You should make a big deal out of this. Arrange for them to be given a tour of the school and to meet teachers and staff if possible. This will ease their nervousness as the first day of school approaches.

Deal with Automobile Issues

Finalize your automobile arrangements. If you are leasing a car yourself, make sure you know the lease terms and conditions—particularly your obligations regarding maintenance and lease termination. Confirm the validity of your driver's license in the foreign country (you will probably be allowed to drive under your U.S. license for a specified period of time) and that you have appropriate insurance coverage. Your U.S. automobile insurance policy may or may not cover you while driving outside the United States. Before you get behind the wheel of the car, learn the driving rules and important road signs.

Familiarize Yourself with Local Surroundings

Locate trains, buses, and other local transportation in your area and figure out how to use them. Find grocery stores, gasoline stations, shopping areas, and other businesses that you will frequent. Identify the best routes to work and school. Ascertain the routes you would take in the event of a crisis or other emergency requiring your quick departure.

Figure Out How to Make Things Work

Many common appliances, utilities, and other items operate very differently than do their U.S. counterparts. Telephones, stoves, hot water heaters, irons, lighting fixtures, air conditioners, heaters, and washers and dryers—in fact, almost all the conveniences you use daily—are included in this caveat. Take the time when you first arrive to get operating instructions translated and to learn how to use all of these. Don't wait until the first freezing night to discover that you have no idea how to turn on the heat.

Create a Directory of Important Contacts

Make a list of people and companies that you might need to contact while living in your new country: emergency services (police, fire, ambulance), emergency/crisis contact numbers provided by the U.S. embassy, landlord, maintenance or repair services, telephone company, utilities, taxis, food delivery services. Write down their numbers and keep them close to your telephone. This is especially important if the local telephone book is written in a language that you cannot read.

Getting Adjusted to a New Culture

Stage One of your new life overseas will be one of *excitement and adventure*. Everything will seem new and interesting. You will use the word quaint a lot. You will think that the inconveniences and challenges encountered are fun to conquer. You will feel like an adventurer. Unfortunately, the initial excitement can erode pretty quickly into Stage Two: *discontent and frustration*. Things that were "quaint" the first week—like that tiny refrigerator/freezer in your new U.K. home—will become (initially) a mild irritation and then a source of extreme aggravation. You will find yourself developing a poor attitude about the people and culture of your new country. This is normal for expats on their first foreign assignment (and even on subsequent ones), and you should not become discouraged or worried that it signals an impending failure. However, it is important to work through these problems so that you do not become permanently entrenched in the downward spiral of Stage Two. Otherwise, your Stage Three will no doubt be a *crash and burn* rather than the *successful adjustment* to your new environment desired.

Your Stage Two discontent and frustration is caused by three factors. First, you are overwhelmed by the changes in your environment that require adjustment on your part:

- Language and communication problems
- Unfamiliar customs and lifestyle (i.e., the day-to-day traditions and practices of the local population)
- Absence of U.S.-style conveniences
- Housing that is very different from your U.S. home
- Restrictions on your activities that were a normal part of your life in the United States, such as consumption of alcohol or mode of dress;
- Reduced mobility (e.g., women are not allowed to drive in some countries)
- Weather that is very different from what you are used to

- A different pace and seeming lack of urgency in getting things done
- Increasing frustration that "nothing works right," even after you have paid to have everything fixed

These are common expat refrains, particularly in countries that are very dissimilar to the United States.

A second problem is that you've lost contact with most of the things in your life that provided comfort and security. You find yourself missing these now, even if they seemed trivial back in the United States. This will in turn lead to criticism and resentment of your new environment because it doesn't provide you with those things that make you feel at home.

Your final area of frustration relates to the multitude of day-to-day activities that you performed with almost no thought or effort back home—for example, buying groceries, making a telephone call, taking a bus or subway, ordering food in a restaurant. These formerly mindless activities are suddenly not mindless at all, and you are required to invest too much thought and effort in doing something that was routine and effortless at home. You can wear yourself out just trying to get through a normal day.

In sum, all of these factors will merge to make you forget that you also had to deal with day-to-day problems, albeit of a very different nature, before you moved overseas. You will start to remember your life in the United States as being much better than it really was. (Okay, the TV really *was* much better.)

Family Matters

Your adjustment process should center on your family. If they have accompanied you overseas, your spouse and/or children can be a positive factor in your own adjustment. However, keep in mind that they also have their own issues and needs. In fact, your family's adjustment might be more difficult than your own. You have your job to keep you occupied, and you will benefit from the resulting personal interaction and work-related challenges.

There is often a tendency to become more involved in your work on an overseas assignment than you were back home, particularly in the beginning. This is a mistake. Put aside time for your family to help them adjust to, and enjoy, their new environment. Approach all of the adjustment issues discussed in this chapter as a family project. Do not leave your spouse and children to figure things out on their own. Aside from the fact that your family should be more important than your job in any event, this approach will in the long run benefit your new job much more than overworking yourself.

If your family does not join you on the overseas assignment, you have a different set of issues to face. For much of my expatriate career, I have been assigned to a variety of countries in different parts of the world while my wife and children lived back in America. As a result, I was not there to help my wife deal with day-to-day issues, and I have missed many of the special events and stages in my children's lives. This time can never be recaptured.

This situation creates the need to find other (affordable) ways to communicate and maintain family bonds. Obviously, e-mail correspondence, telephone calls, and special cards and gifts from exotic countries will help to keep you in contact with your family. In addition, keep your family posted regularly as to your whereabouts and activities, particularly new countries visited and interesting or unusual experiences. Invest in the necessary equipment and software so that you can exchange pictures and/or video clips of each other's current activities. When scheduling your business trips back to the United States, try to include a home stopover with several days of free time to spend exclusively with your family. Also, bring your spouse and/or children to see your new country—this is an experience that may not be available later in their lives.

Try to make each visit something special—an adventure or romantic getaway. Take time off during these visits so that you are not splitting your time between job and family. Your company is getting 100 percent of your time while you are on foreign assignment, and is benefiting from your willingness to be apart from your family for extended periods of time. It will not begrudge a few days off here and there and will be able to survive your absence.

If you have chosen to accept a foreign assignment apart from your family, hopefully there was a positive reason for doing so. You and your family members should periodically remind yourselves why you are doing this, and review the progress you are making toward achieving your goals.

Acclimate in Stages

Don't overwhelm yourself by trying to adjust too fast—adjusting to life in a new country is not the same as jumping into a swimming pool filled with cold water. True, there are many aspects of your new environment that you cannot change. They are what they are, and you have no choice but to adjust immediately. For example, if women are required by law or culture to cover themselves or wear specific items of clothing, you cannot choose to comply by gradually adopting a more modest style of dress. Similarly, if automobiles are driven on the opposite side of the road, you have no choice but to immediately conform your driving to the new rules.

However, you *can* exercise a degree of control over how quickly you expose yourself to the various aspects of your new environment: You don't have to immediately start eating the local diet—there are almost always American or other Western restaurants and food available somewhere nearby. Similarly, you don't have to jump in the car for a long drive your first week—you can take some time to get used to the new driving rules. In sum, you don't have to immerse yourself immediately in all aspects of the new culture and language. You can start with selected tasks that need to be done immediately, limiting your exposure at first. Then, gradually become immersed as you start to feel more comfortable in your new environment. Pace yourself and set goals: for example, today, learn how to use local transportation and go to the store to buy groceries; tomorrow, spend the entire afternoon exploring and learning the city center; and on the weekend, learn how to drive and practice by driving around your surrounding area.

Appreciate Your Host Country's Customs and Traditions

Now is the time to follow up on your initial research (see "Learn Before You Go" in Chapter 4) to really learn about your new home. Seek out opportunities to learn about the customs, religion, politics, history, and, most important, *the people*. Take the time to understand the culture and, to the extent things are different from what you are used to, the underlying reasons *why* they are different. The more you understand the people and the way they think, the easier it will be to not only accept and adjust to such differences, but to appreciate their merits as well. As an example, my friend Martha adopted the tradition of washing her face, feet, and hands five times per day when living in Muslim countries. She discovered that not only was this a healthy practice in the desert climate, but that the local citizens respected her for this respectful observance of their religious custom.

Communicate Early and Often

An important part of adjusting is to start communicating with the local people as soon as possible. Show an interest in their culture and customs. Ask questions about anything you don't understand or are curious about. People everywhere are proud of their country, and your expressions of interest will encourage them to open up to you. However, be careful to not phrase your questions as though you are complaining about an inconvenience or a difference from the United States. (Many people in foreign countries have the perception that we think everything is better in the U.S.)

Dealing with Sensitive or Controversial Issues

Generally, I recommend steering clear of locally sensitive or controversial issues until you develop relationships that allow such dialogue. It is definitely better to avoid these discussions with *anyone* unless conducted in a language that all participants understand and speak very well. Unfortunately, the circumstances of the post-9/11 world do not always allow for such discretion, and you will find yourself being queried by friends and strangers alike on a broad range of global and local issues—particularly those relating to the United States and its government policies. Some of these queries will be driven by a sincere desire for understanding and knowledge. Some will seem emotional and potentially confrontational, perhaps evoking your own sense of patriotism in response.

When I find myself in such situations with strangers or people whom I don't know well, I find that the best policy is to politely listen and avoid saying anything controversial or inflammatory. At most, I might inquire as to their own views and express my understanding of their feelings. Then I turn the discussion to a different topic. After all, little or nothing positive can be accomplished in the context of a conversation with a taxi driver on the way to the airport or a stranger seated next to you on the subway (and under no circumstances allow yourself to be drawn into these discussions at the local bar—alcohol and inflammatory conversations do not mix). The best you can do is to ensure that you do not exhibit the Ugly American stereotypes described in the Introduction.

With friends and colleagues with whom you have developed a personal relationship, I recommend a totally different approach. Open, civil, noninflammatory dialogue between citizens of different cultures is the only hope we have for achieving understanding of others and resolution of global issues. However, be prepared to conduct these discussions with an open mind and to learn something—perhaps to even change your own viewpoint. Otherwise, there is no point to having them.

Avoid the Fortress Mentality

Don't languish in your home. Get out. Get some exercise. Meet some people. Join a group activity that interests you. Many expats become frustrated if they can't get things done or they make mistakes. Their response is to then retreat to the relative familiarity of their home. Subject to my previous advice to acclimate in stages, you must force yourself out there and make your mistakes. If you do so, it will get better—soon.

If you hole up in your house or apartment, and/or surround yourself with the negative American expats described in chapter 4, your attitude will only continue to deteriorate. You will develop a siege mentality and an us-versus-them attitude. This is a particular risk in countries in which Western expats customarily live in walled compounds separated from local residents. These compounds usually provide Western lifestyle and conveniences, and it is an easy misstep to find yourself rarely venturing outside of the walled comfort zone. If you allow this to happen, you will have missed the unique and rich experiences that were available outside of your compound.

Obviously, my recommendation that you get out and explore your new country must be tempered by any safety and security risks in that country. See the section entitled "Living in or Traveling to High-Risk Countries" in chapter 6 for a discussion of these issues.

Start (or Continue) Learning the Language

If you can't speak the local language, anticipate this and prepare for it. Don't set yourself up for frustration by ignoring the issue. If you have an appointment—doctor, hair stylist, whatever—make arrangements in advance for access to an English speaker. If you have trouble making yourself understood in a store, ask for someone who speaks English to help you. One of the first phrases you should learn in any foreign country is "Excuse me, can you speak English?" Don't just assume that everyone does—to do so is impolite and may get your conversation off on the wrong foot.

Learning to speak the local language is important and useful. Although English is now widely spoken throughout the world, your experience in a non-English-speaking country will be greatly enhanced if you can communicate in the local language. The residents will appreciate your efforts and respond positively. Life will be less stressful—shopping, taxis, telephone calls, dealing with repairmen, and so forth. This will in turn significantly impact your perception and attitude toward the country and its people. For example, my own feelings toward France, and the French in general, totally changed after I visited France several times with my wife, who could converse with the local residents in French. Everything was so much easier, and I encountered none of the problems that I had previously attributed to stereotypical French rudeness.

Having your family with you overseas provides an opportunity to turn learning a new language into a family project. It is always easier to learn a foreign language in a group setting that provides practice partners. You can study together and designate certain evenings or hours during the week as "no-English" zones. Obviously, professional lessons are a good idea, if you have the time and resources. However, there are also innumerable self-study CDs, tapes, and books available for most languages (at bookstores everywhere) that you can use alone or in conjunction with lessons.

Due to time and scheduling constraints, I have generally used these self-study materials (rather than formal lessons) in my attempts to learn a foreign language. I have also developed various techniques to augment this self-study:

- Repetition is obviously important. My friend Martha, who unlike me has become very skilled in the local language of every country in which she has lived, states that it is necessary to hear a word 200,000 times before it becomes part of your vocabulary. This seems over-the-top to me, but that might explain why she is fluent or near-fluent in several foreign languages while I am barely conversant in only a couple.

- One of the most effective (and enjoyable) techniques I have found is to rent DVDs of movies that are in both English and the local language. I first watch the movie several times in English with the local language subtitles. Doing this allows me to see the local-language translation while I hear the English. Then I reverse the process and watch the movie in the local language with English subtitles. I thus can hear the local-language translation while seeing the original English. This is not perfect, since subtitles are often shortened and thus do not present a word-for-word translation of the original English. However, I have found that it is one of the fastest ways to become conversant in a foreign language.
- I use my "mindless" time to study. For example, I place a small speaker under my pillow and listen to language tapes before falling asleep. While living in Frankfurt, I had a 45-minute commute to my office via train, and I listened to language tapes both ways. This added one-and-a-half hours of study time that I would ordinarily have used reading a newspaper or staring out the window. Alternatively, Martha considers foreign radio stations to be the ideal medium for learning the language. She advises listening to the radio as much as possible, including while sleeping.
- I carry a small dictionary everywhere and make a point of translating signs and looking up words that I hear.

Of course, the only way to really learn a foreign language is to get out and use it—immerse yourself. I establish two or three "workout" periods each week, during which I venture out and speak nothing but the foreign language, no matter how confused I become. (One problem with this is that local residents often speak English better than I speak the foreign language, so they try to rescue me by reverting to English.)

Eventually, these workout periods will no longer feel like workouts, and you will find yourself communicating more and more in the foreign language. Along the way, there will be plateaus in your progress. It may seem as if you are making no improvement at all, and then all of a sudden you will see a significant jump in your abilities. Be prepared to laugh at your mistakes along the way. When my friend Greg moved to the Czech Republic, being an adventurous sort of guy, he jumped right in and attempted to communicate in Czech. On his first trip to the market to purchase cheese, the vendor kept asking in Czech whether Greg wanted more. Greg kept saying "no," not realizing at the time that the word *ano* means "yes" in Czech, and is often abbreviated to *no*. He learned a valuable lesson that day, had a good laugh, and went home with 10 pounds of cheese.

Do not get discouraged, and don't push yourself unnecessarily. Learning a foreign language can be hard work and very tiring. If you treat your study like a hobby rather than a job, it will be more enjoyable and come to you much more naturally.

You can also translate the written text of many major languages at websites such as www.babelfish.altavista.com, www.freetranslation.com, and www.translation1.paralink.com. These websites provide a surprisingly good translation, both from English into the foreign language and vice versa. They are particularly useful for translating a block of text from an e-mail or other correspondence.

Find and Create Some "Home Anchors"

It is not necessary, or advisable, to totally separate yourself from all things American. In chapter 5, I suggested that you bring some "anchors" with you—meaningful items from your U.S. home that provide comfort as you adjust to your new environment. You should also seek out a few anchors in the foreign country that provide a connection to your U.S. home. These can be a source of relaxation and solace for you during and after your adjustment phase. You might try the following:

- Join a local expatriate group such as American Citizens Abroad or an embassy-sponsored organization, and associate yourself with other American expats who have positive, inquisitive attitudes about the country.
- Find a special American restaurant and visit it periodically.
- Attend American sports events, such as World Football League games in Europe, or play in a weekend expatriate softball league.
- Continue to engage in an activity or hobby that you enjoyed back in the United States, such as playing a musical instrument, playing bridge, or collecting stamps.
- Celebrate U.S. holidays abroad, both with other American expats and your new friends in the foreign country.

These activities can all serve as a lifeline between the comfort and familiarity of your old home and the challenges of your new one. They also provide a means for making new friends and introducing them to American culture. For example, when I lived in England, I sponsored a Fourth of July party every year, at which my English friends would have their once-a-year opportunity to eat hot dogs and play softball.The highlight of the event was always the Dress Like an American contest, with prizes for the outfits that most accurately captured the English stereotypical view of American fashion. We had a good laugh, and it gave the English an opportunity to poke some harmless fun at the colonies. I am told that this annual celebration continued long after I left England.

Keep in Touch with the Home Crowd

With today's technology, such as the Internet and e-mail, it is easy and inexpensive to maintain contact with your home base. Also, numerous international call-back services and special rate plans exist that allow you to make telephone calls to friends in the United States at relatively low rates.

The Internet allows you to connect to radio and television stations in your home area via websites such as www.radiotower.com, www.live-radio.net, and www.tv4all.com. Your local U.S. newspaper probably has an Internet edition that you can visit daily or that allows you to sign up and have local news bulletins emailed to you.

All of these are good ways to keep in touch with your friends and stay up to date on events in your home community, thus providing another anchor for your adjustment process. Once you have settled in and are prepared to entertain visitors, another great way to stay in touch is to invite family and close friends to visit you in your new home. It is fun to show them around and might provide them with a once-in-a-lifetime opportunity to visit that country. It will also help you avoid homesickness.

Don't Give Up

If you are suffering through a difficult adjustment process, it is tempting to take an unscheduled trip back home to regroup and pull yourself back together. Do not go back to your U.S. home while you are in a negative mood about your foreign experience. Things will inevitably seem better there than they really were, and you are likely to give up on your foreign assignment. Remember, there was a reason you moved overseas. You have a plan and goals that you want to achieve. Revisit your plan and reconfirm your goals. If you need to get away, take a vacation within the country and have some fun, or visit another country. You will get through this rough period and laugh about it later.

Remember that even if you faithfully apply the principles and techniques discussed in this chapter, your Stage Three *successful adjustment* will not be achieved overnight. It is a gradual process. Furthermore, you will occasionally backslide into Stage Two again, no matter how much experience you have as an expat. This is just human nature and part of the adjustment process. So, I repeat: *don't give up*.

On the Job—Adapting to Foreign Business Practices, Culture, and Traditions

Much is written about the need to understand and adapt to the different business practices, culture, and traditions of any country in which you are working or doing business. Doing so is not merely courteous and politically correct: it is good for business. The better you understand the people with whom you are doing business, and the more comfortable they feel with you, the more likely you are to achieve your business goals. The bottom line is that understanding the people you interact with, and how they think, gives an advantage in your business dealings.

In the next section, I discuss these issues and reiterate the importance of understanding and adapting to these differences, being flexible and open-minded in your dealings, and communicating in a manner that will be understood and appreciated. However, I would like to start by offering three caveats to this advice.

Caveat 1: It's Okay to Act Like an American

First, I sometimes think that the emphasis on cultural adjustment gets overblown and takes on a life of its own. You don't need to develop a totally new "foreign persona" in order to be successful in international business. It is often okay to just act like an American—on balance, most of us have more positive than negative traits.

I was once told by a Japanese businessman that he liked to do business with Americans because of what he viewed as American traits: confidence, forthrightness, and decisiveness. However, it seemed to him that Americans who visit Japan on business trips are always trying to "be Japanese." His point was that we seem to believe that taking on the characteristics of a Japanese businessperson is a prerequisite to success in Japan. In his opinion, we will be *more successful* if we act more like Americans are expected to act.

The literature on cultural adjustment is replete with examples of business and cultural taboos that will bring disaster to any uninformed American so uncouth as to perpetrate them in the presence of his or her foreign hosts. With all due respect, I think these taboos tend to be exaggerated at times. I once asked a Saudi business partner about the conventional wisdom that one must never show the heels of his shoes to a Saudi. He response was that this was extremely important in dealing with his grandfather and perhaps with his father to a lesser extent. "However," he was quick to add, "I am an Oxford man myself." His point was that the world is indeed becoming a smaller place and that international business people in all countries have developed a level of sophistication that transcends old rituals and customs.

So by all means understand, appreciate, and adapt to differences in foreign business environments. However, don't go overboard. You will just end up looking silly if you try too hard to be take on the persona of your foreign hosts.

Caveat 2: Develop an "International Personality"

To the extent that you *do* adjust your "American-ness" to increase your effectiveness in foreign settings, you can't expect yourself to be a chameleon, constantly changing your style and personality to fit each of the different countries in which you travel and do business. I have often visited as many as four different countries on a single business trip. In preparing for these trips, I always *do* take time to study the business practices, culture, and traditions of any country that is new to me. However, rather than constantly adapting my style and personality to each specific country, I have been better served by developing a single "international personality" that I use wherever I am. Its traits include the following:

- Being polite and respectful to every person I meet, regardless of social or business rank
- Listening and never interrupting another speaker

- Being honest to a fault, but not in an aggressive or rude manner
- Not being too informal in my dress or manners
- Working very hard to not talk too fast, too loud, or too much
- Showing a sincere interest in my host's country and its culture
- Not insisting on having things my way—showing a willingness to consider alternative approaches and ideas, even if I feel certain they will not work
- Not hesitating to ask questions or to point out possible problems, so that there can be an open discussion of issues
- Being willing to admit that I might be wrong or have made a mistake
- Always thinking and pausing before responding, to show that I am giving their statement full consideration

You will get along just fine in any country if, in addition to your basic understanding of that country's business practices, culture, and traditions, you apply these principles to your business dealings. In the final analysis, your foreign business colleagues have the same goals as you—to grow revenue, increase market share, and create profits. They will not mind that you act somewhat American as long as you do your part to achieve these goals.

Caveat 3: The Importance of Mutual Adjustment

Lastly, one would hope that your foreign business partners or work colleagues will view this adjustment process as a two-way street—as a mutual exchange of business culture and traditions. Certainly, if assigned to the foreign office of a U.S. company, you have been sent there for a reason—to introduce new ideas or skills, share knowledge and/or procedures from the headquarters, or otherwise make an impact. Any local manager or employee working in that office should expect to adjust to a certain level of U.S. business practices (if not, they shouldn't be working there). However, even when working at or dealing with a foreign company with no U.S. connection, you nevertheless have a lot to offer in the way of positive U.S. business practices and traditions. Most people with whom you are dealing will want to take advantage of these and meet you halfway in the adjustment process.

Understanding and Adapting

What do you need to understand about the business practices, culture, and traditions of a country in which you are working or doing business? I divide this into three areas: business etiquette, work rules and customs, and management style. The depth of understanding required to be successful depends in large part on how much time you spend in the country. Do you live or travel there often, or are you an infrequent visitor? The more time you spend in a country, the greater your need to thoroughly understand and adapt to the local practices, culture, and traditions.

Business etiquette encompasses issues such as

- business entertaining and gifts,
- making appointments and punctuality,
- greeting protocol,
- business card customs,
- titles and forms of address,
- gestures and body language,
- acceptable style of dressing for business meetings,
- protocol and pace of business meetings, and
- customary topics of conversation (i.e., are questions about one's family or other personal matters acceptable?).

Work rules and customs include, for example: interaction and relationships between employees, work ethic, hours of work, vacation practices, attitude toward overtime, formality and style of work environment, attitudes toward gender-related issues, and procedures for maintaining work discipline.

It also is important to understand the *management style* in your new country, namely, negotiation style and techniques, how decisions are made and who makes them, the nature of the relationship between managers and employees, the role of employees in decision making, the level of bureaucracy prevalent in the organization, style of meetings, and the management structure and hierarchy.

Business etiquette, work rules and customs, and management styles vary not only from country to country, but you might find significant differences even between regions of a relatively small country (e.g., the German, French, and Italian areas of Switzerland).

Understanding a country's business practices, culture, and traditions will enhance your business experience even in a country that you are visiting only for a few days. However, this knowledge is indispensable if you are living and working there as an expat. Numerous sources are available to help you reach the needed level of understanding. The best single source I have found that provides a thorough overview for most countries in which you are likely to live or travel on business is *Kiss, Bow, or Shake Hands: How to Do Business in Sixty Countries* (Adams Media Corporation, 1995, 2006). I routinely review the relevant pages of this book before traveling to a country that is unfamiliar to me. Another good book on this topic is *When in Rome or Rio or Riyadh: Cultural Q & As for Successful Business Behavior Around the World* (Intercultural Press, 2004). Other sources that focus on the business practices, culture, and traditions of a specific country or region are listed in appendix E.

Attitude Is Everything

It is not sufficient to simply understand and adapt to the different business environment in your new country. Your *attitude* toward work colleagues, as evidenced by your conduct and style of interaction, is equally important. Openness, tolerance, patience, humility, sense of humor, flexibility, and communication skills (personality traits discussed in the section entitled "The Right Stuff" in Chapter 1) must be developed and applied in your new job. Review this section again and consider each of these traits in relation to specific situations that arise in your job. Your ability to build personal relationships with your work colleagues hinges on how well you integrate each of these traits into your own personality and style of interaction. And, in the final analysis, the strength of such personal relationships will be the single most important factor in achieving success in your foreign job.

Communicating and Being Understood

I discussed techniques to increase your effectiveness in communicating with your local work colleagues in chapter 1. I have also suggested that your foreign experience will be enhanced if you learn to communicate in the local language. However, be very careful about using your nascent language skills in the work environment. Business communication needs to be very precise and accurate. It is one thing to have a misunderstanding when ordering dinner, but quite another when discussing the terms of a deal.

Also, you must remember your role in the organization. Your attempts at using the local language might make you appear inarticulate and indecisive, thus undermining your authority.

Whenever possible, communicate on important issues via e-mail or other written format. You can be much clearer in a written communication, and the recipient can take time to really understand what you are saying. This is not to de-emphasize the importance of face-to-face communication. After all, you will never develop a personal relationship through e-mail. However, complicated issues can be broached in writing ahead of a meeting or summarized afterward. This will make the face-to-face meetings more effective and useful.

A *good* interpreter can be of significant value in your business discussions, particularly in face-to-face negotiations. I emphasize the word good because a bad interpreter can do more harm than good. A good interpreter does much more than translate words and phrases. He or she will make your thoughts and intentions understood by the person to whom you are speaking. It is thus important that the interpreter not only possesses excellent skills in both languages, but also has a general knowledge of the subject matter being discussed and a strong familiarity with the cultural nuances of both speakers.

Obviously, you must never rely on an interpreter provided by the other party, but should use your own interpreter who understands your goals and has been fully briefed by you prior to the meeting.

When working through interpreters, keep in mind the extreme difficulty of their job. Speak slowly, with breaks after every few sentences. Do not use idioms or complex phrases. Do not interrupt or start speaking again until the interpreter has completed interpreting all of your or the other party's previous sentences.

Actually, using an interpreter in important business discussions is not a bad idea even if you feel reasonably comfortable with the foreign language. It gives you time to think before answering and an opportunity to observe the actions and reactions of the other party—both of which can provide a tactical advantage.

You and the Home Office

If you are working in the foreign operation of a U.S. company, you have an opportunity to play a valuable, multilayered role: representative of the U.S. office; liaison and conciliator between the home and foreign offices; and mentor and advisor with respect to headquarters' policies, procedures, structure, and politics.

In so doing, you will be more effective if you maintain a balance between your loyalties to the headquarters operation and your local colleagues. It is quite easy to tilt one way or the other, and even I have found myself sometimes grumbling under my breath about the "damn Americans." However, you will be more credible and provide a better service to both the home office and the foreign operation (not to mention your career) if you remain objective and above the fray.

In your role as the liaison between the foreign office and various groups at headquarters, keep in mind that you will probably be in a different time zone, making direct communication via telephone or teleconference problematic. You will need to be prepared for your conversations and make your points clearly and concisely within the limited time available to you. E-mail will be an important tool, serving both as your primary means of communication and for verifying verbal communications. I have found it wise to document with a follow-up e-mail all verbal discussions with, and instructions and approvals given by, the parent company. This clarifies that everyone understands what was said (memories are sometimes short) and just might save your neck if something goes wrong.

You also will encounter significant scheduling and logistical difficulties in getting things done between the home office and various foreign offices. If you work for a multinational company with offices throughout the world, you may be surprised at the result you get by calculating the common work days for all of these offices:

- Aside from a few common holidays in predominantly Christian countries, every country will have different (and usually more) holidays than the United States.
- Employees in most of the foreign countries will have up to six weeks of vacation per year, and, since summer is in different months depending on the hemisphere, there won't be much overlap between vacations taken in Europe and those taken in, for example, Australia.
- Much of Asia is already well on its way to Saturday by the time the U.S. hits its stride on Friday.

- Depending on the dominant religion in a country, weekends fall anywhere from Thursday through Sunday.

Attempting to complete a project that requires input from all the various foreign offices can be very frustrating: it requires careful planning and a great deal of patience.

A New Legal Regime

The legal systems in foreign countries relating to business can be radically different from what you are used to. Some legal systems are relatively similar to the U.S. system (e.g., that in the United Kingdom), but in many others the concepts and rules governing business relations are very different. Even where the systems are similar in structure (such as in most of Western Europe) you are likely to find significantly different laws governing important legal concepts such as employer/employee relations. In countries in which the legal system evolved from a different foundation (for example, most Muslim and many Asian countries), you might find little in common with traditional Western concepts, except to the extent these have been adopted (voluntarily or otherwise) to conform to the global economic structure.

Different Attitudes Toward the Law and Lawyers

You might find yourself frustrated by the fact that a foreign legal system does not seem to have clear rules regarding many aspects of business relations. This perception stems in large part from the fact that we come from a rules-oriented society, in which we cherish certainty in our business and legal relations. The rest of the world isn't necessarily like this, particularly outside of the Western European–based legal systems. In many countries, business practices and customs develop unfettered by legal strictures. You should not be surprised if there are no legal rules addressing a particular business issue (or, if such rules exist, they won't be nearly as detailed or static as those commonly encountered in the United States).

When laws do exist, you will often find that they are biased in favor of local companies or individuals. Sometimes this bias reflects the view that local companies and individuals need help in dealing with the economic power of a foreign company. For example, laws may restrict your ability to terminate agreements with local partners to offset a perceived unfair economic advantage. Other laws are not so blatantly xenophobic, but merely reflect the society's value system. For example, employee protection laws usually are not specifically directed at foreign employers but instead reflect a societal view of employee rights. Whatever the case, be aware that the law in foreign countries is often not a level playing field.

Furthermore, in no other countries do lawyers play such a dominant and visible role as they often do in the United States. They tend to play more of a behind-the-scenes advisory role (if any at all). For example, it is not as common in most countries as it is in the U.S. for lawyers to be prominently involved in negotiating commercial deals. It is thus usually not a good idea to show up with lawyers for negotiations or business meetings unless this has been agreed to beforehand. Unfortunately (for U.S. lawyers like myself), the presence of a lawyer at a commercial meeting often will immediately raise suspicion on the part of your foreign colleagues and have a chilling effect upon your discussions.

Similarly, your business colleagues in foreign countries are not likely to give as much credence to legal considerations (or even to the language of signed contracts) as is common in the United States. Legalistic solutions are not favored—personal relationships and interactions are often considered more important than the precise wording of a contract. In fact, contracts often are not viewed so much as a binding legal documents, but merely as a nonbinding framework for discussing future issues that may arise in the course of the relationship. Foreign business people are put off by the American tendency to quickly escalate business issues to a legal footing. They might expect a contract's wording to be "massaged" if necessary to achieve what they view as a fair result. If you find yourself relying on legal arguments or maneuvers to resolve a misunderstanding or dispute, chances are that the underlying business relationship has been irreparably damaged.

Even if there is a defined legal system and you clearly have "the law on your side," there is a huge difference between the theoretical legal world and the real world. You may have a contract or other legal right, but you nevertheless must find some way to enforce it. A legal battle can be time-consuming and expensive, even in so-called Western democracies with well-structured court systems. In countries with different or "less developed" legal systems, recourse to the local courts is usually not a viable option at all. Remember, you are there to do business, not to engage in legal wrangling. With few exceptions, the rest of the world is simply not as legalistic (and certainly not as litigious) as we are. You need to adjust to this reality and find other ways to resolve your disputes.

The Sticky Web of Foreign Labor Laws

The aspect of foreign law that is usually most surprising and frustrating to Americans is the legal regime governing employer/employee relations. Many countries take a much more protective view of employees. The traditional U.S. approach of two weeks' termination notice does not work in most of the world. Termination payouts are usually measured in terms of months, and possibly years, rather than weeks. I have been through this in many countries, and it seems that continental Europe always presents the most challenges. Every termination issue I have dealt with in the Netherlands, France, and Germany has resulted in a protracted court hearing and a very expensive settlement. This obviously places a premium on the quality of your hiring process, because termination of an underperforming (or even unnecessary) employee is likely to be a lengthy and expensive process.

Aside from termination issues, many of the clauses we routinely include in U.S. employment contracts might be invalid under the applicable laws of the foreign country (e.g., clauses concerning ownership of an employee's work product or a company's right to view employee e-mails). Inclusion of these clauses in their U.S. form may result in the entire employment contract being considered void.

On a similar note, you will find that unions tend to be stronger and more prevalent than in the United States. Employees in some countries have a legal right to participate in discussions on key issues—and sometimes to even influence management decisions. Such rights are more common in Western European countries (such as France and Germany) that have strong labor movements and usually apply only to companies with more than a specified number of employees.

The rules are different in every country, as are the steps you can take to minimize your exposure to these rules. Take the time to educate yourself sooner rather than later. If there is one area of foreign business for which I would say that competent legal advice is required from the very beginning, it is employer/employee relations. You should neither hire nor attempt to terminate anyone without appropriate guidance on how to do this correctly from a legal standpoint.

Some Other Laws to Watch Out For

Examples of other common types of laws that you may encounter in your new country (that are different from what you were used to in the United States) include the following:

- Regulations that restrict advertising and marketing activities (e.g., advertisements that compare your products to a competitor's, or e-mail–based marketing campaigns)
- Special certifications or approvals that must be secured before your products can be sold in a country
- Laws that protect agents, distributors, and partners in their commercial relations with foreign companies

- Intellectual property and technology laws that either afford insufficient protection (by Western standards) for your proprietary property or that impose a technology-sharing or transfer requirement
- Special packaging or labeling regulations
- Laws that require printed materials to be in the country's local language

U.S. Laws Still Apply

The United States tends to apply its laws on an extra-territorial basis (i.e., outside of U.S. borders) more than most countries. If you are managing a foreign operation of a U.S. company, you should be aware that many U.S. laws will continue to apply to your operation to the same extent as to the U.S. parent. Here are some examples of such laws that are applied extra-territorially, and that you are likely to encounter:

- Export regulations that restrict the countries to which U.S.-origin goods can be exported or reexported
- Laws prohibiting or restricting business or financial transactions with or in certain countries (e.g., Cuba or North Korea)
- For publicly owned companies, securities regulations that govern insider trading, conduct of officers, financial reporting
- The Foreign Corrupt Practices Act, which prohibits the payment of "corrupt payments" (i.e., bribes) to obtain or retain business

Finding the Right Help

If you are in a position to make decisions that have legal consequences, you need access to competent legal advice in order to understand and deal with everything just discussed. You want a lawyer who is fluent in both English and the local language, who is used to working with U.S. clients and thus understands our business style and idiosyncrasies, and who is familiar with the legal issues for your particular industry. If working in the foreign operation of a U.S. company, you should consider selecting a law firm that has an office (or affiliated firm) in both that country and in the United States. It would thus be in a position to advise you on both the U.S. and foreign laws that apply to your business.

Chapter 5 **Summary** and **Checklist**

Settling In, Setting Up, and Adjusting at Home and at Work

1. *Getting Settled In and Set Up:*
   a. Have someone available locally in the foreign country, hopefully arranged by your company, to assist you in getting acclimated and settled in.
   b. Finalize the important tasks that you researched and started before leaving the United States within the first few days of your arrival: housing arrangements, bank accounts, school enrollment, and locating and contacting hospitals and doctors.
   c. Be aware of different standards for telephone plugs, electrical plugs, electrical voltage and frequency, televisions, and other electrical devices. Learn about the applicable standards, and obtain the necessary converters and adapters, before you use any U.S.-origin devices in your new country.
   d. Make sure that you understand how you will be charged for telephone calls back to the United States and for using your mobile telephone in other countries. You might need a special multiband chip for your mobile telephone in order to have coverage in all needed countries.
   e. Register yourself and your family members at the U.S. embassy as soon as you arrive, so that you can be contacted in the event of emergency. The embassy and consulate

also provide many other valuable services to U.S. citizens living abroad.

f. Remember that voting overseas requires advance planning and completion of an application process for an absentee ballot. You cannot just walk into the U.S. embassy on election day and vote.

g. If you are leasing an automobile, make sure you understand the lease terms, particularly your obligations regarding maintenance and lease termination. Familiarize yourself with driving rules and road signs before getting out on the road.

h. Take the time to figure out how to operate appliances, utilities, and other items in your new home while you have a local person available to help you. Do not wait until you actually need to use them. Also, create a directory of important telephone numbers and keep it close to your telephone.

2. *Getting Adjusted:*

a. Be prepared to go through stages in your adjustment to a new country—starting with a sense of excitement and adventure, followed by a (hopefully short) period of discontent and frustration. These shifting attitudes are normal human nature and should not alarm you. However there are steps discussed in this chapter that you should take to help you work through. these stages.

b. Center your adjustment process around your family, if they are with you. Approach all adjustment issues as a family project, and do not leave your spouse and children to figure it out on their own.

c. Acclimate in stages. You do not need to adjust to every aspect of your new life immediately upon arrival. Start with selected tasks that require prompt attention, and then gradually immerse yourself over a period of several weeks.

d. Take time to understand the country's culture and the reasons why things are different. The more you understand the people and the way they think, the easier it will be to accept and adjust to such differences.
e. Communicate with the local people as much as possible. Ask questions and show an interest in their country. This approach will cause them to open up to you, thus making you and them feel more at ease in your interactions.
f. Do not hole up in your house or apartment. Get out, meet people, and join in activities. If you don't, you might develop a siege mentality.
g. Make an effort to develop some level of proficiency in the local language—doing so will greatly enhance your experience in any non-English-speaking country. Use the techniques discussed in this chapter to augment your study time and speed your progress.
h. Seek out "anchors" in your new country that provide a connection to your U.S. home. These will serve as a lifeline between the comfort and familiarity of your old home and the challenges of your new one.
i. Use modern technology to inexpensively keep in touch with your friends and family back in the U.S. and up to date on events in your home community.
j. Don't give up. Do *not* slink back to the United States if you are suffering through a difficult adjustment process. If you go home while in a negative mood, you are likely to not return to your foreign assignment.

3. *Adapting on the Job:*
   a. Understand and adapt to the business practices, culture, and traditions of any country in which you are working or doing business. Doing so is not only courteous and politically correct but, more important, will give you an advantage in your dealings and help you achieve your business goals.

b. Focus on these three areas of your new country's business environment: business etiquette, work rules and customs, and management style. Understanding these will enhance your business experience even in a country that you are visiting for only a few days. It is indispensable in a country where you are living and working as an expat.
c. However, don't feel you need to develop a totally new "foreign persona" to be successful in international business. It is often okay to just act like an American. The adjustment process should be a two-way street and a mutual cultural exchange between you and your foreign business colleagues and coworkers.
d. Rather than constantly adapting your style and personality to fit each different country in which you work or travel, try to develop an "international personality" as described in this chapter.
e. Focus on your *attitude* toward your work colleagues—it's just as important as your ability to understand and adapt to the foreign business environment.
f. Be careful about using your developing language skills in the business environment. Business communication must be very precise. In addition, lack of fluency might make you seem inarticulate and indecisive, thus undermining your authority. Do not hesitate to use your own interpreter in important business discussions to ensure clear communication, as well as to provide you with a tactical advantage in the discussions.
g. Keep in mind that the legal systems in many foreign countries, as they relate to business, are radically different from the U.S. system, and you will likely encounter attitudes about the law and lawyers that are very different from what you are used to. For example, with few exceptions, the rest of the world is not as legalistic or litigious as

we are. In addition, many of the foreign business laws you encounter will seem biased in favor of local companies or employees. Also be aware that numerous U.S. laws continue to apply to and govern the foreign operations of a U.S. company. Find and use a competent legal advisor who has experience representing companies in your industry and who is used to working with U.S. clients.

Chapter Six

Safety and Security in the Post-9/11 World

Safety and security are of increasing concern for U.S. expats, and given the political realities of today's world, this is fully understandable. However, judging the safety of another country from the headlines or the evening news is like judging the safety of a day at the beach by watching Jaws. *Moreover, Americans have the misconception that the United States is one of the safest places in the world, but there are many countries safer than the U.S. Foreign countries tend to feel unsafe primarily because they are unfamiliar to us. I have often been amazed at how little congruence there is between my own perception of a country I have lived in or visited (not just as to its safety, but all aspects) and its portrayal in the U.S media.*

The safety and security issues associated with living overseas have become exaggerated in many people's minds. It's human nature to take what we see and hear in the media—which reports "extraordinary events"—and assume that it is a reflection of everyday life. Therefore, each reported event, whether a demonstration, bombing, or even an "ordinary" crime, takes on the aura of something that is happening daily. Being safe and secure in a foreign country is to a large extent also an attitudinal issue as well as a physical one. When we learn of or experience one of these events in a foreign country, we tend to dwell on the fact that it is "foreign," and this in turn somehow elevates the danger and harm in our minds. For example, if we are the victim of a crime in our home city, we have learned that we must deal with it and move on. Yet if the same crime is committed against us in a foreign country, we use it as a basis for ill feeling toward *all* the people of the country in which the crime is committed.

This same attitude is common among people in other countries in regard to the United States. Some Europeans believe that there are regular carjackings and murders in Florida because of a couple of incidents that received wide publicity in Europe several years ago. Indeed, many foreign citizens consider the United States to be one of the more dangerous countries in which to travel. After all, the city in which the most civilians of any nationality have been killed by terrorist actions and which has been the subject of the most terrorist threats in the twenty-first century is not found in a foreign country. That city is New York City. Still, in terms of overall safety, New York City is one of the safer major cities in the world, and I do not live in daily fear of any kind of terrorist or criminal act. It's important to keep all this in perspective as you read this chapter.

However, the new American expat clearly lives in a different world since September 11, 2001—a world with new risks and dangers. At one time, Saudi Arabia was one of the safest places for Americans to live. As of the writing of this book, it is on the State Department's list of countries to which Americans should not travel. However, the key to dealing with these new risks and dangers is not to isolate yourself from the rest of the world and stay home. Rather, it is to understand what the real dangers are so that you don't allow them to disrupt your everyday life—and at the same time to take reasonable, practical precautions so you don't get caught up in one of those extraordinary events I've just mentioned.

Keep in mind that many of the everyday security and safety steps discussed in this chapter are the same precautions that you should take as a matter of course in any city that is unfamiliar to you, whether in the United States or abroad. However, in a foreign country you are perhaps more likely to be a target than in the U.S. because you stick out as being vulnerable and you can't read the clues indicating your safety is at risk. It is also much more difficult for you to deal with the aftermath of a crime or accident in a foreign country than at home.

Also keep in mind that if you live in or travel regularly to a high-risk country, you and your company should obtain expert advice from a professional security consultant such as those listed in appendix D. You need more assistance and guidance than can be provided in this or any other book, pamphlet, or other written document and you need to receive it from experienced professionals.

U.S. Government Information on Safety and Security

The U.S. State Department publishes two types of safety information sheets for countries that are considered to be potentially dangerous for U.S. citizens. First, the State Department issues Travel Warnings recommending that Americans avoid certain countries. As of the writing of this book, this list includes 26 countries. The second type of safety information sheet, Public Announcements, informs travelers about terrorist threats and other short-term travel conditions that pose a significant risk to the safety of Americans. These announcements have traditionally been issued to deal with coups and government overthrows, specific bomb threats, violence by terrorists, and anniversary dates of specific terrorist events. As of the writing of this book, Public Announcements are in effect for 16 countries. You can access these Travel Warnings and Public Announcements 24 hours a day either through the State Department's travel website (www.travel.state.gov) or by telephone (202-647-5225) or facsimile (202-647-3000). They also are available through the consular offices of U.S. embassies abroad. It is a good idea to stay familiar with these for any country in which you are living or to which you will be traveling. The State Department also publishes Consular Information Sheets regarding every country in the world (containing information on crime and security conditions), as well as useful articles about safety and security. The series includes titles such as *Crisis Abroad; Travel Warning on Drugs Abroad*; *A Safe Trip Abroad*; *Tips for Older Americans*; *Tips for Women Traveling Alone*; *Tips for Travelers with Disabilities*; *Tips for Students*; and *Information on Preparing for a Crisis Abroad*. All of these can be accessed at the previously mentioned website (Click on *International Travel—Travel Brochures*).

Everyday Safety Issues

Plan for Safety

Have a safety and security plan for yourself and your family. Think about how you would respond to specific types of emergencies or dangerous situations, and discuss these with all of your family members.

Know how to ask for help. If you can't speak the local language, at least know how to say a few phrases ("Can you help me?" "Please call the police," etc.) or carry a card written in the local language that tells your name, address, telephone, and number and has key phrases to which you can point. Such a card is particularly important for children.

Carry a small street map of any city that is new to you, and mark the location of your home or hotel on this map. Carry important telephone numbers—for example, those of the police, the U.S. embassy or consulate, and emergency medical assistance. Know how to use local pay telephones and carry either a calling card or the appropriate coins for making an emergency call.

Street Smarts

Be careful about where you are going and, most important, always know the correct route to your destination before you start out. Be aware of your surroundings and the people around you. Good and bad neighborhoods can intermingle and the relative safety of any street can change quickly. Notice signposts and markers that you can remember in case you get lost or turned around. Take note of safe places you can get to if necessary, such as a police station or a public building such as a major hotel. Walk and act confidently, as though you know what you are doing and where you are going (*especially* if you don't). Don't attract attention to yourself by walking around in an apparent daze or looking as if you are lost. Appearing vulnerable invites trouble.

Be Aware of Current Events

Are elections scheduled that could possibly result in protests or violence? Is an anti-American demonstration scheduled to take place? (Never mind anti-American—you don't want to get caught up in any demonstration regardless of the subject). Are there areas of your city in which the majority of residents are known to have anti-American opinions? Is a soccer match scheduled between two teams whose fans are bitter rivals? In general, don't put yourself in a situation in which you could blithely wander into trouble. Your U.S. consulate is a good source of information about events and areas to be avoided. You can also learn about these abroad from the local media in the country, CNN, American newspapers such as *The Herald Tribune* and *USA Today*, and from other American expats who know the local scene.

Don't Draw Attention to Yourself

Don't be "overtly American" in your dress and actions. I don't recommend wearing a shirt sporting a U.S. flag with the words *These Colors Don't Run* on the back. (I've seen this more than once in a foreign country). It is one thing to be proud of your country and quite another to be stupid. Citizens of foreign countries do not understand the outward displays of patriotism that we're used to in the United States. Why wear something that will start you off on the wrong foot in your dealings with them or, in the extreme, increase your chance of becoming a target for hostility or violence? Aside from this, you should dress conservatively and eschew flashy clothes and expensive jewelry. Try to simply blend in and not bring attention to yourself by the way you act and dress.

Stay in Contact

Always make sure that someone in your company knows where you are planning to be, particularly if you are taking an excursion out of the city where you are living or going to another country. If you are going on an extended trip, arrange with someone that you will call in at designated times.

Protect Money and Valuables

People have a lot of different routines and methods for carrying money and valuables to keep them secure from loss or robbery. For example, I never carry anything in my back pockets, or in anything that is open to the view of others, or in a bag that can be reached into. I carry sufficient money for immediate purposes in a small money clip or wallet in one front pocket. I keep all other items—for example, credit cards, airline cards, driver's license, passport—in a wallet in my other front pocket. It is easy to lose a money wallet or leave it behind, so I keep money and these other items separate.

I also keep everything in the same place and return it to that place immediately after I use it. For example, when traveling I keep my passport in the same compartment of the same wallet in the same pocket at all times. When I get it out to show someone for identification, I return it immediately to the same compartment, wallet, and pocket. I follow the same routine for keys, credit cards, and everything else that I carry with me. This helps me keep track of things and decreases my likelihood of losing something because I am in a hurry.

If I am carrying a shoulder bag, I make sure it has a strong strap that cannot be easily ripped away or cut. I carry it over the opposite shoulder and in front of me. I always wear coats with zippers or buttons and inside pockets.

All of these routines may seem obsessive on my part, but then again, I have rarely lost anything and have never had anything stolen while traveling. It is not important that you adopt *my* routines. There are many variations on this theme (belt wallets, waist or neck pouches, etc.) that are readily available at travel shops and airports. The important thing is that you develop your own routines that work for you and that you use them.

You also should have a copy of your passport identification page, driver's license, both sides of all credit cards, and any other documents that would need to be replaced if lost or stolen. Obviously, keep these copies in a safe place where they will be immediately accessible to you if needed. Credit card companies also provide a local number in foreign countries that can be called 24 hours per day in the event a card is lost or stolen. The numbers for those countries in which you will be living and traveling can be obtained from the credit card company's website and should be written down and kept with the copies of other important documents.

Street Scams

Avoid people who approach you offering bargains and those who want to be your guide. Although some are legitimate, others are scam artists. If you are not familiar with the country, you don't have the necessary knowledge to figure out which category they fall into. Just say "No, thank you," and move on. Don't exchange money with money-changers on the street. You may not know enough about the local currency to avoid being swindled. You also don't know the views of the local police on this type of activity, not to mention that it will require you to pull out a wad of cash in a place where you shouldn't.

Using Public Transportation

Be aware that public transportation is often the scene of crimes against foreigners. It is easy for a foreigner to let his or her guard down in such settings. You may be tired, lost, unable to read the signs, naïve, or easily intimidated because of the unfamiliar environment—all of these factors can combine to make you easy pickings.

When taking a train or bus, make sure you know where you are going and how many stops there are before your destination. Avoid leaving your belongings unguarded because you've fallen asleep or gone to the dining car. Note where the emergency buttons are and the telephones or intercoms closest to your seat. Don't consume any food or drink that a stranger offers you, no matter how nice he or she may seem.

Driving in Foreign Countries

Don't drive a car in a foreign country unless you know how to read the road signs. If you are not familiar with the area, map out your course before starting your journey. Avoid renting station wagons or minivans that don't have secure storage areas. It is not a good idea to be driving around or parking your vehicle with your luggage and other items visible through a window. If you are bumped by a car in an apparent accident, don't stop until you are in a lighted, populated area. The same rule applies if someone pulls alongside you in a car that is not clearly marked as a police car and tries to get you to stop, or where someone appears to be signaling you that there is something wrong with your car. Keep driving (if possible) until you are in a lighted and populated area.

By the way, if you are confused by driving vehicles in which the steering wheel is on the opposite side of the car from the United States, just remember the following: no matter what country you are driving in, "keep your body in the middle of the road". In other words, no matter where you are, the body of the driver should be closest to the middle line separating him or her from oncoming vehicles. This rule will help you navigate intersections and roundabouts. Of course, it is totally inapplicable if you choose to drive a right-side vehicle in a left-side country, or vice versa. However, I don't recommend that you do this.

Hotels

From a safety and security standpoint, it usually is preferable to stay in brand-name chains or larger hotels. However, this practice would rule out some of the most enjoyable hotels in a lot of countries, so I admit that I don't follow it all the time myself. Hopefully, however, your hotel will have a secure parking lot, restricted access to room floors, and a safe for your valuables.

Living in or Traveling to High-Risk Countries

In certain countries, you unfortunately face much more extreme safety and security risks. These are countries that are experiencing civil war or internal unrest, and those in which aggressive anti-American attitudes exist among large numbers of people. High-risk countries include those hosting known terrorist groups targeting the United States (and our friends and allies), or those with a high number of political or financially driven kidnappings of U.S. business executives. The State Department publishes good general advice as to how to protect yourself when living or traveling in known high-risk areas and how to act when caught up in a kidnapping situation (See *A Safe Trip Abroad*, available at www.travel.state.gov). Some of these tips are included in the discussion that follows.

Before traveling to any country that is considered a high-risk for Americans, check your life and health insurance policies to make sure that they continue to provide coverage while you are there. Insurance policies sometimes exclude coverage in countries for which State Department Travel Warnings are in effect.

Where and How to Get Information on High-Risk Countries

One of the best sources of current information regarding countries that pose a high safety or security risk to Americans is the U.S. State

Department. Travel Warnings and Public Announcements are updated constantly and can be accessed 24 hours a day via Internet, telephone, or facsimile. These official bulletins are very good sources, although I would add as a caveat that they *do* present a worst-case scenario and, as a result, are viewed by some expats as exaggerating the danger somewhat. You also can check directly with the U.S. consulate in the country where you are living or traveling. You should do so regularly, particularly immediately prior to embarking on any trip to a high-risk country. This official government information can be supplemented with very good, up-to-date information from a professional security consultant if you have engaged the services of such a firm. Several private firms specialize in providing up-to-date information about security and safety conditions in foreign countries (for a fee). Examples of such firms are listed in appendix D.

It goes without saying that if you are living in a high-risk country or spending more than a few days there, you must register with the U.S. embassy. This is the only way that they U.S. government can assist you in the event of emergency or the need for evacuation.

Traveling

When traveling by air, check State Department Travel Warnings and Public Announcements for any information regarding the airlines you are flying and airports you are using during your trip. The local U.S. consulate for the country of transfer or final destination also can assist you in investigating the security history and current rating for airlines and airports in these countries. At airports, move into secure zones as soon as possible—don't loiter outside the security areas. Try to check in at times when the airport is not crowded, thus avoiding large crowds and long lines at the check-in counter and security barriers.

The same rule applies to travel by train and bus: try to avoid getting stuck in large crowds or long lines or standing around for long periods in waiting halls at train or bus stations. Be alert for anyone acting suspicious, carrying anything suspicious, dressed unusually (e.g., appearing to be hiding something under his or her clothing), or walking off and leaving anything behind. Familiarize yourself with exits and emergency procedures for all modes of public transportation on which you are traveling.

Keep your wits about you while at airports and train and bus stations. Don't leave your luggage outside a restaurant, bar, or store while you go inside, unless it is visible to you during the entire time that you are away from it. Even when sitting at a table or in a waiting area, keep all pieces of luggage, briefcases, and other tems in your personal sight at all times. (Keep in mind that in some countries it is illegal to leave your luggage outside of an airport restaurant or store even if it always visible to you. It might be confiscated and "blown up"—a definite downer for your trip.)

Do not get into any automobile (even a marked taxi) at an airport, train or bus station, hotel, or anywhere else in a high-risk country unless you personally know the driver or the car is provided by someone you know. Arrange secure pickups before you arrive. Generally, the less attention your vehicle draws, the better.

Do not use any luggage ID tags (or place any stickers on your luggage) that easily identify you as a U.S. citizen or as an employee of a U.S. company. Use luggage ID tags that conceal your name and address from casual bystanders.

When traveling in a high-risk country, stay out of areas where you would be alone or exposed. Also, stay away from crowded areas near sites of national, political, or religious significance. Any of these could be a target for violent activity.

Be careful about visas and entry stamps that appear in your passport. For example, if two countries are engaged in hostilities or do not have diplomatic relations, it is not always a good idea to enter either country using a passport that shows you recently visited the other one. If possible, it is best not to carry any identification or business card on your person that identifies you as a member of the U.S. military or National Guard or as an employee of the U.S. government. These can attract attention in a hostage situation and cause you to be singled out for hostile treatment. Keep these types of identification in your checked luggage if possible.

Staying in Hotels

In high-risk countries, forget about staying in quaint or charming hotels and opt for large hotels with advanced security systems. Inquire about the security procedures and systems of a hotel before making your reservation. The local U.S. consulate in the country that you are visiting can assist you with this.

As soon as you check in, identify all exits from the hotel, particularly those from the floor on which your room is located. Do not answer your door unless you are certain who is there, and do not accept any package delivered to you unless you know who sent it. Do not hang around the lobby or other areas close to the entrance any longer than necessary. Stay as "packed" as possible, to facilitate a quick exit in an emergency.

If possible, arrange your local meetings at the hotel, which is likely to have better security than the office of the person you are meeting. If you must leave the hotel, make sure you know the exact route to your destination and the length of time that it should take to get there.

Home Safety and Security

Although I do not recommend this for most expats, if you are planning to live in a high-risk area, you should, if possible, live in a walled, guarded compound that has restricted access. Although these may draw attention and thus become a possible target, it is safer on balance to live in this type of housing than in an apartment or house in the city that is not separated from the general population. If you do live among the general population, try to choose an apartment or house with some form of restricted access and a security system.

When you move into a house or apartment, have the locks changed as soon as possible. Create a "safe area" in your home—a reinforced room or a heavy piece of furniture you can get under quickly in case of an emergency. Reinforce your doors and windows against break-in and install a peephole in your front door so that you can always see who is there before opening it. At night, keep your blinds drawn and, in general, keep away from windows. Install outside security lighting if necessary to see at night, and install timing devices on your internal lighting so that you can give the appearance of being home when you are not. Always have a medical kit and emergency supplies of food and water on hand.

Practice safety and security drills with your family members. Make sure that everyone knows what to do and where to go in the event of an emergency. Before hiring any worker who would have access to your family and home (housekeeper, driver, etc.) check the applicant's background and let a professional screen him or her for security risks. Once so vetted, they must also be included in your family security plan.

Driving in High-Risk Countries

Any automobile that you drive in a high-risk country should have a good security system and a hands-free telephone. Gas caps and hoods should not be capable of being opened from the outside without a key. Electric door locks and windows are strongly recommended. It is also possible to consider run-flat tires (i.e., tires that do not go flat when punctured), a vehicle locator beacon, bulletproof glass, and special communications equipment for your vehicle if the threat level in your area justifies extra precautions. You should carry a medical kit, fire extinguisher, flashlight, and road flares in your vehicle at all times.

Avoid routines in your driving habits. Try to vary your times and routes of daily travel. You should consider obtaining professional training for yourself (and your driver if you have one) in emergency driving techniques.

Don't leave your vehicle accessible to strangers at night or in non-public places. Check your vehicle every time before opening any doors to see if there is any evidence of tampering or anything that seems "not right."

While driving, keep your windows up and your doors locked at all times. When entering or leaving compounds, try to do so at a time when you will not get stuck in a long line. You always are safer if you can keep moving.

Work Safety and Security

A professional security and access system with CCTV cameras should be in place and fully functional. Any security guards should be trained and capable of providing the level of protection commensurate with the risk. Try to limit access to your building at the most remote point possible, and install barriers to protect the building. Your employee parking lot should be included within your security measures.

Implement detection and physical search procedures for visitors to the building and for safely inspecting packages and envelopes delivered to the building. Employees should be made aware of procedures to be followed in the event of an emergency and periodic drills should be conducted. Medical kits and emergency supplies should be on hand and someone should be trained to administer medical assistance.

Communications

When you are away from your home and/or workplace, implement check-in procedures at regular intervals. Make sure your telephone will work for all frequencies and networks in all countries and areas you are visiting. Enter emergency numbers into your speed-dial so that you can dial them easily when under stress. Agree upon code words and phrases that you will use with family members or work colleagues to convey information in emergencies when you cannot speak freely.

The Critical Importance of Obtaining Professional Assistance

All of these tips are useful for maintaining safety and security while living and working in high-risk countries. But the truth is that you need far more help than you will get out of this book or any other publication. For these countries and situations, you and your company need to have a professionally developed and implemented security plan (both for everyday life and for emergencies), and you need to be serious about adhering to it. A professional security consultant will be able to assist you with

- crisis management and response,
- training for defensive living,
- training and supplementing your security personnel,
- conducting physical security reviews of your home, workplace, and automobile,
- planning and training for extraordinary emergencies such as evacuations, kidnappings, and bombings, and
- generally providing a full risk analysis and assessment of your safety and security needs.

Support from Your Employer

Your employer should do its utmost to provide a secure living and working environment for you and your family. Whether your employer is a U.S. or foreign company, it should take this responsibility very seriously and provide you all necessary support, training, and security measures to ensure the maximum level of protection possible. These provisions for your safety should include engaging appropriate professional security consultants to advise on security issues and a willingness to fund and implement their recommendations.

In particular, your company should place no impediments on your ability to leave the country quickly in the event of emergency, and it should be prepared to assist you in this regard with a realistic evacuation plan that can be put into operation on a moment's notice. Notice periods in employment contracts should be waived in the event of a security emergency (including the issuance of a State Department Travel Warning recommending that U.S. citizens leave the country). It is customary in some countries (and sometimes required by law) for a company to hold the passports of its expatriate employees. I personally do not feel comfortable not being in possession of my passport, particularly in a country that I might need to leave quickly. You should discuss with your employer whether it is possible to retain possession of your passport or to at least have a guarantee that it will be returned to you immediately in the event of emergency.

If your employer is not willing to take the security issues discussed in this chapter seriously and to provide a secure living and working environment that includes an evacuation plan in the event of emergency, you should not be working for that company in that country. Either come home or find a different employer.

Staying Out of Trouble

Contrary to popular belief, Americans do not have any special "get of out jail" card when it comes to breaking laws in foreign countries. In fact, asserting your rights as a U.S. citizen may actually do you more harm than good. Violating the laws of a foreign country may get you deported or worse—possibly much worse. If you are arrested, you will soon learn that (1) your ignorance of a law is no more relevant in a foreign country than in the United States, (2) laws in foreign countries are often quite different from those you are accustomed to, (3) some actions you thought were legal in a foreign country are in fact illegal, and, finally, (4) penalties for violating many laws can be draconian compared to those in the U. S. If you are convicted, you will then learn that the U.S. government can do nothing to keep you from going to prison in a foreign country and that prisons in many countries are extremely hazardous to your health (and life).

Following are some suggestions as to places you *should not* go, things you *should not* do, and general advice as to how you can stay out of trouble in foreign countries.

Sex

Many countries in the world have more liberal laws regarding sex than does the United States. Prostitution is common and open in many countries. Red-light districts, such as those in Amsterdam and Hamburg, are magnets for American males, many of whom have never experienced such an intense concentration of these establishments. Putting aside for a moment the risk of sexually transmitted diseases, the likelihood of being ripped off, and the moral implications of how young women (and men) are procured for the sex trade, you should also be aware that many of these establishments are not nearly as legal as you might think. They may be tolerated, but that doesn't mean they are actually legal. There is no better way to screw up and lose a lucrative expat assignment than by not being able to show up for work because you are sitting in jail. If this is the only local cultural activity that appeals to you, just stay in your room, order room service, and watch CNN all night long. When the sun comes up, you'll be glad you did.

Drugs

Ditto for drugs. Stay away from drugs, even if you feel absolutely certain that they are legal in a particular country (and they almost never are). *Even if they are legal*, you may not be used to them so you are likely to lose control and hurt yourself or someone else. A seeming acceptance of drug use by local police can quickly turn into a disaster if you are arrested for possession of illegal drugs. The penalties for illegal drug possession in many countries (including some very advanced countries like Singapore) are horrendous, and there are no exceptions. Every legal right that you take for granted is nonexistent in many countries when it comes to illegal drugs. You run a real risk of rotting away in jail for years, or of being tied to a post and shot.

Also be careful regarding prescription or otherwise legal drugs that you purchase in one country and expect to take into another country. Even if you have a copy of the prescription, that won't make the drugs legal. This can be a particular problem when entering some Middle Eastern countries. Check with the appropriate U.S. consulate if you have any doubts.

If you simply can't resist and do partake of the local drug culture, such as "coffee shops" in Amsterdam, make sure you check all of your clothes before you leave that country. Obviously, only an absolute idiot (or a drug smuggler) would knowingly cross any international border in possession of an illegal drug. However, it is also easy to forget that you stuffed something in your shirt pocket. In fact, it is better to wash all your clothes before you return. The scent of illegal drugs such as marijuana can stay on your clothing even if you are not currently possessing. There is nothing more embarrassing than the experience of a sniffer dog at the baggage claim zeroing in on your suitcase.

Rock and Roll

Actually, rock and roll—meaning your choice of music, fashion, films, books, and so forth—is usually okay. In most countries, you can indulge in your chosen diversions (within reason, of course) to your heart's content. Don't take this for granted, however. You need to be familiar with basic social and religious customs of any country you are planning to visit. In some countries, certain conduct that is common in the United States is viewed not only as being in poor taste or objectionable—it is also illegal. For example, inappropriate or immodest dress is not only offensive to many Muslims, it can result in your arrest in some Muslim countries.

In addition, some forms of Western music and films are prohibited in certain Muslim countries as well as parts of Asia. For example, customs officials view in their entirety any videos brought into Singapore and seize those that contain prohibited sexual content or offensive language. (Standards in Singapore are stricter than you might think—most movies shown in Singapore theaters are heavily edited and often run several minutes shorter than the original version.)

Even books and magazines that seem innocuous (*Time*, for instance) might get you in trouble. Pictures of women in Western attire and advertisements for alcohol can result in a magazine's being seized by customs officers in certain Arab countries. When I moved to Saudi Arabia, someone painstakingly cut all pictures and references to Israel out of an atlas that was in my air-freight shipment. Carrying a picture of the Dalai Lama into Chinese-occupied Tibet can be grounds for immediate deportation. Books dealing with political, social, or cultural issues are often seized by Canadian customs officials under the broad Canadian regulations banning so-called hate propaganda.

The rules are different in every country and may surprise you. Check before you go.

Taking Pictures

Be careful about taking pictures. You never know which building might be a military or government installation. Certainly, do not take pictures of airports and other transportation facilities. With terrorism a genuine threat, the mere act of taking a picture of even an office building, shopping center, or commercial area can attract the attention of the police. In addition, people in many other countries do not like to have their picture taken, and it doesn't take long for a crowd to gather if they start protesting. In general, ask before you take pictures of anyone or of anything that is not an obvious tourist attraction.

Foreign Purchases

When you purchase something in a foreign country, make sure that it can be legally taken out of that country. In some countries it is possible for you to purchase items deemed to be "national treasures" within the country without restriction, yet it may be illegal to take them out of that country. These proscribed items can include antiques, historical objects, and even animal or plant products. It is not always obvious to the uninformed visitor that they would be subject to such a restriction (or why they are). For example, when I visited the Seychelles, I discovered that Coco-de-Mer nuts may not be removed from that country, which is the only place in the world where they grow naturally. To me, they were just nuts that grew on a tree. But to the people of the Seychelles, they are a national treasure that makes their country unique in a small but important way.

Also, make sure that whatever you purchase can be legally brought into the United States. Numerous items that may be purchased legally, and removed from, a foreign country cannot be brought back into the U.S. These include items made from sea turtle shells, certain reptile skins, crocodile leather, ivory, coral, and furs from endangered species.

What to Do If Stopped or Arrested by the Police

If stopped by the police, try not to panic. Keep a clear head. It is common in some countries for the police to stop people, particularly foreigners, to ask questions. Do not assume that you have done something wrong. Show your identification and answer questions politely and truthfully about who you are and what you are doing. Ask if you have done something wrong, and apologize if you have made a mistake.

If detained or arrested, do not resist or run. Be calm, passive but confident, and polite. Do not be rude, aggressive, or confrontational. Do not complain or demand anything, or make sudden movements. Do not allow yourself to be provoked by the questions that are being asked or by any physical contact by the police officer.

Be careful about answering questions, particularly if you are being asked questions by a nonnative-English speaker, or if you are answering in a language other than English. If in doubt or uncertain about a particular question, or if you are clearly suspected or being accused of a specific crime, try to avoid answering. Request that questions be put in writing so that you can understand them, and ask for a translator to be provided. Unless you are certain of the questions being asked and of your answers, the best thing to do is to politely ask what your rights are and to speak to a U.S. consular official. You may not have an absolute right to an attorney as you do in the United States, but you do have a right to speak to a U.S. consular official in almost all countries (pursuant to international treaty). Be polite but persistent about getting access to a U.S. consular official.

If you think you are being harassed by a corrupt police officer, try to stay in a public area and in sight of other people. Do your best to delay or pretend that you don't understand what is being said or expected of you. The longer you can delay, the better chance that the police officer will go away and move on to the next victim. If you are harassed or abused by a police officer, try to remember the name, badge number, license plate, or some identifying feature of the officer. Report the incident to the U.S. Consulate before reporting it to the local police. If you are hit by a police officer, do not strike back. Just cover up and protect yourself as best you can.

Penalties can be severe compared to the penalties for the same crime in the United States. In many countries, you'll find that the rights you're entitled to in the U.S. won't be extended to you. Bail is less common and not guaranteed, so you may sit in jail for a long time just waiting for a trial. Jury trials are not common. Foreign prisons are no Club Med and conditions in them are often much worse than in U.S. prisons. They sometimes don't even provide meals to prisoners, who must arrange for family members or friends to deliver food. The death penalty is applied in some Asian countries for crimes, such as the possession of relatively small amounts of drugs, that are not capital offenses in the U.S.—or in some Shariah (Muslim) law countries, for acts such as adultery that aren't even illegal in the U.S.

Assistance Available from the U.S. Consulate

If you are arrested in a foreign country, it is well worth your effort to try to make immediate contact with the U.S. consulate in that country. The consulate officer will visit you, provide information on local attorneys and judicial procedures, notify your family or friends, and serve as a contact point for requests for money or other assistance. He or she can arrange for food to be provided to you by your family or friends, and arrange for medical examination and treatment. He or she will also attend your trial, monitor your living conditions, and protest mistreatment for unfair or discriminatory procedures. Unfortunately, the consular officer has no ability at all to do what you really want done, which is to get you out of there.

Chapter 6 **Summary** and **Checklist**

Safety and Security in Foreign Countries

1. Keep up to date with U.S. government Travel Warnings and Public Announcements.
2. Have a safety plan for you and your family.
3. Be aware of your surroundings and of current events in your area. Always act confident, but don't draw attention to yourself.
4. Inform family, work colleagues, and/or friends of your whereabouts.
5. Have an established routine and safety procedures for carrying your valuables and personal items.
6. Avoid street people offering bargains, to be your guide, or to change money.
7. Be particularly vigilant when using any public transportation.
8. When driving in a foreign country, know how to read signs and know where you are going. Do not stop for anyone except in lighted, populated areas.
9. When living or traveling in a high-risk country, follow the security procedures for home, travel, hotels, automobiles, work, and communication as discussed in this chapter. Get professional help in developing and implementing such security procedures,

and in training all employees and family members how to respond to security risks and emergencies.

10. Make sure your employer takes the security issues discussed in this chapter seriously and provides a secure living and working environment that includes an evacuation plan in the event of emergency. If not, either come home or find a different employer.
11. Stay away from places that can get you in trouble. Research laws in foreign countries before you go there. Be cooperative with local police, but do not incriminate yourself or answer questions you do not understand. Always request a U.S. consular official if arrested in a foreign country.

Chapter Seven

Coming Home: Returning to Your U.S. Home and Job

You can *go home again, but it's not always easy. Living overseas can change your outlook on virtually every aspect of life. Returning expats often encounter culture shock in reverse. You may also have the issue of finding a job back in the United States, hopefully one that will leverage your foreign work experience to your advantage.*

Returning to the United States after a foreign assignment involves many of the same issues you faced when moving overseas, but in reverse and with a few twists (including new ones related to post-9/11 issues). This chapter explores the twists.

My own experiences returning home after foreign assignments have been a mixed bag. Most of my overseas stints have been for relatively long periods of time (ranging from three to six years), so I have become more "internationalized" than many expats who spend a shorter time overseas. This has provided me with certain advantages vis-à-vis adapting to the challenges of life in other countries, but this degree of expatriate longevity can make returning to the United States more problematic.

I find, for example, that I have a decidedly different view of the world than my colleagues who have never lived outside of the United States. My interests tend to be different from those of someone who has spent his or her life living only in the U.S. This sense of difference is sometimes interpreted as arrogance or elitism, which is absolutely not the case. However, the personality traits that I developed in order to be successful overseas (see "The Right Stuff" in chapter 1) have stayed with me, and they color and affect my worldview. In addition, I continue to be very interested in what is happening in other parts of the world, even in countries that are rarely mentioned in the U.S. media.

I also often find myself out of sync with American pop culture. (While living overseas, I generally haven't kept up to date with American TV, music, sports, and so forth, particularly during my earlier assignments that predated CNN and the Internet.) As an example, when I watch a TV show like *Jeopardy*, I can answer all sorts of difficult questions about international issues, but I'm stumped by the simplest question about a well-known U.S. television show from the past.

In sum, I have often felt more like a foreigner in my own country than I do overseas, and I usually start planning my next foreign assignment as soon as my previous one is over. This is a choice, however, not an inevitability. With a little preparation and understanding of the issues you will face, you can do much better than I have at repatriating and reintegrating into American society and culture.

On the Road Again

As I mentioned earlier, many of the logistics of moving back to the United States are the same as those involved with your original move overseas. You can apply the same techniques discussed in chapter 4 (in reverse) to make your return more successful and less stressful. In addition, you should be aware of the following:

- In general, you need to understand the U.S. Customs regulations that apply to returning expats, so that you know what you can legally bring into the United States and your potential liability for duties. Special rules apply to household goods and personal effects imported into the U.S. by a returning expat. See the U.S. Customs document entitled *Moving Household Goods to the U.S.* at www.cbp.gov (click on *Publications—Travel*) for details of such rules. Have receipts available for any items that cannot be brought into the U.S. duty-free under the household goods and personal effects exemption. I have never been required to pay a duty on any items purchased abroad when returning to the U.S. after a foreign assignment, but the requisite forms and procedures had to be followed in order to assure a timely and trouble-free delivery of such items to my U.S. home.
- If you purchased a car overseas and want to bring it back with you, you will need to check this out very carefully. U.S. automobile standards are different from those of most countries, particularly with respect to emissions. You will very likely have to modify your car in order bring it into the U.S. Also, make sure you understand whether any duties will be assessed on the importation of the car into the U.S.
- If you married a foreign national while overseas or if your children were born overseas, you must make sure that you have the necessary documentation for these family members to enter the U.S. See www.travel.state.gov for information on passports, visas, and other documentation required in order for spouses and foreign-born children to return with you to the U.S. (click on *Children and Family*).

- Depending on where you have been living, don't assume that your pet can return to the U.S. without further documentation and possible immunizations. In fact, you need to check these issues for yourself and family members as well, particularly if you were living in an area in which infectious diseases were prevalent. See www.cdc.gov for requirements on bringing pets back into the U.S. and for general information about health requirements for persons entering the U.S. from specific regions or countries.
- Consider whether you can (or should) bring certain items purchased in a foreign country back to the U.S. Some items that are freely available in foreign countries are either prohibited from being imported into the U.S. (e.g., products made from certain endangered species) or would incur a high duty making their importation economically unfeasible. Furthermore, electrical items that you purchased overseas are unlikely to work in the U.S. due to different voltage or other standards.
- A lot of the hassles and stress involved in moving back to the U.S. can be minimized if you secure the services of an international relocation firm that is experienced in dealing with these issues. See appendix D for a listing of such companies.

Culture Shock on the Flip Side

Adjusting to life back in the United States can be difficult. The problem is not even that your home or your friends have changed, but that you have. You and your family have experienced a new life, developed new attitudes and interests, and been exposed to new cultures and ways of thinking. You are returning to your old zone of comfort, but it isn't as comfortable as it was before you left. Longtime friends may not seem as interesting as they used to be, and they aren't all that intrigued by your unfamiliar attitudes and foreign experiences. Let's face it: your expatriate assignment should probably be considered a failure if you haven't significantly changed. In sum, you will miss the diversity and excitement that you experienced in your overseas environment and will strive to find some way not only to hold on to it but to build upon it.

What follows are some tips that will ease the reintegration into your home culture for you and your family:

- Don't become arrogant about your new attitudes and interests, and don't be critical of your old friends just because you may no longer have as much in common. They had no reason, and no opportunity, to experience what you did.
- You might need to adjust your financial thinking. It is likely that your expatriate compensation allowed you to live at a higher standard than you had in the United States. Returning expats are prone to purchasing expensive houses, cars, and luxury items, and to generally living at a higher standard than they can afford until their overseas savings are exhausted. In part, this is due to the fact that they became accustomed to the higher standard of living, and no one likes to go backward. It is also fueled by a desire to show their friends how successful they were overseas. It is easy to get carried away.
- Depending on their ages when you moved overseas and your length of stay, you may find yourself introducing your children to a totally new environment. Even if they attended U.S.-curriculum schools, the U.S. will be a foreign country to them. Treat their adjustment process accordingly.
- Find new anchors in your home community that help you maintain your interest in, and contacts with, the foreign country in which you lived and the international community in general. Join local groups that are focused on that country and/or international issues. Continue to read the *International Herald Tribune* and foreign newspapers—both

to keep up on international events and to be exposed to differing viewpoints from those found in the U.S. media (this can be done online or through subscription). Find ways to continue using your foreign language skills so that you don't lose them. Keep in touch with your foreign friends and work colleagues, and invite them to visit so that you can return their hospitality.

- You might find that you have picked up habits or attitudes during your expat assignment that die hard. My friend Martha returned from living in Africa, where one typically bargains on all purchases, and found it difficult to go into a U.S. store and pay the price stated on the price tag.

You Can't Step into the Same River Twice

Although you're likely to return with a shifted mind-set and an enlarged worldview, things back in Akron or Spokane won't have stood still either. You may be surprised to find that your favorite ma and pa café is now a chain restaurant or that your downtown movie theatre has closed. There may be a different principal at your children's school, a new pastor at your church, or a no-left-turn sign where there wasn't one before (and the cop doesn't care that you're newly back from Timbuktu).

In the post-9/11 world, you're apt to find additional changes in policy and outlook that returning expats in earlier years didn't have to deal with. It is a new era—at home as well as in the places you visited and lived. While you can be rightfully proud of your new man (or woman)-of-the-world status, you'll need to step back and reclaim your American-ness without sacrificing all you've learned as an expat.

Do this by keeping track of national news and politics and becoming involved in the civic affairs in your community. You may be the only one at the town meeting in a shirt purchased at Wimbledon or the Tour de France, but you need to be there—for your own repatriation and for the unique perspective you can offer.

Your friends and neighbors back in the United States will have been affected by 9/11 and its aftermath in a different way than you. They have not had the benefit of your direct exposure to foreign culture and people during this period. To the contrary, their views have largely been molded by what they see, hear, and read in the U.S. media—which is a mixed bag at best and somewhat dependent on what they have access to (or have selected) as their primary source of information. Your reintegration into the home culture can provide a new window to the world for those who are willing to look through it.

Reentering the U.S. Job Market

Hopefully, you included your return to the U.S. job market as part of your planning that was discussed in chapter 4. Ideally, you and your employer have already discussed mutual employment expectations upon successful completion of your foreign assignment, and you are coming home to a position that both incorporates your international experience and offers an opportunity for increased compensation and career advancement.

Unfortunately, this ideal result is too rarely achieved. A very high percentage of returning expats do not continue working for their former U.S. employers. A surprising number are not offered any position at all, or are offered a position that is inappropriate. It may be at a parallel (or even lower) level than the position held before moving overseas, or it does not take advantage of the skills and knowledge developed during the foreign assignment.

Many U.S. companies simply cannot figure out what to do with a returning expat, because (1) the old position is either filled or is now uninteresting to the expat, and (2) it is not always easy to find a way to use a returning expat's skills and knowledge in the U.S. office. In my own experience with U.S. employers and overseas assignments (four different companies), I have returned to a U.S.-based position with the same employer in only one case. Although the companies and I both made an effort, we were unable to identify a U.S. position that was a good fit—a state of affairs with which many returning expats become familiar.

If you do return to your U.S. employer, you may be surprised at how "un-special" you are treated by your former work colleagues. You have likely been in a position where you exercised more management and decision-making authority than you did previously. However, your colleagues remember you from the way things used to be. They don't understand much about where you've been or what you've been doing. They probably haven't seen your growth and development, and they won't know or understand the value that you brought to the company while overseas. They're likely to see you as someone who's just returned from an exotic location after having a good time seeing the world. They, on the other hand, have been trudging through the real work back here. Why should you just be able to waltz back into the office and get deferential treatment?

All in all, while there are numerous (theoretically) good reasons to return to a U.S.-based position with the same employer, in reality this happens much less than you might think. A strong mutual commitment is required to make such a transition successful from the standpoint of both employer and employee. It is an unfortunate truism that an unknown company that has received no benefit from your overseas efforts might be a better fit and place a higher value on you than your original U.S. employer.

It is thus a good idea to consider and plan for this likelihood well in advance of your return to the United States, even if your goal is to indeed return to a position with your former U.S. employer. Test the market and see what is out there for someone possessing your new skills and knowledge. Apply the principles and techniques discussed in chapter 2, "Finding a New Job Overseas," only in reverse. You now have a background and level of experience in a specific country (and in international business in general) that is unique in the U.S. job market. Focus on finding a position in which such background and experience are valued, and in which your new skills will be aptly used and rewarded.

Another Option

You could avoid these challenges of returning home altogether—by simply not doing it. If you have done well overseas and enjoyed your expatriate experience, why not consider staying on or finding a new assignment in a different country? After all, you have gotten through all the tough stuff, and you now possess knowledge and skills that may never be fully utilized if you return to the United States. You are a member of the fraternity of U.S. expats, and it can be difficult to give up this special status.

Whether you come back to the United States, or decide to continue your life as an expat, hold on to what you have experienced and learned from the rest of the world. As a new American expat, apply the principles discussed in this book to your relationships with people from other countries. Become, and always continue to be, an American world citizen...because the United States and the world need as many of these as they can get.

Chapter 7 **Summary** and **Checklist**

Coming Home

1. Be aware that like your initial move overseas, your return to the U.S. will require preparation and adjustment. You will face many of the same issues, but in reverse and with a few twists.
2. Understand that you and your family will encounter reverse culture shock, due primarily to the fact that you have significantly changed during the foreign assignment while your U.S. home and friends may have changed less or in different ways.
3. Check out U.S. customs regulations carefully before shipping or bringing your belongings back to the United States. Special rules apply to returning expats that should allow you to bring most household goods and personal effects back to the U.S. duty-free. However, some items purchased overseas are prohibited from being imported into the U.S., and the requisite forms and procedures must be strictly complied with.
4. Obtain all necessary documentation required for your spouse (if you married a foreign national while overseas) and foreign-born children to be allowed to return with you to the United States.
5. Check any special U.S. health requirements applicable to people entering the United States from the country or region in which you have been living. If you have pets, check on those regulations as well.
6. Consider engaging the services of an experienced international relocation firm. Doing so could make your move back to the United States a great deal easier.

7. Be prepared to adjust your financial thinking. You have possibly become used to a standard of living while overseas that you cannot maintain after returning to the United States.
8. Understand that your friends, your community, and your country have not been in suspended animation. Just as you hope for accommodation of the person you have become, stretch yourself to accept home and friends as you find them. Become politically or civically active to reassert your citizenship and help this country live up to the best that America stands for.
9. Remember that a very high percentage of returning expats do not continue working for their former U.S. employers after returning to the United States. A strong mutual commitment is required to make such a transition successful, and you may be better off finding a position with a new company that more appropriately leverages your new skills and experience. Prepare for this likelihood in advance and test the market using the job-finding techniques discussed in chapter 2.
10. Find new anchors in your home community that help you maintain your interest in, and contacts with, the country in which you lived and the international community in general. Continue to expose yourself to international events and differing viewpoints from around the world.

Appendix A:

Recommended Reading on General Expatriate Issues

- Adams, John. 1998. *U.S Expatriate Handbook: Guide to Living & Working Abroad*. Morgantown, WV: West Virginia University.
- Black, J. Stewart, and Hal Gregersen. 1998. *So You're Going Overseas: A Handbook for Personal and Professional Success*. San Diego, CA: Global Business Publishers (accompanying workbooks are available).
- Hess, Melissa, and Patricia Linderman. 2002. *The Expert Expatriate: Your Guide to Successful Relocation Abroad*. Yarmouth, ME: Nicholas Brealey Publishing/Intercultural Press.
- Kohls, L. Robert. 2001. *Survival Kit for Overseas Living: For Americans Planning to Live and Work Abroad,* 4th ed. Yarmouth, ME: Nicholas Brealey Publishing/Intercultural Press.
- Kruemplemann, Elizabeth. 2002. *The Global Citizen: A Guide to Creating an International Life and Career*. Berkeley, CA: Ten Speed Press.
- Pollock, David, and Ruth Van Reken. *Third Culture Kids: The Experience of Growing Up Among Worlds*. Yarmouth, ME: Nicholas Brealey Publishing/Intercultural Press.
- Storti, Craig. 2001. *The Art of Coming Home*. Yarmouth, ME: Intercultural Press.
- Storti, Craig. 2001. *The Art of Crossing Cultures*, 2nd ed. Yarmouth, ME: Intercultural Press.

Appendix B:

Expatriate Internet Resources with General Information and Advice on Expatriate Issues

- www.anamericanabroad.com (chat room available)
- www.escapeartist.com
- www.expatexchange.com
- www.goingglobal.com (fee-based)
- www.livingabroad.com (fee-based)
- www.overseasdigest.com
- www.transitionsabroad.com

Appendix C:

Resources and Organizations to Contact for Information Regarding International Jobs

Recommended Reading

- Gliozzo, Charles, ed. 2004. *Directory of International Internships,* 5th ed. East Lansing, MI: Michigan State University. (Available for order at www.isp.msu.edu.)
- Griffith, Susan. 2003. *Work Your Way Around the World,* 11th. ed. Oxford, UK: Vacation Work Publications.
- Kocher, Eric, and Nina Segal. 2003. *International Jobs: Where They Are, How to Get Them,* 6th ed., Cambridge, MA: Perseus Publishing.
- Lauber, Daniel. 2002. *International Job Finder: Where the Jobs Are Worldwide*. Gardena, CA: SCB Distributors.
- Mohammed, Jeff. 2003. *Teaching English Overseas: A Job Guide for Americans and Canadians.* Spring, TX: English International Inc.
- Mueller, Nancy. 2000. *Work Worldwide: International Career Strategies for the Adventurous Job Seeker*. Emeryville, CA: Avalon Travel Publishing.
- Oldman, Mark, and Sarah Hamadeh. 2004. *The Internship Bible.* Princeton, NJ: Princeton Review.

Job-Finding Internet Resources

- www.backdoorjobs.com (student positions, internships, and short-term jobs)
- www.english-international.com (teaching English as a second language)

- www.escapeartist.com
- www.eslcafe.com (teaching English as a second language)
- www.gradschools.com (teaching English as a second language)
- www.monster.com
- www.overseasdigest.com
- www.overseasjobs.com
- www.quintcareers.com
- www.rileyguide.com
- www.tefl.com (teaching English as a second language)
- www.theeducationportal.com (teacher-exchange portal)
- www.transitionsabroad.com

International Job Bulletins

- International Employment Gazette (www.intemployment.com)
- International Employment Hotline and International Career Employment Weekly (www.internationaljobs.org)

Organizations

Internships

- AIESEC (www.aiesec.org)
- Association for International Practical Training (www.aipt.org)
- BUNAC (www.bunac.org)
- Council on International Education Exchange (www.ciee.org)
- International Association for the Exchange of Students for Technical Experience (www.iaeste.org)
- International Cooperative Education Program (www.icemenlo.com)
- Internship opportunities at government agencies and international organizations may be accessed through www.studentjobs.com

(portal cosponsored by U.S. Office of Personnel Management and the U.S. Department of Education)

- Internship opportunities at UN organizations may be accessed through www.unsystem.org

Teaching Overseas

- Association of American Schools in South America (www.aassa.com)
- Council for International Exchange of Scholars (www.cies.org)
- European Council of International Schools (www.ecis.org)
- Fulbright Teacher and Administrator Exchange Program (www.fulbrightexchanges.org)
- International Schools Services (www.iss.edu)
- Search Associates (www.search-associates.com)

Volunteer

- Citizens Democracy/MBA Corps (www.cdc.org)
- United Nations Volunteer Program (www.unv.org)
- U.S. Peace Corps (www.peacecorps.gov)
- Volunteers for the International Executive Corps (www.iesc.org).

Directories of U.S. and Foreign Companies

- *America's Corporate Families and International Affiliates*, Dun & Bradstreet (www.dnb.com).
- *Directory of American Firms Operating in Foreign Countries*, Uniworld Business Publications (www.uniworldbp.com).
- *Directory of Foreign Firms Operating in the United States*, Uniworld Business Publications (www.uniworldbp.com).
- Information regarding U.S. companies with overseas operations can also be obtained through the membership directories of local international and world trade clubs (see www.fita.org for a listing of such clubs).

Appendix D:

Government and Private Organizations that Provide Information and Assistance for Expatriates and for Companies Employing Expatriates

American Expatriate Organizations

- American Citizens Abroad (www.aca.ch)
- Association of Americans Resident Overseas (www.aaro.org)

Government and Commercial Agencies

- Federal Voting Assistance Program (www.fvap.gov)
- Foreign chamber of commerce offices located in the United States (www.chamberofcommerce.com)
- Foreign embassies and consulates located in the United States (www.embassyworld.com or www.travel.state.gov)
- United States State Department, Embassies and Consulates (www.travel.state.gov)

Health Organizations and Information Sources

- American Society of Tropical Medicine (www.astmh.org)
- International Association of Medical Assistance to Travelers (www.iamat.org)
- International Society of Travel Medicine (www.istm.org)
- U.S. Center for Disease Control (www.cdc.gov)
- World Health Organization (www.who.org)

Selected International Relocation and Human Resource Outsourcing Firms

- Berlitz Cross-Cultural/Relocation Services (www.berlitz.com)

- Cendant Mobility (www.cendantmobility.com)
- GMAC Global Relocation Services (www.gmac-relocation.com)
- Primacy Relocation Services (www.primacy.com)
- Prudential Relocation (www.prudential.com)

Selected Tax Advisors Specializing in Expatriate Taxation Issues

- Donald Walter (CPA) (www.globaltaxhelp.com)
- Global Tax Network, LLC (www.gtntax.com)
- Jane Bruno JD, LLM, Gutta, Koutoulas and Relis LLC (www.americantaxhelp.com)
- Steven Y. C. Kang CPA and Associates (www.taxsav.com)

International Transport of Pets

- Independent Pet and Animal Transportation Association (www.ipata.com)

Information on Electrical, Telephone, and TV Standards in Foreign Countries

- www.travel-advisor-online.com

Information and Assistance Regarding Safety and Security While Living and Traveling Overseas

- International security and safety consulting and management services: Marcus Group Co. Ltd. (www.marcus-group.com), Marc Bradshaw, President
- Private firms that provide international security and safety information and consultancy services: Control Risks Group (www.crg.com), iJet Travel Risk Management (www.ijet.com), International SOS (www.internationalsos.com)
- United States State Department, Embassies and Consulates: Travel Warnings, Public Announcements, Consular Information Sheets and publications (www.travel.state.gov)

Appendix E:

Resources for Learning More about Specific Foreign Countries

- *Culture Shock!* Series, Portland, OR: Graphic Arts Center Publishing (series of books about many countries by different authors).
- Foreign chamber of commerce offices located in the U.S. (www.chamberofcommerce.com).
- Foreign embassies and consulates located in the U.S. (www.embassy-world.com or www.travel.state.gov).
- Lewis, Richard. 2005. *When Cultures Collide: Managing Successfully Across Cultures*. Yarmouth, ME: Nicholas Brealey Publishing/Intercultural Press.
- Morrison, Terri, Wayne Conaway, and George Borden. 1995. *Kiss, Bow, or Shake Hands: How to Do Business in Sixty Countries*. Holbrook, MA: Adams Media Corporation (new edition scheduled 2006).
- Morrison, Terri, Wayne Conaway, and Joseph Douress. 2000. *Dun & Bradstreet's Guide to Doing Business Around the World*. Upper Saddle River, NJ: Prentice Hall.
- Olofsson, Gwyneth. 2004. *When in Rome or Rio or Riyadh: Cultural Q & As for Successful Business Behavior Around the World*. Yarmouth, ME: Intercultural Press.
- U.S. State Department Background Notes available through U.S. embassies and consulates worldwide and at www.travel.state.gov.

Internet Resources

- www.expatexchange.com
- www.escapeartist.com
- www.goingglobal.com (fee-based)
- www.livingabroad.com (fee-based)
- www.overseasdigest.com
- www.transitionsabroad.com

Index

About the Author

William Russell Melton has more than 25 years experience specializing in international business, both as an international lawyer and executive manager. During this period, he has lived and worked in seven different countries—the United Kingdom, the Netherlands, Saudi Arabia, Germany, Singapore, Switzerland and Bahrain—and has served both as International General Counsel for U.S corporations and as the Managing Director for operations based in these countries.

In his role as Managing Director, Melton was vested with overall responsibility for sales, marketing, human resource issues, finance, administration and other aspects of the operations he was managing. His experience also includes setting up new operations in more than 20 countries, recruiting and managing distributors and other partners globally, and restructuring and turning around under-performing foreign operations. The companies he has worked for have ranged from small start-ups to Fortune 500 companies, and he thus understands the special issues faced by both small and large businesses in the international business arena.

The author's experience working and living overseas, including the first two and a half years following the 9/11 attacks and subsequent international upheaval, has provided him with a unique perspective on the issues facing American expatriates in today's world. Melton holds a B.S., J.D. in law, and M.A. in Soviet and East European Studies. His website is www.wrm-international.com and he can be reached via email at wrm@wrm-international.com.